T0344293

# Galois' Theory
## of Algebraic Equations

### Second Edition

# Galois' Theory
## of Algebraic Equations

## Second Edition

## Jean-Pierre Tignol
*Université Catholique de Louvain, Belgium*

 **World Scientific**

NEW JERSEY · LONDON · SINGAPORE · BEIJING · SHANGHAI · HONG KONG · TAIPEI · CHENNAI · TOKYO

*Published by*

World Scientific Publishing Co. Pte. Ltd.

5 Toh Tuck Link, Singapore 596224

*USA office:* 27 Warren Street, Suite 401-402, Hackensack, NJ 07601

*UK office:* 57 Shelton Street, Covent Garden, London WC2H 9HE

**Library of Congress Cataloging-in-Publication Data**
Names: Tignol, Jean-Pierre, author.
Title: Galois' theory of algebraic equations / by Jean-Pierre Tignol
    (Université Catholique de Louvain, Belgium).
Other titles: Leçons sur la théorie des équations. English | Lectures on the theory of equations
Description: 2nd World Scientific edition. | New Jersey : World Scientific, 2016. |
    Originally published in English in 1988 jointly by: Harlow, Essex, England :
    Longman Scientific & Technical; and, New York : Wiley. |
    Includes bibliographical references and index.
Identifiers: LCCN 2015038537 | ISBN 9789814704694 (hardcover : alk. paper)
Subjects: LCSH: Equations, Theory of. | Galois theory.
Classification: LCC QA211 .T5413 2016 | DDC 512/.32--dc23
LC record available at http://lccn.loc.gov/2015038537

**British Library Cataloguing-in-Publication Data**
A catalogue record for this book is available from the British Library.

Printed in Singapore

à Paul

For inquire, I pray thee, of the former age,
and prepare thyself to the search of their fathers:
For we are but of yesterday, and know nothing,
because our days upon earth are a shadow.

Job 8, 8–9.

# Preface to the Second Edition

After the first edition of this book was published, the bicentennial of Galois' birth (in 2011) occasioned a renewal of interest in his highly atypical oeuvre. Scholarship on the theory of equations and Galois theory has been significantly expanded, notably by Stedall's and Ehrhardt's monographs [68] and [27], and through the publication of a new edition of Galois' mathematical writings by Neumann [55]. While these circumstances influenced my decision to prepare a second edition of this book, the crucial factor in this regard stems from an uncanny experience that I had while working on a completely different project: as Max Knus and I were studying constructions on étale algebras inspired by the representation theory of linear algebraic groups, it dawned on me that some of the constructions were *exact analogues* of those that Galois had used to attach groups to equations. Revisiting Galois' memoir with this analogy in mind, I was awed by the efficiency of this perspective in elucidating Galois' statements and enabling the presentation of full proofs of his propositions.

In this new edition, the chapter on Galois has been completely rewritten to take advantage of this viewpoint. The exposition is now much closer to his memoir, and remarkably elementary.[1] It is also mostly free of anachronisms, although I did not refrain from using a few modern notions when I felt they could illuminate Galois' words without overextending his vision. Thus, for the reader in search of a precise idea of what Galois wrote, my account is no substitute for the original,[2] but I hope it will foster a better

---

[1] Keep in mind that "elementary" is not the same as "easy"; in this case (as often), it is in fact the opposite: Galois' elementary arguments are remarkably ingenious and sometimes quite intricate.

[2] Galois' memoir is now readily available on the web site of the *Bibliothèque nationale de France*, http://gallica.bnf.fr/ or the *Bibliothèque de l'Institut de France*, http://www.bibliotheque-institutdefrance.fr/, or in translation [26] and [55, Ch. IV].

understanding of this difficult text. To the epilogue (Chapter 15), I have added a new appendix to indicate the close relation between Galois' definition of the group of an equation and the modern notion of torsor that inspired my new analysis of his memoir.

Chapters 1–13 are mostly unchanged, except for a few slight revisions in the wording, which I hope are improvements. The more significant alterations occur in the discussion of elimination theory in §5.6, and of radical extensions in §13.1.

This second edition also gives me the opportunity to acknowledge input from several people who gave me feedback on various aspects of the first edition. I am especially grateful to Benjamin Barras, Oscar Luis Palacios-Vélez, and Robert Perlis for lists of typographical errors and valuable comments, which have been taken into account in this new edition. I am also indebted to Karim-Johannes Becher and James O'Shea, who kindly read and commented the revised version of Chapter 14. Their numerous constructive suggestions allowed me to eliminate mistakes and to improve the exposition in several places.

# Preface to the First Edition (2001)

In spite of the title, the main subject of these lectures is not algebra, even less history, as one could conclude from a glance over the table of contents, but methodology. Their aim is to convey to the audience, which originally consisted of undergraduate students in mathematics, an idea of how mathematics is made. For such an ambitious project, the individual experience of any but the greatest mathematicians seems of little value, so I thought it appropriate to rely instead on the collective experience of generations of mathematicians, on the premise that there is a close analogy between collective and individual experience: the problems over which past mathematicians have stumbled are most likely to cause confusion to modern learners, and the methods which have been tried in the past are those which should come to mind naturally to the (gifted) students of today. The way in which mathematics is made is best learned from the way mathematics has been made, and that premise accounts for the historical perspective on which this work is based.

The theme used as an illustration for general methodology is the theory of equations. The main stages of its evolution, from its origins in ancient times to its completion by Galois around 1830 will be reviewed and discussed. For the purpose of these lectures, the theory of equations seemed like an ideal topic in several respects: first, it is completely elementary, requiring virtually no mathematical background for the statement of its problems, and yet it leads to profound ideas and to fundamental concepts of modern algebra. Secondly, it underwent a very long and eventful evolution, and several gems lie along the road, like Lagrange's 1770 paper, which brought order and method to the theory in a masterly way, and Vandermonde's visionary glimpse of the solution of certain equations of high degree, which hardly unveiled the principles of Galois theory sixty years before Galois' memoir. Also instructive from a methodological point of view is the relationship between the general theory, as developed by Cardano, Tschirnhaus, Lagrange and Abel, and the attempts by Viète, de Moivre, Vandermonde and Gauss at significant examples, namely the so-called cyclotomic equations, which arise from the division of the circle into equal parts. Works in these two directions are closely intertwined like themes in a counterpoint, until their resolution in Galois' memoir. Finally, the algebraic theory of equations is now a closed subject, which reached complete maturity a long time ago; it is therefore possible to give a fair assessment of its various aspects. This is of course not true of Galois theory, which still provides inspiration for original research in numerous directions, but these lectures are concerned with the theory of equations and not with the Galois theory of fields. The evolution from Galois' theory to modern Galois theory falls beyond the scope of this work; it would certainly fill another book like this one.

As a consequence of emphasis on historical evolution, the exposition of mathematical facts in these lectures is genetic rather than systematic, which means that it aims to retrace the concatenation of ideas by following (roughly) their chronological order of occurrence. Therefore, results which are logically close to each other may be scattered in different chapters, and some topics are discussed several times, by little touches, instead of being given a unique definitive account. The expected reward for these circumlocutions is that the reader could hopefully gain a better insight into the inner workings of the theory, which prompted it to evolve the way it did.

Of course, in order to avoid discussions that are too circuitous, the works of mathematicians of the past—especially the distant past—have been somewhat modernized as regards notation and terminology. Although considering sets of numbers and properties of such sets was clearly alien to the patterns of thinking until the nineteenth century, it would be futile to ignore the fact that (naive) set theory has now pervaded all levels of mathematical education. Therefore, free use will be made of the definitions of some basic algebraic structures such as field and group, at the expense of lessening some of the most original discoveries of Gauss, Abel and Galois. Except for those definitions and some elementary facts of linear algebra which are needed to clarify some proofs, the exposition is completely self-contained, as can be expected from a genetic treatment of an elementary topic.

It is fortunate to those who want to study the theory of equations that its long evolution is well documented: original works by Cardano, Viète, Descartes, Newton, Lagrange, Waring, Gauss, Ruffini, Abel, Galois are readily available through modern publications, some even in English translations. Besides these original works and those of Girard, Cotes, Tschirnhaus and Vandermonde, I relied on several sources, mainly on Bourbaki's Note historique [7] for the general outline, on van der Waerden's "Science Awakening" [79] for the ancient times and on Edwards' "Galois theory" [26] for the proofs of some propositions in Galois' memoir. For systematic expositions of Galois theory, with applications to the solution of algebraic equations by radicals, the reader can be referred to any of the fine existing accounts, such as Artin's classical booklet [2], Kaplansky's monograph [42], the books by Morandi [54], Rotman [63] or Stewart [70], or the relevant chapters of algebra textbooks by Cohn [17], Jacobson [40], [41] or van der Waerden [77], and presumably to many others I am not aware of. In the present lectures, however, the reader will find a thorough treatment of cyclotomic equations after Gauss, of Abel's theorem on the impossibility of solving the general equation of degree 5 by radicals, and of the conditions for solvability of algebraic equations after Galois, with complete proofs. The point of view differs from the one in the quoted references in that it is strictly utilitarian, focusing (albeit to a lesser extent than the original papers) on the concrete problem at hand, which is to solve equations. Incidentally, it is striking to observe, in comparison, what kind of acrobatic tricks are needed to apply modern Galois theory to the solution of algebraic equations.

The exercises at the end of some chapters point to some extensions of the theory and occasionally provide the proof of some technical fact which is alluded to in the text. They are never indispensable for a good understanding of the text. Solutions to selected exercises are given at the end of the book.

This monograph is based on a course taught at the Université catholique de Louvain from 1978 to 1989, and was first published by Longman Scientific & Technical in 1988. It is a much expanded and completely revised version of my "Leçons sur la théorie des équations" published in 1980 by the (now vanished) Cabay editions in Louvain-la-Neuve. The wording of the Longman edition has been recast in a few places, but no major alteration has been made to the text.

I am greatly indebted to Francis Borceux, who invited me to give my first lectures in 1978, to the many students who endured them over the years, and to the readers who shared with me their views on the 1988 edition.

Their valuable criticism and encouraging comments were all-important in my decision to prepare this new edition for publication. Through the various versions of this text, I was privileged to receive help from quite a few friends, in particular from Pasquale Mammone and Nicole Vast, who read parts of the manuscript, and from Murray Schacher and David Saltman, for advice on (American-) English usage. Hearty thanks to all of them. I owe special thanks also to T.S. Blyth, who edited the manuscript of the Longman edition, to the staffs of the Centre général de Documentation (Université catholique de Louvain) and of the Bibliothèque Royale Albert 1$^{er}$ (Brussels) for their helpfulness and for allowing me to reproduce parts of their books, and to Nicolas Rouche, who gave me access to the riches of his private library.

On the TEXnical side, I am grateful to Suzanne D'Addato (who also typed the 1988 edition) and to Béatrice Van den Haute, and also to Camille Debiève for his help in drawing the figures.

Finally, my warmest thanks to Céline, Paul, Ève and Jean for their infectious joy of living and to Astrid for her patience and constant encouragement. The preparation of the 1988 edition for publication spanned the whole life of our little Paul. I wish to dedicate this book to his memory.

# Contents

*Preface to the Second Edition*      vii

*Preface to the First Edition (2001)*      ix

1.   Quadratic Equations      1

     1.1    Babylonian algebra . . . . . . . . . . . . . . . . . . . . . .   2
     1.2    Greek algebra . . . . . . . . . . . . . . . . . . . . . . . . .   5
     1.3    Arabic algebra . . . . . . . . . . . . . . . . . . . . . . . .   9

2.   Cubic Equations      13

     2.1    Priority disputes on the solution of cubic equations . . . .   13
     2.2    Cardano's formula . . . . . . . . . . . . . . . . . . . . . . .   15
     2.3    Developments arising from Cardano's formula . . . . . . .   16

3.   Quartic Equations      21

     3.1    The unnaturalness of quartic equations . . . . . . . . . . .   21
     3.2    Ferrari's method . . . . . . . . . . . . . . . . . . . . . . . .   22

4.   The Creation of Polynomials      25

     4.1    The rise of symbolic algebra . . . . . . . . . . . . . . . . .   25
         4.1.1    L'Arithmetique . . . . . . . . . . . . . . . . . . . . .   26
         4.1.2    In Artem Analyticem Isagoge . . . . . . . . . . . . .   29
     4.2    Relations between roots and coefficients . . . . . . . . . .   30

5.   A Modern Approach to Polynomials      41

     5.1    Definitions . . . . . . . . . . . . . . . . . . . . . . . . . . .   41
     5.2    Euclidean division . . . . . . . . . . . . . . . . . . . . . . .   43

5.3  Irreducible polynomials . . . . . . . . . . . . . . . . . . . .  48

5.4  Roots . . . . . . . . . . . . . . . . . . . . . . . . . . . . .  51

5.5  Multiple roots and derivatives . . . . . . . . . . . . . . .  53

5.6  Common roots of two polynomials . . . . . . . . . . . .  56

Appendix: Decomposition of rational functions into sums of
partial fractions . . . . . . . . . . . . . . . . . . . . . . .  60

6.  Alternative Methods for Cubic and Quartic Equations  63

6.1  Viète on cubic equations . . . . . . . . . . . . . . . . . .  63

6.1.1  Trigonometric solution for the irreducible case . .  63

6.1.2  Algebraic solution for the general case . . . . . . .  64

6.2  Descartes on quartic equations . . . . . . . . . . . . . .  66

6.3  Rational solutions for equations with rational coefficients .  67

6.4  Tschirnhaus' method . . . . . . . . . . . . . . . . . . . .  68

7.  Roots of Unity  73

7.1  The origins of de Moivre's formula . . . . . . . . . . . .  73

7.2  The roots of unity . . . . . . . . . . . . . . . . . . . . .  80

7.3  Primitive roots and cyclotomic polynomials . . . . . . .  85

Appendix: Leibniz and Newton on the summation of series . . .  89

Exercises . . . . . . . . . . . . . . . . . . . . . . . . . . . . . .  90

8.  Symmetric Functions  93

8.1  Waring's method . . . . . . . . . . . . . . . . . . . . . .  96

8.2  The discriminant . . . . . . . . . . . . . . . . . . . . . .  101

Appendix: Euler's summation of the series of reciprocals of
perfect squares . . . . . . . . . . . . . . . . . . . . . . .  105

Exercises . . . . . . . . . . . . . . . . . . . . . . . . . . . . . .  107

9.  The Fundamental Theorem of Algebra  109

9.1  Girard's theorem . . . . . . . . . . . . . . . . . . . . . .  110

9.2  Proof of the fundamental theorem . . . . . . . . . . . . .  113

10.  Lagrange  117

10.1  The theory of equations comes of age . . . . . . . . . . .  117

10.2  Lagrange's observations on previously known methods . .  121

10.3  First results of group theory and Galois theory . . . . . .  131

Exercises . . . . . . . . . . . . . . . . . . . . . . . . . . . . . .  142

11. Vandermonde 143

    11.1 The solution of general equations . . . . . . . . . . . . . 144
    11.2 Cyclotomic equations . . . . . . . . . . . . . . . . . . . . 148
    Exercises . . . . . . . . . . . . . . . . . . . . . . . . . . . . . 154

12. Gauss on Cyclotomic Equations 155

    12.1 Number-theoretic preliminaries . . . . . . . . . . . . . . 156
    12.2 Irreducibility of the cyclotomic polynomials of prime index 162
    12.3 The periods of cyclotomic equations . . . . . . . . . . . 169
    12.4 Solvability by radicals . . . . . . . . . . . . . . . . . . . 178
    12.5 Irreducibility of the cyclotomic polynomials . . . . . . . . 182
    Appendix: Ruler and compass construction of regular polygons 185
    Exercises . . . . . . . . . . . . . . . . . . . . . . . . . . . . . 192

13. Ruffini and Abel on General Equations 193

    13.1 Radical extensions . . . . . . . . . . . . . . . . . . . . . 195
    13.2 Abel's theorem on natural irrationalities . . . . . . . . . 203
    13.3 Proof of the unsolvability of general equations of degree
         higher than 4 . . . . . . . . . . . . . . . . . . . . . . . 209
    Exercises . . . . . . . . . . . . . . . . . . . . . . . . . . . . . 211

14. Galois 215

    14.1 Arrangements and permutations . . . . . . . . . . . . . 220
    14.2 The Galois group of an equation . . . . . . . . . . . . . 225
    14.3 The Galois group under base field extension . . . . . . . 236
    14.4 Solvability by radicals . . . . . . . . . . . . . . . . . . . 246
    14.5 Applications . . . . . . . . . . . . . . . . . . . . . . . . . 256
        14.5.1 Irreducible equations of prime degree . . . . . . . 256
        14.5.2 Abelian equations . . . . . . . . . . . . . . . . . . 265
    Exercises . . . . . . . . . . . . . . . . . . . . . . . . . . . . . 268

15. Epilogue 271

    Appendix 1: The fundamental theorem of Galois theory . . . . 274
    Appendix 2: Galois theory *à la* Grothendieck . . . . . . . . . 283
        Étale algebras . . . . . . . . . . . . . . . . . . . . . . . . 283
        Galois algebras . . . . . . . . . . . . . . . . . . . . . . . 285
        Galois groups . . . . . . . . . . . . . . . . . . . . . . . . 287
    Exercises . . . . . . . . . . . . . . . . . . . . . . . . . . . . . 290

*Selected Solutions*                                                                            291

*Bibliography*                                                                                  299

*Index*                                                                                         305

## Chapter 1

# Quadratic Equations

Since the solution of a linear equation $ax = b$ does not use anything more than a division, it hardly belongs to the algebraic theory of equations; it is therefore appropriate to begin our discussion with quadratic equations

$$ax^2 + bx + c = 0 \qquad (a \neq 0).$$

Dividing each side by $a$, we reduce to the case where the coefficient of $x^2$ is 1:

$$x^2 + px + q = 0.$$

The solution of this equation is well-known: when $\left(\frac{p}{2}\right)^2$ is added to each side, the square of $x + \frac{p}{2}$ appears and the equation can be written

$$\left(x + \frac{p}{2}\right)^2 + q = \left(\frac{p}{2}\right)^2.$$

(This procedure is called "completion of the square.") The values of $x$ easily follow:

$$x = -\frac{p}{2} \pm \sqrt{\left(\frac{p}{2}\right)^2 - q}.$$

This formula is so well-known that it may be rather surprising to note that the solution of quadratic equations could not have been written in this form before the seventeenth century.[1] Nevertheless, mathematicians had been solving quadratic equations for about 40 centuries before. The purpose of this first chapter is to give a brief outline of this "prehistory" of the theory of quadratic equations.

---

[1] The first uniform solution for quadratic equations (regardless of the signs of coefficients) is due to Simon Stevin in "L'Arithmetique" [69, p. 595], published in 1585. However, Stevin does not use literal coefficients, which were introduced some years later by François Viète: see Chapter 4, §4.1.

## 1.1   Babylonian algebra

The first known solution of a quadratic equation dates from about 2000 B.C.; on a Babylonian tablet, one reads (see van der Waerden [79, p. 69])

> I have subtracted from the area the side of my square: 14.30. Take 1, the coefficient. Divide 1 into two parts: 30. Multiply 30 and 30: 15. You add to 14.30, and 14.30.15 has the root 29.30. You add to 29.30 the 30 which you have multiplied by itself: 30, and this is the side of the square.

This text obviously provides a procedure for finding the side of a square (say $x$) when the difference between the area and the side (i.e., $x^2 - x$) is given; in other words, it gives the solution of $x^2 - x = b$.

However, one may be puzzled by the strange arithmetic used by Babylonians. It can be explained by the fact that their base for numeration is 60; therefore 14.30 really means $14 \cdot 60 + 30$, i.e., 870. Moreover, they had no symbol to indicate the absence of a number or to indicate that certain numbers are intended as fractions. For instance, when 1 is divided by 2, the result which is indicated as 30 really means $30 \cdot 60^{-1}$, i.e., 0.5. The square of this 30 is then 15 which means 0.25, and this explains why the sum of 14.30 and 15 is written as 14.30.15: in modern notation, the operation is $870 + 0.25 = 870.25$.

After clearing the notational ambiguities, it appears that the author correctly solves the equation $x^2 - x = 870$, and gets $x = 30$. The other solution $x = -29$ is neglected, since the Babylonians had no negative numbers.

This lack of negative numbers prompted Babylonians to consider various types of quadratic equations, depending on the signs of coefficients. There are three types in all:

$$x^2 + ax = b, \quad x^2 - ax = b, \quad \text{and} \quad x^2 + b = ax,$$

where $a$, $b$ stand for positive numbers. (The fourth type $x^2 + ax + b = 0$ obviously has no (positive) solution.)

Babylonians could not have written these various types in this form, since they did not use letters in place of numbers, but from the example above and from other numerical examples contained on the same tablet, it clearly appears that the Babylonians knew the solution of

$$x^2 + ax = b \quad \text{as} \quad x = \sqrt{\left(\frac{a}{2}\right)^2 + b} - \frac{a}{2}$$

and of

$$x^2 - ax = b \quad \text{as} \quad x = \sqrt{\left(\frac{a}{2}\right)^2 + b} + \frac{a}{2}.$$

How they argued to get these solutions is not known, since in every extant example, only the procedure to find the solution is described, as in the example above. It is very likely that they had previously found the solution of geometric problems, such as to find the length and the breadth of a rectangle, when the excess of the length on the breadth and the area are given. Letting $x$ and $y$ respectively denote the length and the breadth of the rectangle, this problem amounts to solving the system

$$\begin{cases} x - y = a \\ xy = b. \end{cases} \tag{1.1}$$

By elimination of $y$, this system yields the following equation for $x$:

$$x^2 - ax = b. \tag{1.2}$$

If $x$ is eliminated instead of $y$, we get

$$y^2 + ay = b. \tag{1.3}$$

Conversely, equations (1.2) and (1.3) are equivalent to system (1.1) after setting $y = x - a$ or $x = y + a$.

They probably deduced their solution for quadratic equations (1.2) and (1.3) from their solution of the corresponding system (1.1), which could be obtained as follows: let $z$ be the arithmetic mean of $x$ and $y$.

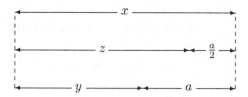

In other words, $z$ is the side of the square which has the same perimeter as the given rectangle:

$$z = x - \frac{a}{2} = y + \frac{a}{2}.$$

Compare then the area of the square (i.e., $z^2$) to the area of the rectangle ($xy = b$). We have

$$xy = \left(z + \frac{a}{2}\right)\left(z - \frac{a}{2}\right)$$

hence $b = z^2 - \left(\frac{a}{2}\right)^2$. Therefore, $z = \sqrt{\left(\frac{a}{2}\right)^2 + b}$ and it follows that

$$x = \sqrt{\left(\frac{a}{2}\right)^2 + b} + \frac{a}{2} \quad \text{and} \quad y = \sqrt{\left(\frac{a}{2}\right)^2 + b} - \frac{a}{2}.$$

This solves at once the quadratic equations $x^2 - ax = b$ and $y^2 + ay = b$.

Looking at the various examples of quadratic equations solved by Babylonians, one notices a curious fact: the third type $x^2 + b = ax$ does not explicitly appear. This is even more puzzling in view of the frequent occurrence in Babylonian tablets of problems such as to find the length and the breadth of a rectangle when the perimeter and the area of the rectangle are given, which amounts to the solution of

$$\begin{cases} x + y = a \\ xy = b. \end{cases} \tag{1.4}$$

By elimination of $y$, this system leads to $x^2 + b = ax$. So, why did Babylonians solve the system (1.4) and never consider equations like $x^2 + b = ax$?

A clue can be discovered in their solution of system (1.4), which is probably obtained by comparing the rectangle with sides $x$, $y$ to the square with perimeter $\frac{a}{2}$:

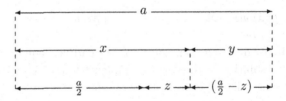

One then sets $x = \frac{a}{2} + z$, hence $y = \frac{a}{2} - z$, and derives the value of $z$ from the equation $b = \left(\frac{a}{2}\right)^2 - z^2$.

Whatever their method, the solution they obtain is

$$x = \frac{a}{2} + \sqrt{\left(\frac{a}{2}\right)^2 - b},$$

$$y = \frac{a}{2} - \sqrt{\left(\frac{a}{2}\right)^2 - b},$$

thus assigning one value for $x$ and one value for $y$, while it is clear to us that $x$ and $y$ are interchangeable in the system (1.4): we would have given two values for each one of the unknown quantities, and found

$$x = \frac{a}{2} \pm \sqrt{\left(\frac{a}{2}\right)^2 - b}, \quad y = \frac{a}{2} \mp \sqrt{\left(\frac{a}{2}\right)^2 - b}.$$

In the Babylonian wording, however, $x$ and $y$ are not interchangeable: they are the length and the breadth of a rectangle, so there is an implicit condition that $x \geq y$. According to Gandz [29, §9], the type $x^2 + b = ax$ was systematically and purposely avoided by Babylonians because, unlike the two other types, it has *two* positive solutions (which are the length $x$ and the breadth $y$ of the rectangle). The idea of two values for one quantity was probably very embarrassing to them, it would have struck Babylonians as an illogical absurdity, as sheer nonsense.

However, this observation that algebraic equations of degree higher than 1 have several interchangeable solutions is of fundamental importance: it is the corner-stone of Galois theory, and we shall have the opportunity to see to what clever use it will be put by Lagrange and later mathematicians.

As André Weil commented in relation with another topic [87, p. 104]:

> This is very characteristic in the history of mathematics. When there is something that is really puzzling and cannot be understood, it usually deserves the closest attention because some time or other some big theory will emerge from it.

## 1.2 Greek algebra

The Greeks deserve a prominent place in the history of mathematics, for being the first to perceive the usefulness of proofs. Before them, mathematics were rather empirical. Using deductive reasoning, they built a huge mathematical monument, which is remarkably illustrated by Euclid's celebrated masterwork "The Elements" (c. 300 B.C.).

The Greeks' major contribution to algebra during this classical period is foundational. They discovered that the naive idea of number (i.e., integer or rational number) is not sufficient to account for geometric magnitudes. For instance, there is no line segment that could be used as a length unit to measure the diagonal and the side of a square by integers: the ratio of the diagonal to the side (i.e., $\sqrt{2}$) is not a rational number, or in other words, the diagonal and the side are incommensurable.

The discovery of irrational numbers was made among followers of Pythagoras, probably between 430 and 410 B.C. (see Knorr [46, p. 49]). It is often credited to Hippasus of Metapontum, who was reportedly drowned at sea for producing a downright counterexample to the Pythagoreans' doctrine that "all things are numbers." However, no direct account is extant, and how the discovery was made is still a matter of conjecture. It is widely

believed that the first magnitudes that were shown to be incommensurable
are the diagonal and the side of a square, and the following reconstruction
of the proof has been proposed by Knorr [46, p. 27].

Assume the side $AB$ and the diagonal $AC$ of the square $ABCD$ are
both measured by a common segment; then $AB$ and $AC$ both represent
numbers (= integers) and the squares on them, which are $ABCD$ and
$EFGH$, represent square numbers. From the figure, it is clear (by counting
triangles) that $EFGH$ is the double of $ABCD$, so $EFGH$ is an even square
number and its side $EF$ is therefore even. It follows that $EB$ also represents
a number, hence $EBKA$ is a square number.

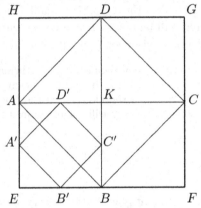

Since the square $ABCD$ clearly is the double of the square $EBKA$, the
same arguments show that $AB$ is even, hence $A'B'$ represents a number.

We now see that $A'B'$ and $A'C'$ (= $EB$), which are the halves of $AB$
and $AC$, both represent numbers; but $A'B'$ and $A'C'$ are the side and the
diagonal of a new (smaller) square, so we may repeat the same arguments
as above.

Iterating this process, we see that the numbers represented by $AB$ and
$AC$ are indefinitely divisible by 2. This is obviously impossible, and this
contradiction proves that $AB$ and $AC$ are incommensurable.

This result clearly shows that integers are not sufficient to measure
lengths of segments. The right level of generality is that of ratios of lengths.
Prompted by this discovery, the Greeks developed new techniques to oper-
ate with ratios of geometric magnitudes in a logically coherent way, avoiding
the problem of assigning numerical values to these magnitudes. They thus
created a "geometric algebra," which is methodically taught by Euclid in
"The Elements."

By contrast, Babylonians seem not to have been aware of the theoretical difficulties arising from irrational numbers, although these numbers were of course unavoidable in the treatment of geometric problems: they simply replaced them by rational approximations. For instance, the following approximation of $\sqrt{2}$ has been found on some Babylonian tablet: 1.24.51.10, i.e., $1 + 24 \cdot 60^{-1} + 51 \cdot 60^{-2} + 10 \cdot 60^{-3}$ or $1.41421296296296\ldots$, which is accurate up to the fifth place. (See van der Waerden [79, p. 45].)

Although Euclid does not explicitly deal with quadratic equations, the solution of these equations can be detected under a geometric garb in some propositions of the Elements. For instance, Proposition 5 of Book II states [37, v. I, p. 382]:

> If a straight line be cut into equal and unequal segments, the rectangle contained by the unequal segments of the whole together with the square on the straight line between the points of section is equal to the square on the half.

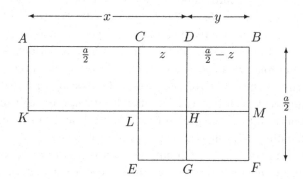

On the figure above, the straight line $AB$ has been cut into equal segments at $C$ and unequal segments at $D$, and the proposition asserts that the rectangle $AH$ together with the square $LG$ (which is equal to the square on $CD$) is equal to the square $CF$. (This is clear from the figure, since the rectangle $AL$ is equal to the rectangle $DF$.)

If we understand that the unequal segments in which the given straight line $AB = a$ is cut are unknown, it appears that this proposition provides us with the core of the solution of the system

$$\begin{cases} x + y = a \\ \quad xy = b. \end{cases}$$

Indeed, setting $z = x - \frac{a}{2}$ "the straight line between the points of section," it states that $b + z^2 = \left(\frac{a}{2}\right)^2$. It then readily follows that

$$z = \sqrt{\left(\frac{a}{2}\right)^2 - b}$$

hence

$$x = \frac{a}{2} + \sqrt{\left(\frac{a}{2}\right)^2 - b} \quad \text{and} \quad y = \frac{a}{2} - \sqrt{\left(\frac{a}{2}\right)^2 - b},$$

as in Babylonian algebra. In subsequent propositions, Euclid also teaches the solution of

$$\begin{cases} x - y = a \\ xy = b \end{cases}$$

which amounts to $x^2 - ax = b$ or $y^2 + ay = b$. He returns to the same type of problems, but in a more elaborate form, in propositions 28 and 29 of book VI. (Compare Kline [45, pp. 76–77] and van der Waerden [79, p. 121].)

The Greek mathematicians of the classical period thus reached a very high level of generality in the solution of quadratic equations, since they considered equations with (positive) *real* coefficients. However, geometric algebra, which was the only rigorous method of operating with real numbers before the nineteenth century, is very difficult. It imposes tight limitations that are not natural from the point of view of algebra; for instance, a great skill in the handling of proportions is required to go beyond degree three.

To progress in the theory of equations, it was necessary to think more about formalism and less about the nature of coefficients. Although later Greek mathematicians such as Hero and Diophantus took some steps in that direction, the really new advances were brought by other civilizations. Hindus, and Arabs later, developed techniques of calculation with irrational numbers, which they treated unconcernedly, without worrying about their irrationality. For instance, they were familiar with formulas like

$$\sqrt{a} + \sqrt{b} = \sqrt{a + b + 2\sqrt{ab}}$$

which they obtained from $(u + v)^2 = u^2 + v^2 + 2uv$ by extracting the square root of each side and replacing $u$ and $v$ by $\sqrt{a}$ and $\sqrt{b}$ respectively. Their notion of mathematical rigor was rather more relaxed than that of Greek mathematicians, but they paved the way to a more formal (or indeed algebraic) approach to quadratic equations (see Kline [45, Ch. 9, §2]).

## 1.3   Arabic algebra

The next landmark in the theory of equations is the book "Al-jabr w'al muqabala" (c. 830 A.D.), due to Mohammed ibn Musa al-Khowarizmi [5], [43].

The title refers to two basic operations on equations. The first is *al-jabr* (from which the word "algebra" is derived) which means "the restoration" or "making whole." In this context, it stands for the restoration of equality in an equation by adding to one side a negative term that is removed from the other. For instance, the equation $x^2 = 40x - 4x^2$ is converted into $5x^2 = 40x$ by *al-jabr* [5, p. 36]. The second basic operation *al muqabala* means "the opposition" or "balancing"; it is a simplification procedure by which like terms are removed from both sides of an equation. For example (see [5, p. 40]), *al muqabala* changes $50 + x^2 = 29 + 10x$ into $21 + x^2 = 10x$.

In this work, al-Khowarizmi initiates what might be called the classical period in the theory of equations, by reducing the old methods for solving equations to a few standardized procedures. For instance, in problems involving several unknowns, he systematically sets up an equation for one of the unknowns, and he solves by completion of the square the three types of quadratic equations

$$x^2 + ax = b, \qquad x^2 + b = ax, \qquad x^2 = ax + b,$$

giving the two (positive) solutions for the type $x^2 + b = ax$.

Al-Khowarizmi first explains the procedure, as a Babylonian would have done:

> *Roots and Squares are equal to Numbers;* for instance, "one square, and ten roots of the same, amount to thirty-nine dirhems;" that is to say, what must be the square which, when increased by ten of its own roots, amounts to thirty-nine? The solution is this: you halve the number of the roots, which in the present instance yields five. This you multiply by itself; the product is twenty-five. Add this to thirty-nine; the sum is sixty-four. Now take the root of this, which is eight, and subtract from it half the number of the roots, which is five; the remainder is three. This is the root of the square which you sought for; the square itself is nine [5, p. 8].

However, after explaining the procedure for solving each of the six types $mx^2 = ax$, $mx^2 = b$, $ax = b$, $mx^2 + ax = b$, $mx^2 + b = ax$ and $mx^2 = ax + b$, he adds:

We have said enough, says Al-Khowarizmi, so far as numbers
are concerned, about the six types of equations. Now, however,
it is necessary that we should demonstrate geometrically the
truth of the same problems which we have explained in numbers
[43, p. 77].

He then gives geometric justifications for his rules for the last three
types, using completion of the square as in the following example for $x^2 +
10x = 39$:

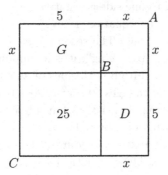

Let $x^2$ be the square $AB$. Then $10x$ is divided into two rectangles $G$
and $D$, each being $5x$ and being applied to the side $x$ of the square $AB$.
By hypothesis, the value of the shape thus produced is $x^2 + 10x = 39$.
There remains an empty corner of value $5^2 = 25$ to complete the square
$AC$. Therefore, if 25 is added, the square $(x + 5)^2$ is completed, and its
value is $39 + 25 = 64$. It then follows that $(x + 5)^2 = 64$, hence $x + 5 = 8$
and $x = 3$ (see [5, p. 15]).

It should be observed that the geometry behind this construction is
much more elementary than in Euclid's Elements, since it is not logically
connected by deductive reasoning to a small number of axioms, but relies
instead on intuitive geometric evidence. From the point of view of algebra,
on the other hand, al-Khowarizmi's work is largely ahead of Euclid's, and
it set the stage for the later development of algebra as an independent
discipline.

Another remarkable achievement of the Arab period is a geometric solution of cubic equations due to Omar Khayyam (c. 1079). For instance, the
solution of $x^3 + b^2x = b^2c$ is obtained by intersecting the parabola $x^2 = by$
with the circle of diameter $c$ that is tangent to the axis of the parabola at
its vertex.

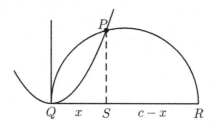

To prove that the segment $x$ as shown on the figure satisfies $x^3 + b^2x = b^2c$, we start from the relation $x^2 = b \cdot PS$, which yields

$$\frac{b}{x} = \frac{x}{PS}.$$

On the other hand, since the triangles $QSP$ and $PSR$ are similar, we have

$$\frac{x}{PS} = \frac{PS}{c - x},$$

hence

$$\frac{b}{x} = \frac{PS}{c - x}.$$

As $PS = b^{-1}x^2$, this equation yields

$$\frac{b}{x} = \frac{x^2}{b(c - x)},$$

hence $x^3 = b^2c - b^2x$, as required.

Omar Khayyam also gives geometric solutions for the other types of cubic equations by intersection of conics (see van der Waerden [80, Ch. 1 C]), but these brilliant solutions are of little use for practical purposes, and an algebraic solution was still longed for.

In 1494, Luca Pacioli closes his influential treatise "Summa de Arithmetica Geometria Proportioni et Proportionalita" (one of the first printed books in mathematics) with the remark that the solutions of $x^3 + mx = n$ and $x^3 + n = mx$ (in modern notations) are as impossible as the quadrature of the circle. (See Kline [45, p. 237], Cardano [13, p. 8].) However, unexpected developments were soon to take place.

# Chapter 2

# Cubic Equations

## 2.1 Priority disputes on the solution of cubic equations

The algebraic solution of $x^3 + mx = n$ was first obtained around 1515 by Scipione del Ferro, professor of mathematics in Bologna. Not much is known about his solution as, for some reason, he decided not to publicize his result. After his death in 1526, his method passed to some of his pupils.

The second discovery of the solution is much better known, through the accounts of its author himself, Niccolò Tartaglia (c. 1499–1557), from Brescia. In 1535, Tartaglia, who had dealt with some very particular cases of cubic equations, was challenged to a public problem-solving contest by Antonio Maria Fior, a former pupil of Scipione del Ferro. When he heard that Fior had received the solution of cubic equations from his master, Tartaglia threw all his energy and his skill into the struggle. He succeeded in finding the solution just in time to inflict upon Fior a humiliating defeat.

The news that Tartaglia had found the solution of cubic equations reached Girolamo Cardano (1501–1576), a very versatile scientist, who wrote a number of books on a wide variety of subjects, including medicine, astrology, astronomy, philosophy, and mathematics. Cardano then asked Tartaglia to give him his solution, so that he could include it in a comprehensive treatise on mathematics, but Tartaglia flatly refused, because he was himself planning to write a book on this topic. It turns out that Tartaglia later changed his mind upon Cardano's insistence, since in 1539 he handed on to Cardano the solution of $x^3 + mx = n$, $x^3 = mx + n$ and a very brief indication on $x^3 + n = mx$ in verses[1] (see Tartaglia [73, p. 120]):

*Quando chel cubo con le cose apresso*

---

[1] As pointed out by Boorstin (cited by Weeks [84, p. lx]), verses were a useful memorization aid at a time when paper was expensive.

*Se agualia à qualche numero discreto*
*Trouan dui altri differenti in esso*
*Dapoi terrai questo per consueto*
*Ch'el lor produtto sempre sia eguale*
*Al terzo cubo delle cose neto*
*El residuo poi suo generale*
*Delli lor lati cubi bene sottratti*
*Varra la tua cosa principale.*

...

This excerpt gives the formula for $x^3 + mx = n$. The equation is indicated in the first two verses: the cube and the things equal to a number. *Cosa* (= thing) is the word for the unknown. To express the fact that the unknown is multiplied by a coefficient, Tartaglia simply uses the plural form *le cose*. He then gives the following procedure: find two numbers which differ by the given number and such that their product is equal to the cube of the third of the number of things. Then the difference between the cube roots of these numbers is the unknown.

With modern notation, we would write that, to find the solution of

$$x^3 + mx = n,$$

we only need to find $t$, $u$ such that

$$t - u = n \quad \text{and} \quad tu = \left(\frac{m}{3}\right)^3;$$

then

$$x = \sqrt[3]{t} - \sqrt[3]{u}.$$

The values of $t$ and $u$ are easily found (see the system (1.1), p. 3):

$$t = \sqrt{\left(\frac{n}{2}\right)^2 + \left(\frac{m}{3}\right)^3} + \frac{n}{2},$$

$$u = \sqrt{\left(\frac{n}{2}\right)^2 + \left(\frac{m}{3}\right)^3} - \frac{n}{2}.$$

Therefore, a solution of $x^3 + mx = n$ is given by the following formula:

$$x = \sqrt[3]{\sqrt{\left(\frac{n}{2}\right)^2 + \left(\frac{m}{3}\right)^3} + \frac{n}{2}} - \sqrt[3]{\sqrt{\left(\frac{n}{2}\right)^2 + \left(\frac{m}{3}\right)^3} - \frac{n}{2}}. \qquad (2.1)$$

However, the poem does not provide any justification for this formula. Of course, it "suffices" to check that the value of $x$ in (2.1) satisfies $x^3 + mx = n$, but this approach was far beyond the reach of mathematicians of the time, who were using rudimentary notation (see Ch. 4). Even an easy formula like

$$(a - b)^3 = a^3 - 3a^2b + 3ab^2 - b^3$$

could be properly proved only by dissection of a cube in three-dimensional space (see Cardano [13, Ch. VI, Cor. 1]).

Having received Tartaglia's poem, Cardano set to work; he not only found justifications for the formulas[2] but he also solved all the other types of cubics. He then published his results in 1545, giving due credit to Tartaglia and to del Ferro, in the epoch-making book "Artis Magnae, sive de regulis algebraicis" (The Great Art, or the Rules of Algebra [13]), usually referred to as "Ars Magna." A bitter quarrel then erupted between Tartaglia and Cardano, the former claiming that Cardano had solemnly sworn never to publish Tartaglia's solution, while the latter countered that the formula was not Tartaglia's exclusive property, as it had previously been found by del Ferro. (See Toscano [74] for an entertaining account of the disputes between Fior, Tartaglia, and Cardano.)

## 2.2  Cardano's formula

Although Cardano lists 13 types of cubic equations and gives a detailed solution for each of them, we shall use modern notation in this section, and explain Cardano's method for the general cubic equation

$$x^3 + ax^2 + bx + c = 0.$$

First, the change of variable $y = x + \frac{a}{3}$ converts the equation into one that lacks the second degree term:

$$y^3 + py + q = 0, \tag{2.2}$$

where

$$p = b - \frac{a^2}{3} \quad \text{and} \quad q = c - \frac{a}{3}b + 2\left(\frac{a}{3}\right)^3. \tag{2.3}$$

If $y = \sqrt[3]{t} + \sqrt[3]{u}$, then

$$y^3 = t + u + 3\sqrt[3]{tu}(\sqrt[3]{t} + \sqrt[3]{u})$$

and equation (2.2) becomes

$$(t + u + q) + (3\sqrt[3]{tu} + p)(\sqrt[3]{t} + \sqrt[3]{u}) = 0.$$

---

[2]To get an impression of the difficulty of this endeavor, see the laborious proof of (2.1) in [13, Ch. XI] or [71, Ch. II, 3].

This equation clearly holds if the rational part $t + u + q$ and the irrational part $(\sqrt[3]{t} + \sqrt[3]{u})(3\sqrt[3]{tu} + p)$ both vanish or, in other words, if

$$\begin{cases} t + u = -q \\ tu = -\left(\frac{p}{3}\right)^3. \end{cases}$$

This system has the solution

$$t, u = -\frac{q}{2} \pm \sqrt{\left(\frac{p}{3}\right)^3 + \left(\frac{q}{2}\right)^2}$$

(see (1.4), p. 4); hence a solution for equation (2.2) is

$$y = \sqrt[3]{-\frac{q}{2} + \sqrt{\left(\frac{p}{3}\right)^3 + \left(\frac{q}{2}\right)^2}} + \sqrt[3]{-\frac{q}{2} - \sqrt{\left(\frac{p}{3}\right)^3 + \left(\frac{q}{2}\right)^2}} \qquad (2.4)$$

and a solution for the initial equation $x^3 + ax^2 + bx + c = 0$ easily follows by substituting for $p$ and $q$ the expressions given by (2.3). Equation (2.4) is known as *Cardano's formula* for the solution of the cubic (2.2).

## 2.3   Developments arising from Cardano's formula

The solution of cubic equations was a remarkable achievement, but Cardano's formula is far less convenient than the corresponding formula for quadratic equations because it has some drawbacks which undoubtedly baffled sixteenth century mathematicians (its discoverers in the first place).

(a) First, when some solution is expected, it is not always yielded by Cardano's formula. This could have struck Cardano when he was devising examples for illustrating his rules, such as

$$x^3 + 16 = 12x$$

(see Cardano [13, p. 12]) which is constructed to give 2 as an answer. Cardano's formula yields

$$x = \sqrt[3]{-8} + \sqrt[3]{-8} = -4.$$

Why does it yield $-4$ and not 2?

It is likely that the above observation had first prompted Cardano to investigate a question much more interesting to him: How many solutions does a cubic equation have? He was thus led to observe that cubic equations may have three solutions (including the negative ones, which Cardano terms "false" or "fictitious," but not the imaginary ones) and to investigate the relations between these solutions (see Cardano [13, Chapter I]).

(b) Next, when there is a rational solution, its expression according to Cardano's formula can be rather awkward. For instance, it is easily seen that 1 is solution of

$$x^3 + x = 2,$$

but Cardano's formula yields

$$x = \sqrt[3]{1 + \tfrac{2}{3}\sqrt{\tfrac{7}{3}}} + \sqrt[3]{1 - \tfrac{2}{3}\sqrt{\tfrac{7}{3}}}.$$

Now, the equation above has only one real root, since the function $f(x) = x^3 + x$ is monotonically increasing (as it is the sum of two monotonically increasing functions) and, therefore, takes the value 2 only once. We are thus compelled to conclude

$$1 = \sqrt[3]{1 + \tfrac{2}{3}\sqrt{\tfrac{7}{3}}} + \sqrt[3]{1 - \tfrac{2}{3}\sqrt{\tfrac{7}{3}}}, \tag{2.5}$$

a rather surprising result.

Already in 1540, Tartaglia tried to simplify the irrational expressions arising in his solution of cubic equations (see Hankel [35, p. 373]). More precisely, he tried to determine under which condition an irrational expression like $\sqrt[3]{a + \sqrt{b}}$ could be simplified to $u + \sqrt{v}$. This problem can be solved as follows (in modern notation): starting with

$$\sqrt[3]{a + \sqrt{b}} = u + \sqrt{v} \tag{2.6}$$

and taking the cube of both sides, we obtain

$$a + \sqrt{b} = u^3 + 3uv + (3u^2 + v)\sqrt{v},$$

hence, equating separately the rational and the irrational parts (which is licit if $a$, $b$, $u$ and $v$ are rational numbers),

$$\begin{cases} a = u^3 + 3uv, \\ \sqrt{b} = (3u^2 + v)\sqrt{v}. \end{cases} \tag{2.7}$$

Subtracting the second equation from the first, we then obtain

$$a - \sqrt{b} = (u - \sqrt{v})^3$$

hence

$$\sqrt[3]{a - \sqrt{b}} = u - \sqrt{v}. \tag{2.8}$$

Multiplying (2.6) and (2.8), we obtain

$$\sqrt[3]{a^2 - b} = u^2 - v \tag{2.9}$$

which can be used to eliminate $v$ from the first equation of system (2.7). We thus get

$$a = 4u^3 - 3(\sqrt[3]{a^2 - b})u.$$

Therefore, if $a$ and $b$ are rational numbers such that $\sqrt[3]{a^2 - b}$ is rational and if the equation

$$4u^3 - 3(\sqrt[3]{a^2 - b})u = a \qquad (2.10)$$

has a rational solution $u$, then

$$\sqrt[3]{a + \sqrt{b}} = u + \sqrt{v} \qquad \text{and} \qquad \sqrt[3]{a - \sqrt{b}} = u - \sqrt{v},$$

where $v$ is given by equation (2.9):

$$v = u^2 - \sqrt[3]{a^2 - b}.$$

This effectively provides a simplification in the irrational expressions

$$\sqrt[3]{-\frac{q}{2} \pm \sqrt{\left(\frac{p}{3}\right)^3 + \left(\frac{q}{2}\right)^2}}$$

that appear in Cardano's formula for the solution of $x^3 + px + q = 0$, but this simplification is useless as far as the solution of cubic equations is concerned. Indeed, if $a = -\frac{q}{2}$ and $b = \left(\frac{p}{3}\right)^3 + \left(\frac{q}{2}\right)^2$, one has to find a rational solution of equation (2.10):

$$4u^3 + pu = -\frac{q}{2},$$

and this exactly amounts to finding a rational solution of the initial equation $x^3 + px + q = 0$, since these equations are related by the change of variable $x = 2u$. However, this process can be used to show, for instance, that

$$\sqrt[3]{1 + \tfrac{2}{3}\sqrt{\tfrac{7}{3}}} = \tfrac{1}{2} + \tfrac{1}{2}\sqrt{\tfrac{7}{3}} \quad \text{and} \quad \sqrt[3]{1 - \tfrac{2}{3}\sqrt{\tfrac{7}{3}}} = \tfrac{1}{2} - \tfrac{1}{2}\sqrt{\tfrac{7}{3}}$$

from which formula (2.5) follows.

**(c)** The most serious drawback of Cardano's formula appears when one tries to solve an equation like

$$x^3 = 15x + 4.$$

It is easily seen that $x = 4$ is a solution, but Cardano's formula yields a very embarrassing expression:

$$x = \sqrt[3]{2 + \sqrt{-121}} + \sqrt[3]{2 - \sqrt{-121}},$$

in which square roots of negative numbers are extracted. The case where $\left(\frac{p}{3}\right)^3 + \left(\frac{q}{2}\right)^2 < 0$ is known as the "casus irreducibilis" of cubic equations. For a long time, the validity of Cardano's formula in this case has been a matter of debate, but the discussion of this case had a very important by-product: it prompted the use of complex numbers.

Complex numbers had been, up to then, brushed aside as absurd, nonsensical expressions. A remarkably explicit example of this attitude appears in the following excerpt from Chapter 37 of the Ars Magna [13, p. 219], [71, Ch. II, 3]:

> If it should be said, Divide 10 into two parts the product of which is 30 or 40, it is clear that this case is impossible. Nevertheless, we will work thus: ...

Cardano then applies the usual procedure with the given data, which amounts to solving $x^2 - 10x + 40 = 0$, and comes up with the solution: these parts are $5 + \sqrt{-15}$ and $5 - \sqrt{-15}$. He then justifies his result:

> Putting aside the mental tortures involved,[3] multiply $5 + \sqrt{-15}$ by $5 - \sqrt{-15}$, making $25 - (-15)$ which is $+15$. Hence this product is 40. [...] So progresses arithmetic subtlety the end of which, as is said, is as refined as it is useless.

However, with the "casus irreducibilis" of cubic equations, complex numbers were imposed upon mathematicians. The operations on these numbers are clearly taught, in a nearly modern way, by Rafaele Bombelli (c. 1526–1573), in his influential treatise: "Algebra" (1572). In this book, Bombelli boldly applies to cube roots of complex numbers the same simplification procedure as in (b) above, and he obtains, for instance

$$\sqrt[3]{2 + \sqrt{-121}} = 2 + \sqrt{-1} \quad \text{and} \quad \sqrt[3]{2 - \sqrt{-121}} = 2 - \sqrt{-1},$$

from which it follows that Cardano's formula gives indeed 4 for a solution of $x^3 = 15x + 4$.

Complex numbers thus appeared, not to solve quadratic equations that lack solutions (and do not need any), but to explain why Cardano's formula, efficient as it may seem, fails in certain cases to provide expected solutions to cubic equations.

---

[3] In the original text: "dismissis incruciationibus." Perhaps Cardano played on words here, since another translation for this passage is: "the cross-multiples having canceled out," referring to the fact that in the product $(5 + \sqrt{-15})(5 - \sqrt{-15})$, the terms $5\sqrt{-15}$ and $-5\sqrt{-15}$ cancel out.

# Chapter 3

# Quartic Equations

## 3.1 The unnaturalness of quartic equations

The solution of quartic equations was found soon after that of cubic equations. It is due to Ludovico Ferrari (1522–1565), a pupil of Cardano, and it first appeared in Cardano's "Ars Magna."

Ferrari's method is very ingenious, relying mainly on transformation of equations, but it aroused less interest than the solution of cubic equations. This is clearly shown by its place in the "Ars Magna": while Cardano spends thirteen chapters to discuss the various cases of cubic equations, Ferrari's method is briefly sketched in the penultimate chapter.

The reason for this relative disregard may be found in the introduction of the "Ars Magna" [13, p. 9]:

> Although a long series of rules might be added and a long dis-
> course given about them, we conclude our detailed consider-
> ation with the cubic, others being merely mentioned, even if
> generally, in passing. For as *positio* [the first power] refers to a
> line, *quadratum* [the square] to a surface, and *cubum* [the cube]
> to a solid body, it would be very foolish for us to go beyond
> this point. Nature does not permit it.

This passage illustrates the equivocal status of algebra in the sixteenth century. Its logical foundations were still geometric, as in the classical Greek period; in this framework, each quantity has a dimension and only quantities of the same dimension can be added or equated. For instance, an equation like $x^2 + b = ax$ makes sense only if $x$ and $a$ are line segments and $b$ is an area, and equations of degree higher than three don't make any sense at all.[1]

---

[1] A way out of this difficulty was eventually found by Descartes. In "La Geometrie" [23,

However, from an arithmetical point of view, quantities are regarded as dimensionless numbers, which can be raised to any power and equated unconcernedly. This way of thought was clearly prevalent among Babylonians, since the very statement of the problem: "I have subtracted from the area the side of my square: 14.30" is utter nonsense from a geometric point of view. The Arabic algebra also stresses arithmetic, although al-Khowarizmi provides geometric proofs of his rules (see §1.3).

In the "Ars Magna," both the geometric and the arithmetic approaches to equations are present. On one hand, Cardano tries to base his results on Euclid's "Elements," and on the other hand, he gives the solution of equations of degree 4. He also solves some equations of higher degree, such as $x^9 + 3x^6 + 10 = 15x^3$ [13, p. 159], in spite of his initial statement that it would be "foolish" to go beyond degree 3. However, the arithmetic approach, which would eventually predominate, still suffered from its lack of a logical base until the early seventeenth century (see §4.1).

## 3.2   Ferrari's method

In this section, we use modern notation to discuss Ferrari's solution of quartic equations. Let
$$x^4 + ax^3 + bx^2 + cx + d = 0$$
be an arbitrary quartic equation. By the change of variable $y = x + \frac{a}{4}$ the cubic term cancels out, and the equation becomes
$$y^4 + py^2 + qy + r = 0, \tag{3.1}$$
with
$$p = b - 6\left(\frac{a}{4}\right)^2,$$
$$q = c - \frac{a}{2}b + \left(\frac{a}{2}\right)^3, \tag{3.2}$$
$$r = d - \frac{a}{4}c + \left(\frac{a}{4}\right)^2 b - 3\left(\frac{a}{4}\right)^4.$$
Moving the linear terms of (3.1) to the right-hand side and completing the square on the left-hand side, we obtain
$$\left(y^2 + \frac{p}{2}\right)^2 = -qy - r + \left(\frac{p}{2}\right)^2.$$

p. 5], published in 1637, he introduces the following convention: if a unit line segment $e$ is chosen, then the square $x^2$ of a line segment $x$ is the side of the rectangle constructed on $e$ which has the same area as the square with side $x$. Thus, $x^2$ is a line segment, and arbitrary powers of $x$ can be interpreted as line segments in a similar way.

If we add a quantity $u$ to the expression squared in the left-hand side, we get

$$\left(y^2 + \frac{p}{2} + u\right)^2 = -qy - r + \left(\frac{p}{2}\right)^2 + 2uy^2 + pu + u^2. \qquad (3.3)$$

The idea is to determine $u$ in such a way that the right-hand side also becomes a square. Looking at the terms in $y^2$ and in $y$, it is easily seen that if the right-hand side is a square, then it is the square of $\sqrt{2u}y - \frac{q}{2\sqrt{2u}}$; therefore, we should have

$$-qy - r + \left(\frac{p}{2}\right)^2 + 2uy^2 + pu + u^2 = \left(\sqrt{2u}y - \frac{q}{2\sqrt{2u}}\right)^2. \qquad (3.4)$$

Equating the constant terms, we see that this equation holds if and only if

$$-r + \left(\frac{p}{2}\right)^2 + pu + u^2 = \frac{q^2}{8u}$$

or equivalently, after clearing the denominator and rearranging terms,

$$8u^3 + 8pu^2 + (2p^2 - 8r)u - q^2 = 0. \qquad (3.5)$$

Therefore, by solving this cubic equation, we can find a quantity $u$ for which equation (3.4) holds. Returning to equation (3.3), we then have

$$\left(y^2 + \frac{p}{2} + u\right)^2 = \left(\sqrt{2u}y - \frac{q}{2\sqrt{2u}}\right)^2,$$

hence

$$y^2 + \frac{p}{2} + u = \pm\left(\sqrt{2u}y - \frac{q}{2\sqrt{2u}}\right).$$

The values of $y$ are then obtained by solving the two quadratic equations above (one corresponding to the sign $+$ for the right-hand side, the other to the sign $-$).

To complete the discussion, it remains to consider the case where $u = 0$ is a root of equation (3.5), since the calculations above implicitly assume $u \neq 0$. But this case occurs only if $q = 0$ and then the initial equation (3.1) is

$$y^4 + py^2 + r = 0.$$

This equation is easily solved, since it is a quadratic equation in $y^2$.

In summary, the solutions of

$$x^4 + ax^3 + bx^2 + cx + d = 0$$

are obtained as follows: let $p$, $q$, and $r$ be defined as before (see (3.2)) and let $u$ be a solution of (3.5). If $q \neq 0$, the solutions of the initial quartic equation are

$$x = \varepsilon\sqrt{\frac{u}{2}} + \varepsilon'\sqrt{-\frac{u}{2} - \frac{p}{2} - \frac{\varepsilon q}{2\sqrt{2u}}} - \frac{a}{4}$$

where $\varepsilon$ and $\varepsilon'$ can be independently $+1$ or $-1$. If $q = 0$, the solutions are

$$x = \varepsilon\sqrt{-\frac{p}{2} + \varepsilon'\sqrt{\left(\frac{p}{2}\right)^2 - r}} - \frac{a}{4}$$

where $\varepsilon$ and $\varepsilon'$ can be independently $+1$ or $-1$.

Equation (3.5), on which the solution of the quartic equation depends, is called the *resolvent cubic* equation (relative to the given quartic equation). Depending on the way equations are set up, one may come up with other resolvent cubic equations. For instance, from equation (3.1) one could pass to

$$(y^2 + v)^2 = (-py^2 - qy - r) + 2vy^2 + v^2$$

where $v$ is an arbitrary quantity (which plays the same role as $\frac{p}{2} + u$ in the preceding discussion). The condition on $v$ for the right-hand side to be a perfect square is then

$$8v^3 - 4pv^2 - 8rv + 4pr - q^2 = 0. \tag{3.6}$$

After having determined $v$ such that this condition holds, one finishes as before.

This second method is clearly equivalent to the previous one, by the change of variable $v = \frac{p}{2} + u$. Therefore, equation (3.6), which is obtained from (3.5) by this change of variable, is also entitled to be called a resolvent cubic.

# Chapter 4

# The Creation of Polynomials

## 4.1 The rise of symbolic algebra

In comparison with the rapid development of the theory of equations around the middle of the sixteenth century, progress during the next two centuries was rather slow. The solution of cubic and quartic equations was a very important breakthrough, and it took some time before the circle of new ideas arising from these solutions was fully explored and understood, and new advances were possible.

First of all, it was necessary to devise appropriate notations for handling equations. In the solution of cubic and quartic equations, Cardano was straining to the utmost the capabilities of the algebraic system available to him. Indeed, his notation was rudimentary: the only symbolism he uses consists in abbreviations such as p: for *plus,* m: for *minus* and R for *radix* (= root).

For instance, the equation $X^2 + 2X = 48$ is written as

$$1. \text{ quad. p: 2 pos. aeq. } 48$$

(quad. is for *quadratum,* pos. for *positiones* and aeq. for *aequatur* or *aequalia*), and

$$(5 + \sqrt{-15})(5 - \sqrt{-15}) = 25 - (-15) = 40$$

is written (see Cajori [12, §140], Struik [71, Ch. II, 3])

$$\begin{array}{l} 5 \text{ p:R m: } 15 \\ \underline{5 \text{ m:R m: } 15} \\ 25 \text{ m:m: } 15 \quad \tilde{q}\text{d. est } 40. \end{array}$$

Using this embryonic notation, transformation of equations was clearly a tour de force, and a more efficient notation had to develop in order to illuminate this new part of algebra.

This development was rather erratic. Advances made by some authors were not immediately taken up by others, and the process of normalization of notations took a long time. For example, the symbols + and − were already used in Germany since the end of the fifteenth century (Cajori [12, §201]), but they were not widely accepted before the early seventeenth century, and the sign = for equality, first proposed by Recorde in 1557, had to struggle with Descartes' symbol ∞ for nearly two centuries (Cajori [12, §267]).

These are relatively minor points, since it may be assumed that p:, m: and aeq. were as convenient to Cardano as +, − and = are to us. There is one point, however, where a new notation was vital. In effect, it helped create a new mathematical object: polynomials. There is indeed a significant step from

$$1. \text{ quad. p: 2 pos. aeq. } 48$$

which is the mere statement of a problem, to the calculation with polynomials like $X^2 + 2X - 48$, and this step was considerably facilitated by a suitable notation.

Significant as it may be, the evolution from equations to polynomials is rather subtle, and leading mathematicians of this period rarely took the pain to clarify their views on the subject; the rise of the concept of polynomial was most often overshadowed by its application to the theory of equations, and it can only be gathered from indirect indications.

Two milestones in this evolution are "L'Arithmetique" (1585) of Simon Stevin (1548–1620) and "In Artem Analyticem Isagoge" (= Introduction to the Analytic Art) (1591) of François Viète (1540–1603).

### 4.1.1  *L'Arithmetique*

This book combines notational advances made by Bombelli and earlier authors (see Cajori [12, §296]) with theoretical advances made by Pedro Nunes (1502–1578) (see Bosmans [6, p. 165]) to present a comprehensive treatment of polynomials. Stevin's notation for polynomials, which he terms "multinomials" [69, p. 521] or "integral algebraic numbers" [69, p. 518, pp. 570 ff] has a surprising touch of modernity: the indeterminate is denoted by ①, its square by ②, its cube by ③, etc., and the constant term is indicated by ⓪ (sometimes omitted), so that a "multinomial" appears as an expression like

$$3③ + 5② - 4① + 6⓪ \qquad (\text{or } 3③ + 5② - 4① + 6).$$

Such an expression could be regarded (from a modern point of view) as a finite sequence of real numbers, or, better, as a sequence of real numbers which are all zero except for a finite number of them, or as a function from $\mathbb{N}$ to $\mathbb{R}$ with finite support (compare §5.1).

Variants of Stevin's economical notation for the powers of the unknown had in fact already been used in a few isolated texts: as early as 1484, Nicolas Chuquet writes $.12.^2$ m̃. $8^0$ for $12x^2 - 8$ in his "Triparty en la science des nombres" [16, p. 154] (see Cajori [12, §131], Struik [71, Ch. II, 2]). Also, in Bombelli's "Algebra" of 1572,

$$4 \underset{\smile}{2} \text{m.16.} \underset{\smile}{1} \text{p.16}$$

stands for $4x^2 - 16x + 16$ (see Cajori [12, §144]). This notational advance eventually abolished the psychological barrier of the third degree (see §3.1), by placing all the powers of the unknown on an equal footing. It is however rather impractical for polynomials with several unknowns.

Most important is Stevin's observation that the operations on "integral algebraic numbers" share many features with those on "integral arithmetic numbers" (= integers). In particular, he shows [69, Probleme 53, p. 577] that Euclid's algorithm for determining the greatest common divisor of two integers applies nearly without change to find the greatest common divisor of two polynomials (see §5.2, and particularly p. 45).

Although the concept of polynomial is quite clear, the way equations are set up is rather awkward in "L'Arithmetique," because equations are replaced by proportions and the solution of equations is called by Stevin "the rule of three of quantities." In modern notation, the idea is to replace the solution of an equation like

$$X^2 - aX - b = 0$$

by the following problem: find the fourth proportional $u$ in

$$\frac{X^2}{aX + b} = \frac{X}{u}$$

or, more generally, find $P(u)$ in

$$\frac{X^2}{aX + b} = \frac{P(X)}{P(u)},$$

where $P(X)$ is an arbitrary polynomial in $X$, see [69, p. 592]. (Of course, the solutions that are equal to zero must then be rejected.)

In Stevin's words [69, p. 595]:

> Given three terms, of which the first ②, the second ① ⓪, the third an arbitrary algebraic number: To find their fourth proportional term.

This fancy approach to equations may have been prompted by Stevin's methodical treatment of polynomials: an equality like

$$X^2 = aX + b$$

would mean that the *polynomials* $X^2$ and $aX + b$ are equal; but polynomials are equal if and only if the coefficients of like powers of the unknown are the same in both polynomials, and this is clearly not the case here, since $X^2$ appears on the left-hand side but not on the right-hand side. To solve the equation is really to find a *value* $u$ of the "algebraic number" $X$ such that $u^2 = au + b$. Stevin's own explanation [69, pp. 581–582], while not quite illuminating, at least shows that he was fully aware of this notational difficulty and made a clear distinction between a quantity and its value:

> The reason why we call rule of three, or invention of the fourth proportional of quantities, that which is commonly called equation of quantities: [...] Because this word equation has made beginners think that it was some singular matter, which however is common in ordinary arithmetic, since we seek to three given terms a fourth proportional. As that, which is called equation, does not consist in the equality of absolute quantities, but in the equality of their values, so this proportion is concerned with the value of quantities [...].

This approach met with little success. Even Albert Girard (1595–1632), the first editor of Stevin's works, did not follow Stevin's set up in his own work, and it was soon abandoned.

Stevin's formal treatment of polynomials is rather isolated too; in later works, polynomials were most often considered as functions, although formal operations like Euclid's algorithm were performed. For instance, here is the definition of an equation according to René Descartes (1596–1650) [23, p. 156]:

> An equation consists of several terms, some known and some unknown, some of which are together equal to the rest; or rather, all of which taken together are equal to nothing; for this is often the best form to consider.

A polynomial then appears as "the sum of an equation" [23, p. 159].

In practical terms, the concept of polynomial in the seventeenth century is not very different from the modern notion, and the need for a more formal definition was not felt for a long time, but one can get some idea of the difference between Descartes' view and ours from the following excerpt of "La Geometrie" (1637) [23, p. 159]: (the emphasis is mine)

> Multiplying together the two *equations* $x - 2 = 0$ and $x - 3 = 0$, we have $x^2 - 5x + 6 = 0$, or $x^2 = 5x - 6$. This is an equation in which $x$ has the value 2 and *at the same time* the value 3.

### 4.1.2 *In Artem Analyticem Isagoge*

A major advance in notation with far-reaching consequences was Viète's idea, put forward in his "Introduction to the Analytic Art" (1591) [83], of designating by letters all the quantities, known or unknown, occurring in a problem. Although letters were occasionally used for unknowns as early as the third century A.D. (by Diophantus of Alexandria, see Cajori [12, §101]), the use of letters for known quantities was very new. It also proved to be very useful, since for the first time it was possible to replace various numerical examples by a single "generic" example, from which all the others could be deduced by assigning values to the letters. However, it should be observed that this progress did not reach its full extent in Viète's works, since Viète completely disregarded negative numbers; therefore, his letters always stood for *positive* numbers only.

This slight limitation notwithstanding, the idea had another important consequence: by using symbols as his primary means of expression and showing how to calculate with those symbols, Viète initiated a completely formal treatment of algebraic expressions, which he called *logistice speciosa* [83, p. 17] (as opposed to the *logistice numerosa*, which deals with numbers). This "symbolic logistic" gave some substance, some legitimacy to algebraic calculations, which allowed Viète to free himself from the geometric diagrams used so far as justifications.

On the other hand, Viète's calculations are somewhat hindered by his insistence that each coefficient in an equation be endowed with a dimension, in such a way that all the terms have the same dimension: the "prime and perpetual law of equations" is that "homogeneous terms must be compared with homogeneous terms" [83, p. 15]. Moreover, Viète's notation is not as advanced as it could be, since he does not use numerical exponents. For instance, instead of

$$\text{Let } A^3 + 3BA = 2Z,$$

Viète writes (see Cajori [12, §177])

$$\text{Proponatur } A \text{ cubus} + B \text{ plano 3 in } A \text{ aequari } Z \text{ solido 2,}$$

where the qualifiers *plano* and *solido* indicate that $B$ and $Z$ have degree 2 and 3 respectively (for homogeneity).

These minor flaws were soon corrected. Thomas Harriot in "Artis analyticae praxis" (1631) (see Stedall [68, Ch. 2]) and René Descartes in "La Geometrie" (1637) [23] shaped the notation that is still in use today (except for Descartes' sign $\infty$ for equality). Thus, in less than one century, algebraic notation had dramatically improved, reaching the same level of generality and the same versatility as ours. These notational advances fostered a deeper understanding of the nature of equations, and the theory of equations was soon advanced in some important points, such as the number of roots and the relations between roots and coefficients of an equation.

## 4.2  Relations between roots and coefficients

Cardano's observations on the number of roots of cubic and quartic equations (see §2.3) were substantially generalized during the next century. Progress in plane trigonometry brought rather unexpected insights into this question.

In 1593, at the end of the preface of his book "Ideae Mathematicae" [61] (see also Goldstine [34, §1.6]), Adriaan van Roomen[1] (1561–1615) issued the following challenge to "all the mathematicians throughout the whole world":

Find the solution of the equation

$$45X - 3795X^3 + 95634X^5 - 1138500X^7 + 7811375X^9 - 34512075X^{11}$$
$$+ 105306075X^{13} - 232676280X^{15} + 384942375X^{17} - 488494125X^{19}$$
$$+ 483841800X^{21} - 378658800X^{23} + 236030652X^{25} - 117679100X^{27}$$
$$+ 46955700X^{29} - 14945040X^{31} + 3764565X^{33} - 740259X^{35}$$
$$+ 111150X^{37} - 12300X^{39} + 945X^{41} - 45X^{43} + X^{45} = A.$$

He gave the following examples, the second of which is erroneous:

---

[1]then professor at the University of Louvain.

(a) if $A = \sqrt{2 + \sqrt{2 + \sqrt{2 + \sqrt{2}}}}$, then $X = \sqrt{2 - \sqrt{2 + \sqrt{2 + \sqrt{2 + \sqrt{3}}}}}$;

(b) if $A = \sqrt{2 + \sqrt{2 - \sqrt{2 - \sqrt{2 - \sqrt{2 - \sqrt{2}}}}}}$

$$\text{[it should be } A = \sqrt{2 - \sqrt{2 - \sqrt{2 + \sqrt{2 + \sqrt{2}}}}}\text{]},$$

then $X = \sqrt{2 - \sqrt{2 + \sqrt{2 + \sqrt{2 + \sqrt{2 + \sqrt{2}}}}}}$

$$\text{[it should be } X = \sqrt{2 - \sqrt{2 + \sqrt{2 + \sqrt{2 + \sqrt{2 + \sqrt{3}}}}}}\text{]};$$

(c) if $A = \sqrt{2 + \sqrt{2}}$

$$= \sqrt{3.41421356237309504880168872420969807856969671875375},$$

then

$$X = \sqrt{2 - \sqrt{2 + \sqrt{\frac{3}{16}} + \sqrt{\frac{15}{16}} + \sqrt{\frac{5}{8} - \sqrt{\frac{5}{64}}}}}$$

$$= \sqrt{0.00274093049085225243101588312112683881 80},$$

and he asked for a solution when

$$A = \sqrt{\frac{7}{4} - \sqrt{\frac{5}{16}} - \sqrt{\frac{15}{8} - \sqrt{\frac{45}{64}}}}.$$

Of course, this was not just any 45th degree equation; its coefficients had been very carefully chosen.

# PROBLEMA MATHEMATICVM
*omnibus totius orbis Mathematicis ad conſtruendū propoſitum.*

S I duorum terminorum prioris ad poſteriorem proportio ſit, ut 1 (1) ad 45 (1) -- 3795 (3) + 9, 5634 (5) -- 113, 8500 (7) + 781, 1375 (9) -- 3451, 2075 (11) + 1, 0530, 6075 (13) -- 2, 3267, 6280 (15) + 3, 8494, 2375 (17) -- 4, 8849, 4125 (19) + 4, 8384, 1800 (21) -- 3, 7865, 8800 (23) + 2, 3603, 0652 (25) -- 1, 1767, 9100 (27) + 4695, 5700 (29) -- 1494, 5040 (31) + 376, 4565 (33) -- 74, 0259 (35) + 11, 1150 (37) -- 1, 2300 (39) + 945 (41) -- 45 (43) + 1 (45), decurque terminus poſterior, invenire priorem.

### Exemplum primum datum.

S It terminus poſterior r bin. 2 + r bin. 2 + r bin 2 + r 2. quæritur terminus prior. SoLVTIO. Dico terminū priorem eſſe r bin. 2 - r bin. 2 + r bin, 2 + r bin. 2 + r 3.

### Exemplum secundum datum.

Sit terminus poſterior r bin. 2 + r bin. 2 - r bin. 2 - r bin. 2 - r bin. 2 - r 2. quæritur terminus prior. SOLVTIO. Terminus prior eſt r bin. 2 - r bin. 2 + r bin. 2 + r bin. 2 + r bin. 2 + r 2.

### Exemplum tertium datum.

Sit terminus poſterior r bin. 2 + r 2, quæritur terminus prior.
SoL. Terminus prior eſt r bin. 2 - r quadrin. 2 + $r\frac{3}{16}$ + $r\frac{15}{16}$ + r bin. $\frac{5-r\ 5}{8\ 64}$

Si in numeris abſolutis ſolinomijs id proponere libuerit : Sit poſterior terminus r 3 $\frac{4142,1356,2373,0950,4880,1688,7242,0969,8078,5696,7187,5375.}{10000,0000,0000,0000,0000,0000,0000,0000,0000,0000,0000,0000.}$
Quæritur terminus prior. SOLVTIO. Terminus prior erit
r $\frac{27,4093,0490,8522,5245,1015,8831,2112,6838,8180.}{10000,0000,0000,0000,0000,0000,0000,0800,0000,0000.}$

### EXEMPLVM QVÆSITVM.

S It poſterior terminus r trinomia 1 $\frac{3}{4}$ -- r $\frac{5}{16}$ -- r bin. 1 $\frac{7}{8}$ -- r $\frac{45}{64}$. quæritur terminus prior . Hoc exemplum omnibus Mathematicis ad conſtruendum ſit propoſitum . Non dubito quin *Ludolf van Cöllen* ejus ſolutionem, ſaltem in numeris ſolinomijs ſit inventurus.

M E-

When this problem was submitted to Viète, he recognized that the left-hand side of the equation is the polynomial by which $2\sin 45\alpha$ is expressed as a function of $2\sin\alpha$ (see equation (4.2) below, p. 34). Therefore, it suffices to find an arc $\alpha$ such that $2\sin 45\alpha = A$, and the solution of Van Roomen's equation is $X = 2\sin\alpha$. In Van Roomen's examples:

(a) $A = 2\sin\frac{15\pi}{25}$, and $X = 2\sin\frac{\pi}{2^5 \cdot 3}$,

(b) $A$ should be $2\sin\frac{15\pi}{2^6}$, and $X = 2\sin\frac{\pi}{2^6\cdot3}$,

(c) $A = 2\sin\frac{3\pi}{2^5}$, and $X = 2\sin\frac{\pi}{2^3\cdot3\cdot5}$,

and in the proposed problem $A = 2\sin\frac{\pi}{3\cdot5}$, hence $X = 2\sin\frac{\pi}{3^3\cdot5^2}$. That Van Roomen's examples correspond to these arcs can be verified by the formulas

$$2\sin\tfrac{\alpha}{2} = \pm\sqrt{2 - 2\cos\alpha} \qquad 2\cos\tfrac{\alpha}{2} = \pm\sqrt{2 + 2\cos\alpha}$$

$$2\cos\tfrac{\pi}{3} = 1 \qquad 2\sin\tfrac{\pi}{3} = \sqrt{3}$$

$$2\cos\tfrac{2\pi}{5} = \tfrac{\sqrt{5}-1}{2} \qquad 2\sin\tfrac{2\pi}{5} = \sqrt{\tfrac{5+\sqrt{5}}{2}}$$

(see also Remark 7.6(b)). From these last results, the value of $\sin\frac{\pi}{15}$ and of $\cos\frac{\pi}{15}$ can be calculated by the addition formulas, since $\frac{\pi}{15} = \frac{2\pi}{5} - \frac{\pi}{3}$.

It turns out that the solution $2\sin\frac{\pi}{3^3\cdot5^2}$ of Van Roomen's equation could also be expressed by radicals, but this expression, which does not involve square roots only, is of little use for the determination of its numerical value since it requires the extraction of roots of complex numbers. Only the numerical value, suitably approximated to the ninth place, is given by Viète. But Viète does not stop there. While Van Roomen asked for *the* solution of his 45th degree equation, Viète shows that this equation has 23 positive solutions, and, in passing, he points out that it also has 22 negative solutions [82, Cap. 6]. Indeed, if $\alpha$ is an arc such that $2\sin45\alpha = A$, then, letting

$$\alpha_k = \alpha + k\frac{2\pi}{45},$$

one also has $2\sin45\alpha_k = A$ for all $k = 0, 1, \ldots, 44$, so that $2\sin\alpha_k$ is a solution of Van Roomen's equation. If $A \geq 0$ (and $A \leq 2$), then one can choose $\alpha$ between 0 and $\frac{\pi}{90}$, hence

$$2\sin\alpha_k \begin{cases} \geq 0 & \text{for } k = 0, \ldots, 22, \\ < 0 & \text{for } k = 23, \ldots, 44. \end{cases}$$

Another interesting feature of Viète's brilliant solution [82] is that, instead of solving directly Van Roomen's equation, which amounts, as we have seen, to the division of an arc into 45 parts, Viète decomposes the problem: since $45 = 3^2 \cdot 5$, the problem can be solved by the trisection of the arc, followed by the trisection of the resulting arc and the division into five parts of the arc thus obtained. As Viète shows, $2\sin n\alpha$ is given as a function of $2\sin\alpha$ by an equation of degree $n$, for $n$ odd (see equation (4.2) below), hence the solutions of Van Roomen's equation of degree 45 can be obtained by solving successively two equations of degree 3 and one equation of degree 5. This idea of solving an equation step by step was to play a central role in Lagrange's and Gauss' investigations, two hundred years later (see Chapters 10 and 12).

In modern language, Viète's results on the division of arcs can be stated as follows: for any integer $n \geq 1$, let $\lfloor \frac{n}{2} \rfloor$ be the largest integer which is less than (or equal to) $\frac{n}{2}$, and define[2]

$$f_n(X) = \sum_{i=0}^{\lfloor \frac{n}{2} \rfloor} (-1)^i \frac{n}{n-i} \binom{n-i}{i} X^{n-2i}$$

where $\binom{n-i}{i} = \frac{(n-i)!}{i!(n-2i)!}$ is the binomial coefficient. Then, for all $n \geq 1$,

$$2\cos n\alpha = f_n(2\cos\alpha) \tag{4.1}$$

and for all *odd* $n \geq 1$,

$$2\sin n\alpha = (-1)^{(n-1)/2} f_n(2\sin\alpha). \tag{4.2}$$

Formula (4.1) can be proved by induction on $n$, using

$$2\cos(n+1)\alpha = (2\cos\alpha)(2\cos n\alpha) - 2\cos(n-1)\alpha \tag{4.3}$$

and (4.2) is easily deduced from (4.1), by applying (4.1) to $\beta = \frac{\pi}{2} - \alpha$, which is such that $\cos\beta = \sin\alpha$. (The original formulation is not quite so general, but Viète shows how to compute recursively the coefficients of $f_n$, see [82, Cap. 9], [83, pp. 432 ff] or Goldstine [34, §1.6].)

For each integer $n \geq 1$, the equation

$$f_n(X) = A \qquad \cdot$$

has degree $n$, and the same arguments as for Van Roomen's equation show that this equation has $n$ solutions (at least when $|A| \leq 2$). These examples, which are quite explicit for $n = 3, 5, 7$ in Viète's works [82, Cap. 9], [83, pp. 445 ff], may have been influential in the progressive emergence of the idea that equations of degree $n$ have $n$ roots, although this idea was still somewhat obscured by Viète's insistence on considering positive roots only (see also §6.1).

In later works, such as "De Recognitione Aequationum" (On Understanding Equations), published posthumously in 1615, Viète also stressed the importance of understanding the structure of equations, meaning by

---

[2]The polynomials $f_n$ are related to the Chebyshev polynomials of the first kind $T_n$ by $f_n(2X) = 2T_n(X)$. (The polynomials $f_n$ are all monic, whereas $T_n$ is not monic for $n \geq 2$.)

this the relations between roots and coefficients. However, the theoretical tools at his disposal were not sufficiently developed, and he failed to grasp these relations in their full generality. For example, he shows [83, pp. 210–211] that if an equation[3]

$$B^p A - A^3 = Z^s \qquad (4.4)$$

(in the indeterminate $A$) has two roots $A$ and $E$, then assuming $A > E$, one has

$$B^p = A^2 + E^2 + AE,$$
$$Z^s = A^2 E + E^2 A.$$

The proof is as follows: since

$$B^p A - A^3 = Z^s \qquad \text{and} \qquad B^p E - E^3 = Z^s,$$

one has $B^p A - A^3 = B^p E - E^3$, hence

$$B^p (A - E) = A^3 - E^3$$

and, dividing both sides by $A - E$,

$$B^p = A^2 + E^2 + AE.$$

The formula for $Z^s$ is then obtained by substituting for $B^p$ in the initial equation (4.4).

The structure of equations was eventually discovered in its proper generality and its simplest form by Albert Girard (1595–1632), and published in "Invention nouvelle en l'algebre" (1629) [33].

> Since the theorem which will follow requires some new expressions, we shall begin with definitions [71, p. 82], [33, p. E2 v°].

Girard calls an equation *incomplete* if it lacks at least one term (i.e., if at least one of the coefficients is zero); the various terms are called *mixed* ("meslés") and the last is called the *closure*. The *first faction* of the solutions is their sum, the *second faction* is the sum of their products two by two, the *third* is the sum of their products three by three and the *last* is their product. Finally, an equation is in the *alternate order* when the odd powers of the unknown are on one side of the equality and the even powers on the other side, and when moreover the coefficient of the highest power is 1.

---

[3]The superscripts of $B$ and $Z$ indicate the dimensions: $p$ is for *plano* and $s$ for *solido*.

Girard's main theorem is then

> All equations of algebra receive as many solutions as the denomination of the highest term shows, except the incomplete, and the first faction of the solutions is equal to the number of the first mixed, their second faction is equal to the number of the second mixed; their third to the third mixed, and so on, so that the last faction is equal to the closure, and this according to the signs that can be observed in the alternate order [71, p. 85], [33, p. E4].

The restriction to complete equations is not easy to explain. Half a page later, Girard points out that incomplete equations have not always as many solutions, and that in this case some solutions are imaginary ("impossible" is Girard's own word). However, it is clear that even complete equations may have imaginary solutions (consider for instance $x^2 + x + 1 = 0$), and this fact could not have escaped Girard.

At any rate, Girard claims that the relations between roots and coefficients also hold in this case, provided that the equation be completed by adding powers of the unknown with coefficient 0. Therefore, the theorem asserts that each equation

$$X^n + s_2 X^{n-2} + s_4 X^{n-4} + \cdots = s_1 X^{n-1} + s_3 X^{n-3} + s_5 X^{n-5} + \cdots$$

or

$$X^n - s_1 X^{n-1} + s_2 X^{n-2} - s_3 X^{n-3} + \cdots + (-1)^n s_n = 0$$

has $n$ roots $x_1, \ldots, x_n$ such that

$$s_1 = \sum_{i=1}^{n} x_i, \quad s_2 = \sum_{i<j} x_i x_j, \quad s_3 = \sum_{i<j<k} x_i x_j x_k, \quad \ldots, \tag{4.5}$$

$$s_n = x_1 x_2 \ldots x_n.$$

It should be observed that this "theorem" is not nearly as precise as the modern formulation of the fundamental theorem of algebra, since Girard does not explicitly assert that the roots are of the form $a + b\sqrt{-1}$. It is therefore more a *postulate* than a theorem: it claims the existence of "impossible" roots of polynomials, but it is essentially unprovable[4] since nothing else is said about these roots, except (implicitly) that one can calculate with them as if they were numbers. Of course, in all the examples, it turns out that the impossible roots are of the form $a + b\sqrt{-1}$, but Girard nowhere explains what he has in mind.

---

[4] At least, the proof and, for that matter, even a correct statement of the theorem were very far beyond the reach of seventeenth century mathematicians: see §9.1.

> If someone were to ask what is the purpose of the solutions that
> are impossible, then I answer in three ways: for the certitude of
> the general rule, and the fact that there are no other solutions,
> and for its use [71, p. 86], [33, p. F r°].

Girard elaborates as follows on the utility of impossible roots: if one
seeks the (positive) values of $(x + 1)^2 + 2$, where $x$ is such that $x^4 = 4x - 3$, then since the solutions of this last equation are $1$, $1$, $-1 + \sqrt{-2}$
and $-1 - \sqrt{-2}$, one gets $(x + 1)^2 + 2 = 6$, $6$, $0$ or $0$, so that $6$ is the unique
result. One would never have been so sure of that without the impossible
roots.

Of course, Girard does not provide the faintest hint of a proof of his
theorem. It would have been very interesting to see at least how he found
the relations (4.5) between the roots and the coefficients of an equation.
These relations readily follow from the identification of coefficients in

$$X^n - s_1 X^{n-1} + s_2 X^{n-2} - \cdots + (-1)^n s_n = (X - x_1)(X - x_2) \cdots (X - x_n)$$

but this equation was probably not known to Girard. (It may have been first
considered by Harriot around 1605; see Stedall [68, Ch. 2].) Indeed, Girard
does not seem to have been aware of the fact that a number $a$ is a root of
a polynomial $P(X)$ if and only if $X - a$ divides $P(X)$: this observation is
usually credited to Descartes [23, p. 159], although Nunes may have been
aware of it as early as 1567, and perhaps earlier (see Bosmans [6, pp. 163 ff]).

On the subject of impossible roots, Descartes first seems more cautious
than Girard (the emphasis is mine):

> Every equation *can* have as many distinct roots as the number
> of dimensions of the unknown quantity in the equation [23,
> p. 159].

This at least can be proved by Descartes' preceding observation (see
Theorem 5.15, p. 52). However, Descartes further states:

> Neither the true nor the false roots are always real; sometimes
> they are imaginary; that is, while we can always conceive of as
> many roots for each equation as I have already assigned, yet
> there is not always a definite quantity corresponding to each
> root so conceived of [23, p. 175].

Around the middle of the seventeenth century, the fact that the number
of solutions of an equation is equal to the degree was becoming common
knowledge, like a piece of mathematical "folklore," accepted without proof
and never questioned. At least, it was a good working hypothesis, and

mathematicians started to calculate formally with roots of equations without worrying about their nature, pushing further Girard's results to discover what kind of information can be obtained rationally from the coefficients of an equation.

Girard himself had shown (omitting the details of his calculations) that the sum of the squares of the solutions, the sum of their cubes, and the sum of their fourth powers, can be calculated from the coefficients [33, p. F2 r°]: if $x_1, \ldots, x_n$ are the solutions of

$$X^n - s_1 X^{n-1} + s_2 X^{n-2} - s_3 X^{n-3} + \cdots + (-1)^n s_n = 0,$$

let, for any integer $k$,

$$\sigma_k = \sum_{i=1}^{n} x_i^k;$$

then

$$\sigma_1 = s_1,$$
$$\sigma_2 = s_1^2 - 2s_2,$$
$$\sigma_3 = s_1^3 - 3s_1 s_2 + 3s_3,$$
$$\sigma_4 = s_1^4 - 4s_1^2 s_2 + 4s_1 s_3 + 2s_2^2 - 4s_4.$$

Around 1666, general formulas for the sum of any power of the solutions were found by Isaac Newton (1642–1727) [56, p. 519] (who was probably unaware of Girard's work: see footnote (12) in [56, p. 518]). Newton's clever observation is that, while the formulas for $\sigma_1, \sigma_2, \ldots$ in terms of $s_1, \ldots, s_n$ do not seem to follow a simple pattern, yet there are simple formulas expressing $\sigma_k$ in terms of $s_1, \ldots, s_n$ and $\sigma_1, \ldots, \sigma_{k-1}$. These formulas can be used to calculate recursively the various $\sigma_k$ in terms of $s_1, \ldots, s_n$. Newton's formulas are

$$\sigma_1 = s_1,$$
$$\sigma_2 = s_1 \sigma_1 - 2s_2,$$
$$\sigma_3 = s_1 \sigma_2 - s_2 \sigma_1 + 3s_3,$$
$$\sigma_4 = s_1 \sigma_3 - s_2 \sigma_2 + s_3 \sigma_1 - 4s_4,$$
$$\sigma_5 = s_1 \sigma_4 - s_2 \sigma_3 + s_3 \sigma_2 - s_4 \sigma_1 + 5s_5,$$

and, generally,

$$\sigma_k = \begin{cases} \sum_{i=1}^{k-1}(-1)^{i+1} s_i \sigma_{k-i} + (-1)^{k+1} k\, s_k & \text{for } k \leq n, \\ \sum_{i=1}^{n}(-1)^{i+1} s_i \sigma_{k-i} & \text{for } k > n. \end{cases} \tag{4.6}$$

These formulas (for $k \leq n$) were published without proof in "Arithmetica Universalis" (1707) [57, p. 107] (see also [56, p. 519], [59, p. 361]). Various proofs were subsequently proposed (see Stedall [68, Ch. 6]). At least, the formula for the sum of $n$-th powers of $x_1$, ..., $x_n$ (i.e., the case $k = n$ of (4.6)) is easy to prove: we have to show

$$\sigma_n - s_1\sigma_{n-1} + s_2\sigma_{n-2} - s_3\sigma_{n-3} + \cdots + (-1)^n n\, s_n = 0. \qquad (4.7)$$

Now, since $x_1$, ..., $x_n$ are solutions of the given equation, we have for $i = 1$, ..., $n$

$$x_i^n - s_1 x_i^{n-1} + s_2 x_i^{n-2} - s_3 x_i^{n-3} + \cdots + (-1)^n s_n = 0. \qquad (4.8)$$

Formula (4.7) follows by summing the equations (4.8) for $i = 1$, ..., $n$. Likewise, the cases where $k > n$ of (4.6) can be obtained by multiplying (4.8) by $x_i^{k-n}$ and summing for $i = 1$, ..., $n$. The challenge is to prove Newton's formulas (4.6) for $k < n$. Euler [28] suggested a clever reduction[5] to the easy case where $k$ is the degree of the given equation: consider the following equation of degree $k$ $(< n)$

$$Y^k - s_1 Y^{k-1} + s_2 Y^{k-2} - \cdots + (-1)^k s_k = 0.$$

Letting $y_1$, ..., $y_k$ denote its solutions, we have

$$s_1 = \sum_{i=1}^{k} y_i, \quad s_2 = \sum_{i<j} y_i y_j, \quad \ldots, \quad s_k = y_1 \ldots y_k.$$

Euler contended without explanation that we must therefore also have

$$\sum_{i=1}^{n} x_i^\ell = \sum_{j=1}^{k} y_j^\ell \qquad \text{for all } \ell = 1, \ldots, k.$$

Newton's formula (4.6) for $\sigma_k$ then follows from the easy case (4.7) for the sums of powers of $y_1$, ..., $y_k$.

The following argument, taken from Bourbaki [7, App. 1 n° 3], captures the spirit of Euler's proof while avoiding his perplexing contention. Regarding $x_1$, ..., $x_n$ as independent indeterminates over $\mathbb{Z}$, consider the following homogeneous polynomial of degree $k$ $(< n)$ with integral coefficients:

$$F(x_1, \ldots, x_n) = \sigma_k - s_1\sigma_{k-1} + s_2\sigma_{k-2} - s_3\sigma_{k-3} + \cdots + (-1)^k k\, s_k.$$

If we set to 0 any $n - k$ indeterminates among $x_1$, ..., $x_n$, the polynomial vanishes as we see by applying the easy case of Newton's formulas to the

---

[5]In the same paper, Euler gives another ingenious proof based on power series expansions, which is reproduced in Bourbaki [7, App. 1 n° 3].

remaining $k$ indeterminates, which yields equation (4.7) with $k$ substituted for $n$. But as $F$ is homogeneous of degree $k$, it is a sum of terms of the form $cx_{i_1}\ldots x_{i_k}$ with $c \in \mathbb{Z}$ and $i_1 \leq \cdots \leq i_k$. If $F \neq 0$, then there must be some choice of indices $i_1 \leq \cdots \leq i_k$ for which the coefficient $c$ is not 0. But then $F$ does not vanish when we set to 0 the indeterminates $x_j$ for $j \notin \{i_1, \ldots, i_k\}$. Therefore, we must have $F = 0$, which proves Newton's formulas (4.6) for $k < n$.

Newton's formulas (4.6) were of course not his most prominent achievement, even if we restrict to the theory of equations. Indeed, Newton's contributions to numerical aspects were more influential (see for instance Goldstine [34, Ch. 2], Stedall [68, Ch. 9]).

Numerical methods to find the roots of polynomials were an important aspect of the theory of equations, developed by some authors who also contributed to the more algebraic side: see for instance Cardano's "golden rule" [13, Ch. 30], Stevin's "Appendice algebraique" [69, pp. 740–745] or Viète's "De Numerosa Potestatum ad Exegesim Resolutione" (On the numerical resolution of powers by exegetics) [83, pp. 311–370]. These numerical methods were much more successful for the solution of explicit numerical equations than the algebraic formulas "by radicals," because algebraic formulas are available only for degrees up to 4 and they are by no means more accurate than the numerical methods (see also §2.3). Therefore, the numerical solution of equations soon developed into a new branch of mathematics, growing more accurate and powerful while the algebraic theory of equations was progressively stalled (see however Stedall [68, Ch. 4] for an account of the discussions surrounding Descartes' "rule of signs" for the number of positive and negative roots, and Newton's rule for the number of imaginary roots).

Since Descartes' and Newton's rules have little relevance to Galois theory, and since the discussion of numerical methods falls beyond the scope of this book, we take the occasion of this period of relatively low activity in algebra to justify a pause in the historical exposition. We turn in the next chapter to a modern exposition of the above-mentioned results on polynomials in one indeterminate, in order to show on which mathematical base later results were grounded.

# Chapter 5

# A Modern Approach to Polynomials

## 5.1 Definitions

In modern terminology, a polynomial in one indeterminate with coefficients in a ring $A$ can be defined as a map

$$P \colon \mathbb{N} \to A$$

such that the set $\operatorname{supp} P = \{n \in \mathbb{N} \mid P_n \neq 0\}$, called the *support* of $P$, is finite. The addition of polynomials is the usual addition of maps,

$$(P + Q)_n = P_n + Q_n$$

and the product is the convolution product

$$(PQ)_n = \sum_{i+j=n} P_i \cdot Q_j.$$

Every element $a \in A$ is identified with the polynomial $a \colon \mathbb{N} \to A$ that maps $0$ to $a$ and $n$ to $0$ for $n \neq 0$. Denoting by

$$X \colon \mathbb{N} \to A$$

the polynomial that maps $1$ onto the unit element $1 \in A$ and the other integers to $0$, it is then easily seen that every polynomial $P$ can be uniquely written as

$$P = \sum_{i \in \mathbb{N}} P_i \cdot X^i.$$

Therefore, we henceforth write

$$a_0 + a_1 X + \cdots + a_n X^n \tag{5.1}$$

(as is usual!) for the polynomial that maps $i \in \mathbb{N}$ to $a_i$ for $i = 0, \ldots, n$ and to $0$ for $i > n$. Accordingly, the set of all polynomials with coefficients

in $A$ (or polynomials *over* $A$) is denoted $A[X]$. Straightforward calculations show that $A[X]$ is a ring, which is commutative if and only if $A$ is
commutative.

The ring of polynomials in any number $m$ of indeterminates over $A$ can
be similarly defined as the ring of maps from $\mathbb{N}^m$ to $A$ with finite support,
with the convolution product.

Of course, the definition above looks somewhat artificial. The naive
approach to polynomials is to consider expressions like (5.1), where $X$ is an
undefined object, called an *indeterminate*, or a *variable*. While this terminology will be retained in the sequel, it should be observed that, without
any other proper definition, to say something is an indeterminate or a
variable is hardly a definition. Moreover, it fosters confusion between the
polynomial

$$P(X) = a_0 + a_1 X + \cdots + a_n X^n$$

and the associated polynomial function

$$P(\cdot)\colon A \to A$$

which maps $x \in A$ onto $P(x) = a_0 + a_1 x + \cdots + a_n x^n$. This same confusion
has prompted the use of the term *constant* polynomials for the elements of
$A$, considered as polynomials. While this confusion is not so serious when
$A$ is a field with infinitely many elements (see Corollary 5.16 below), it
could be harmful when $A$ is finite. For instance, if $A = \{a_1, \ldots, a_n\}$ (with
$n \geq 2$), then the polynomial $(X - a_1) \ldots (X - a_n)$ is not the zero polynomial
since the coefficient of $X^n$ is 1, but its associated polynomial function maps
every element of $A$ to 0.

The *degree* of a non-zero polynomial $P$ is the largest integer $n$ for which
the coefficient of $X^n$ in the expression of $P$ is not zero; this coefficient is
called the *leading coefficient* of $P$, and $P$ is said to be *monic* if its leading
coefficient is 1.

The degree of $P$ is denoted by $\deg P$. One also sets $\deg 0 = -\infty$, so
that the following relations hold without restriction if $A$ is a domain (i.e.,
a ring $A \neq \{0\}$ in which $ab = 0$ implies $a = 0$ or $b = 0$):

$$\deg(P + Q) \leq \max(\deg P, \deg Q),$$
$$\deg(PQ) = \deg P + \deg Q.$$

When $A$ is a (commutative) field, the ring $A[X]$ has a field of fractions
$A(X)$, constructed as follows: the elements of $A(X)$ are the equivalence

classes of couples $P/Q$, where $P$, $Q \in A[X]$ and $Q \neq 0$, under the equivalence relation

$$P/Q = P'/Q' \quad \text{if} \quad Q'P = P'Q.$$

The addition of equivalence classes is defined by

$$(P_1/Q_1) + (P_2/Q_2) = (P_1Q_2 + P_2Q_1)/Q_1Q_2$$

and the multiplication by

$$(P_1/Q_1) \cdot (P_2/Q_2) = (P_1P_2)/(Q_1Q_2).$$

It is then easily verified that $A(X)$ is a field, called the field of *rational functions*[1] in one indeterminate $X$ over the field $A$. The same construction can be applied to the ring of polynomials in $m$ indeterminates and yields the field of rational functions in $m$ indeterminates over the field $A$.

However, the ring $A[X]$ of polynomials in one indeterminate over a field $A$ has particularly nice properties, which follow from the Euclidean division algorithm. These properties are reviewed in the next section.

## 5.2 Euclidean division

The following Euclidean Division Property actually holds over arbitrary commutative rings, provided the leading coefficient of the divisor is invertible (which is of course the case when the ring of coefficients is a field). In Chapter 14 we will use the next theorem for polynomials with coefficients in a ring of polynomials: see the proof of Lemma 14.19 (p. 238).

**Theorem 5.1 (Euclidean Division Property).** *Let $A$ be a commutative ring and let $P_1$, $P_2 \in A[X]$. If the leading coefficient of $P_2$ is invertible in $A$ (which implies $P_2 \neq 0$), then there exist polynomials $Q$, $R \in A[X]$ such that*

$$P_1 = P_2Q + R \qquad and \qquad \deg R < \deg P_2.$$

*Moreover, the polynomials $Q$ and $R$ are uniquely determined by these properties.*

The polynomials $Q$ and $R$ are called respectively the *quotient* and the *remainder* of the division of $P_1$ by $P_2$.

---

[1] Just like polynomials, the elements in $A(X)$ should be viewed as formal expressions, and not as maps. For instance, if $A$ is finite, $A = \{a_1, \ldots, a_n\}$, the rational function $1/((X - a_1) \ldots (X - a_n))$ cannot be evaluated at any point in $A$.

**Proof.** The existence of $Q$ and $R$ is proved by induction on $\deg P_1$. If $\deg P_1 < \deg P_2$, we set $Q = 0$ and $R = P_1$. If $\deg P_1 - \deg P_2 = d \geq 0$, then letting $c \in A$ be the quotient of the leading coefficient of $P_1$ by that of $P_2$ ($c$ exists because the leading coefficient of $P_2$ is invertible), we have

$$\deg(P_1 - cX^d P_2) < \deg P_1.$$

Therefore, by the induction hypothesis, there exist $Q$ and $R \in A[X]$ such that

$$P_1 - cX^d P_2 = P_2 Q + R \quad \text{and} \quad \deg R < \deg P_2.$$

This equation yields

$$P_1 = P_2(Q + cX^d) + R$$

hence $Q + cX^d$ and $R$ satisfy the required properties.

To prove the uniqueness of $Q$ and $R$, assume

$$P_1 = P_2 Q_1 + R_1 = P_2 Q_2 + R_2$$

with $\deg R_1 < \deg P_2$ and $\deg R_2 < \deg P_2$. Then $R_1 - R_2 = P_2(Q_2 - Q_1)$ and this equality is impossible if both sides are non-zero, since the degree of the right-hand side is then at least equal to $\deg P_2$, while the degree of the left-hand side is strictly less than $\deg P_2$. $\qquad \square$

From now on, we only consider polynomials over a field $F$.

**Definitions 5.2.** Let $P_1$, $P_2 \in F[X]$. We say $P_2$ *divides* $P_1$ if

$$P_1 = P_2 Q \qquad \text{for some } Q \in F[X]$$

or, equivalently when $P_2 \neq 0$, if the remainder of the division of $P_1$ by $P_2$ is 0.

A *greatest common divisor* (GCD) of $P_1$ and $P_2$ is a polynomial $D \in F[X]$ that has the following two properties:

(a) $D$ divides $P_1$ and $P_2$,

(b) if $S$ is a polynomial that divides $P_1$ and $P_2$ then $S$ divides $D$.

If 1 is a GCD of $P_1$ and $P_2$, then $P_1$ and $P_2$ are said to be *relatively prime*.

Since it is by no means obvious that any two polynomials $P_1$, $P_2$ have a GCD, our first objective is to devise a method of finding such a GCD, thereby proving its existence. We closely follow Euclid's algorithm for finding the GCD of two integers or the greatest common measure of two line segments, assuming (without loss of generality) that $\deg P_1 \geq \deg P_2$.

If $P_2 = 0$, then $P_1$ is a GCD of $P_1$ and $P_2$. Otherwise, we divide $P_1$ by $P_2$:

$$P_1 = P_2 Q_1 + R_1. \tag{E.1}$$

Next, we divide $P_2$ by the remainder $R_1$, provided that it is not zero,

$$P_2 = R_1 Q_2 + R_2 \tag{E.2}$$

then we divide the first remainder by the second, and so on, as long as the remainders are not zero:

$$R_1 = R_2 Q_3 + R_3, \tag{E.3}$$
$$R_2 = R_3 Q_4 + R_4, \tag{E.4}$$
$$\cdots \qquad\qquad \cdots$$
$$R_{n-2} = R_{n-1} Q_n + R_n, \tag{E.n}$$
$$R_{n-1} = R_n Q_{n+1} + R_{n+1}. \tag{E.n + 1}$$

Since $\deg P_2 > \deg R_1 > \deg R_2 > \ldots$, this sequence of integers cannot extend indefinitely. Therefore, $R_{n+1} = 0$ for some $n$.

*Claim:* If $R_{n+1} = 0$, then $R_n$ is a GCD of $P_1$ and $P_2$. (If $n = 0$, then set $R_n = P_2$.)

To see that $R_n$ divides $P_1$ and $P_2$, observe that equation $(E.n + 1)$, together with $R_{n+1} = 0$, implies that $R_n$ divides $R_{n-1}$. It then follows from equation $(E.n)$ that $R_n$ also divides $R_{n-2}$. Going up in the sequence of equations $(E.n)$, $(E.n - 1)$, $(E.n - 2)$, $\ldots$, $(E.2)$, $(E.1)$, we conclude recursively that $R_n$ divides $R_{n-3}$, $\ldots$, $R_2$, $R_1$, $P_2$ and $P_1$.

Assume next that $P_1$ and $P_2$ are both divisible by some polynomial $S$. Then equation $(E.1)$ shows that $S$ also divides $R_1$. Since it divides $P_2$ and $R_1$, $S$ also divides $R_2$, by equation $(E.2)$. Going down in the above sequence of equations $(E.2)$, $(E.3)$, $\ldots$, $(E.n)$, we finally see that $S$ divides $R_n$. This completes the proof that $R_n$ is a GCD of $P_1$ and $P_2$.

We next observe that the GCD of two polynomials is not unique (except over the field with two elements), as the following theorem shows.

**Theorem 5.3.** *Any two polynomials $P_1, P_2 \in F[X]$ which are not both zero have a unique monic greatest common divisor $D_1$, and a polynomial $D \in F[X]$ is a greatest common divisor of $P_1$ and $P_2$ if and only if $D = cD_1$ for some $c \in F^\times$ $(= F - \{0\})$. Moreover, if $D$ is a greatest common divisor of $P_1$ and $P_2$, then*

$$D = P_1 U_1 + P_2 U_2 \qquad \text{for some } U_1, U_2 \in F[X].$$

**Proof.** Euclid's algorithm already yields a greatest common divisor $R_n$ of $P_1$ and $P_2$. Dividing $R_n$ by its leading coefficient, we get a monic GCD of $P_1$ and $P_2$. Now, assume $D$ and $D'$ are GCDs of $P_1$ and $P_2$. Then $D$ divides $D'$ since $D$ satisfies condition (a) and $D'$ satisfies condition (b). The same argument, with $D$ and $D'$ interchanged, shows that $D'$ divides $D$. Let $D' = DQ$ and $D = D'Q'$ for some $Q$, $Q' \in F[X]$. It follows that $QQ' = 1$, so that $Q$ and $Q'$ are constants, which are inverse of each other. This proves at once the second statement and the uniqueness of the monic GCD of $P_1$ and $P_2$, since $D$ and $D'$ cannot be both monic unless $Q = Q' = 1$.

Suppose $D$ is any GCD of $P_1$ and $P_2$; then $D$ and the greatest common divisor $R_n$ found by Euclid's algorithm are related by $D = cR_n$ for some $c \in F^\times$. Therefore, it suffices to prove the last statement for $R_n$. For this, we consider again the sequence of equations $(E.1), \ldots, (E.n)$. From equation $(E.n)$, we get

$$R_n = R_{n-2} - R_{n-1}Q_n.$$

We then use equation $(E.n-1)$ to eliminate $R_{n-1}$ in this expression of $R_n$, and we obtain

$$R_n = -R_{n-3}Q_n + R_{n-2}(1 + Q_{n-1}Q_n).$$

This shows that $R_n$ is a sum of multiples of $R_{n-2}$ and $R_{n-3}$. Now, $R_{n-2}$ can be eliminated using $(E.n-2)$. We thus obtain an expression of $R_n$ as a sum of multiples of $R_{n-3}$ and $R_{n-4}$. Going up in the sequence of equations $(E.n-3), (E.n-4), \ldots, (E.1)$, we end up with an expression of $R_n$ as a sum of multiples of $P_1$ and $P_2$,

$$R_n = P_1U_1 + P_2U_2$$

for some $U_1, U_2 \in F[X]$.

The argument above can be made more transparent with only a sprinkle of matrix algebra. We may rewrite equation $(E.1)$ in the form

$$\begin{pmatrix} P_1 \\ P_2 \end{pmatrix} = \begin{pmatrix} Q_1 & 1 \\ 1 & 0 \end{pmatrix} \begin{pmatrix} P_2 \\ R_1 \end{pmatrix},$$

equation $(E.2)$ in the form

$$\begin{pmatrix} P_2 \\ R_1 \end{pmatrix} = \begin{pmatrix} Q_2 & 1 \\ 1 & 0 \end{pmatrix} \begin{pmatrix} R_1 \\ R_2 \end{pmatrix},$$

and so on, finishing with equation $(E.n+1)$ (with $R_{n+1} = 0$) in the form

$$\begin{pmatrix} R_{n-1} \\ R_n \end{pmatrix} = \begin{pmatrix} Q_{n+1} & 1 \\ 1 & 0 \end{pmatrix} \begin{pmatrix} R_n \\ 0 \end{pmatrix}.$$

Combining the matrix equations above, we get

$$\begin{pmatrix} P_1 \\ P_2 \end{pmatrix} = \begin{pmatrix} Q_1 & 1 \\ 1 & 0 \end{pmatrix} \begin{pmatrix} Q_2 & 1 \\ 1 & 0 \end{pmatrix} \cdots \begin{pmatrix} Q_{n+1} & 1 \\ 1 & 0 \end{pmatrix} \begin{pmatrix} R_n \\ 0 \end{pmatrix}.$$

Each matrix $\begin{pmatrix} Q_i & 1 \\ 1 & 0 \end{pmatrix}$ is invertible, with inverse $\begin{pmatrix} 0 & 1 \\ 1 & -Q_i \end{pmatrix}$, hence the preceding equation yields, after multiplying each side on the left successively by $\begin{pmatrix} 0 & 1 \\ 1 & -Q_1 \end{pmatrix}$, $\begin{pmatrix} 0 & 1 \\ 1 & -Q_2 \end{pmatrix}$, ...,

$$\begin{pmatrix} 0 & 1 \\ 1 & -Q_{n+1} \end{pmatrix} \cdots \begin{pmatrix} 0 & 1 \\ 1 & -Q_2 \end{pmatrix} \begin{pmatrix} 0 & 1 \\ 1 & -Q_1 \end{pmatrix} \begin{pmatrix} P_1 \\ P_2 \end{pmatrix} = \begin{pmatrix} R_n \\ 0 \end{pmatrix}.$$

If $U_1, \ldots, U_4 \in F[X]$ are such that

$$\begin{pmatrix} 0 & 1 \\ 1 & -Q_{n+1} \end{pmatrix} \cdots \begin{pmatrix} 0 & 1 \\ 1 & -Q_2 \end{pmatrix} \begin{pmatrix} 0 & 1 \\ 1 & -Q_1 \end{pmatrix} = \begin{pmatrix} U_1 & U_2 \\ U_3 & U_4 \end{pmatrix},$$

it follows that

$$U_1 P_1 + U_2 P_2 = R_n$$

(and $U_3 P_1 + U_4 P_2 = 0$). □

Because of its repeated use in the sequel, the following special case seems to be worth pointing out explicitly:

**Corollary 5.4.** *If $P_1$, $P_2$ are relatively prime polynomials in $F[X]$, then there exist polynomials $U_1$, $U_2 \in F[X]$ such that*

$$P_1 U_1 + P_2 U_2 = 1.$$

**Remarks 5.5.** **(a)** The proof given above is effective: Euclid's algorithm yields a procedure for constructing the polynomials $U_1$, $U_2$ in Theorem 5.3 and Corollary 5.4.

**(b)** Since the GCD of two polynomials $P_1$, $P_2 \in F[X]$ can be found by rational calculations (i.e., calculations involving only the four basic operations of arithmetic), it does not depend on any particular property of the field $F$. The point of this observation is that if the field $F$ is embedded in a larger field $K$, then the polynomials $P_1$, $P_2 \in F[X]$ can be regarded as polynomials over $K$, but their monic GCD in $K[X]$ is the same as their monic GCD in $F[X]$. This is noteworthy in view of the fact that the *irreducible factors* of $P_1$ and $P_2$ depend on the base field $F$, as is clear from Example 5.7(b) below.

(c) Euclid's algorithm was originally designed to find the greatest common measure of two line segments (which is not always possible, as we saw in §1.2), or the greatest common divisor of two integers. The latter is always possible, because of the Euclidean division property in $\mathbb{Z}$, i.e., the analogue of Theorem 5.1 with $P_1$, $P_2 \in \mathbb{Z}$, where the remainder $R$ is subject to the condition that $|P_2| > R \geq 0$. Mimicking the proof of Theorem 5.3, we obtain the following analogue of Corollary 5.4, which will also be useful: *if $n_1$, $n_2 \in \mathbb{Z}$ are relatively prime, then there exist $m_1$, $m_2 \in \mathbb{Z}$ such that*

$$n_1 m_1 + n_2 m_2 = 1.$$

## 5.3 Irreducible polynomials

As in the preceding section, we write $F$ for an arbitrary field.

**Definition 5.6.** A polynomial $P \in F[X]$ is said to be *irreducible in $F[X]$* (or *over $F$*) if $\deg P > 0$ and $P$ is not divisible by any polynomial $Q \in F[X]$ such that $0 < \deg Q < \deg P$.

From this definition, it follows that if a polynomial $D$ divides an irreducible polynomial $P$, then either $D$ is a constant or $\deg D = \deg P$. In the latter case, the quotient of $P$ by $D$ is a constant, hence $D$ is the product of $P$ by a non-zero constant. In particular, for any polynomial $S \in F[X]$, either 1 or $P$ is a GCD of $P$ and $S$. Consequently, either $P$ divides $S$ or $P$ is relatively prime to $S$.

**Examples 5.7.** (a) By definition, it is clear that every polynomial of degree 1 is irreducible. It will be proved later that over the field of complex numbers, only these polynomials are irreducible.

(b) Theorem 5.12 below will show that if a polynomial of degree at least 2 has a root in the base field, then it is not irreducible. The converse is true for polynomials of degree 2 or 3; namely, if a polynomial of degree 2 or 3 has no roots in the base field, then it is irreducible over this field, see Corollary 5.13. It follows that, for instance, the polynomial $X^2 - 2$ is irreducible over $\mathbb{Q}$, but not over $\mathbb{R}$. Thus, the irreducibility of polynomials of degree at least 2 depends on the base field (compare Remark 5.5(b)).

(c) A polynomial (of degree at least 4) may be reducible over a field without having any root in this field. For instance, in $\mathbb{Q}[X]$, the polynomial $X^4 + 4$ is reducible since

$$X^4 + 4 = (X^2 + 2X + 2)(X^2 - 2X + 2).$$

**Remark.** To determine whether a given polynomial with rational coefficients is irreducible or not over $\mathbb{Q}$ may be difficult, although a systematic procedure has been devised by Kronecker, see Van der Waerden [77, §32]. This procedure is not unlike that which is used to find the rational roots of polynomials with rational coefficients, see §6.3.

**Theorem 5.8.** *Every non-constant polynomial* $P \in F[X]$ *is a finite product*

$$P = c\,P_1 \ldots P_n$$

*where* $c \in F^\times$ *and* $P_1, \ldots, P_n$ *are monic irreducible polynomials (not necessarily distinct). Moreover, this factorization is unique, except for the order of the factors.*

**Proof.** The existence of the above factorization is easily proved by induction on $\deg P$. If $\deg P = 1$ or, more generally, if $P$ is irreducible, then $P = c \cdot P_1$ where $c$ is the leading coefficient of $P$ and $P_1 = c^{-1}P$ is irreducible and monic. If $P$ is reducible, then it can be written as a product of two polynomials of degree strictly less than $\deg P$. By induction, each of these two polynomials has a finite factorization as above, and these factorizations multiply up to a factorization of $P$.

To prove that the factorization is unique, we use the following lemma.

**Lemma 5.9.** *If a polynomial divides a product of $r$ factors and is relatively prime to the first $r-1$ factors, then it divides the last one.*

**Proof.** It suffices to consider the case $r = 2$, since the general case then easily follows by induction.

Assume that a polynomial $S$ divides a product $T \cdot U$ and is relatively prime to $T$. By Corollary 5.4 we can find polynomials $V$, $W$ such that $SV + TW = 1$. Multiplying each side of this equality by $U$, we obtain

$$S(UV) + (TU)W = U.$$

Now, $S$ divides the left-hand side, since it divides $TU$ by hypothesis; it then follows that $S$ divides $U$. $\qquad\square$

**End of the proof of Theorem 5.8.** It remains to prove the uniqueness (up to the order of factors) of factorizations. Assume

$$P = c\,P_1 \ldots P_n = d\,Q_1 \ldots Q_m \tag{5.2}$$

where $c, d \in F^\times$ and $P_1, \ldots, P_n, Q_1, \ldots, Q_m$ are monic irreducible polynomials.

First, $c = d$ since $c$ and $d$ are both equal to the leading coefficient of $P$. Therefore, (5.2) yields

$$P_1 \ldots P_n = Q_1 \ldots Q_m. \tag{5.3}$$

Next, since $P_1$ divides the product $Q_1 \ldots Q_m$, it follows from Lemma 5.9 that it cannot be relatively prime to all the factors. Renumbering $Q_1, \ldots, Q_m$ if necessary, we may assume that $P_1$ is not relatively prime to $Q_1$, so that their monic GCD, which we denote by $D$, is not 1. Since $D$ divides $P_1$, which is irreducible, it is equal to $P_1$ up to a constant factor. Since moreover $D$ and $P_1$ are both monic, the constant factor is 1, hence $D = P_1$.

We may argue similarly with $Q_1$, since $Q_1$ is also monic and irreducible, and we thus obtain $D = Q_1$, hence

$$P_1 = Q_1.$$

Canceling $P_1 \ (= Q_1)$ in equation (5.3), we get a similar equality, with one factor less on each side:

$$P_2 \ldots P_n = Q_2 \ldots Q_m.$$

Using inductively the same argument, we obtain (renumbering $Q_2, \ldots, Q_m$ if necessary)

$$P_2 = Q_2, \quad P_3 = Q_3, \quad \ldots, \quad P_n = Q_n,$$

and it follows that $n \leq m$. If $n < m$, then comparing the degrees of both sides of (5.3) we get

$$\deg Q_{n+1} = \cdots = \deg Q_m = 0$$

and this is absurd since $Q_i$ is irreducible for $i = 1, \ldots, m$. Therefore, $n = m$ and the proof is complete. $\qquad\square$

Lemma 5.9 has another consequence which is worth noting in view of its repeated use in the sequel:

**Proposition 5.10.** *If a polynomial is divisible by pairwise relatively prime polynomials, then it is divisible by their product.*

**Proof.** Let $P_1, \ldots, P_r$ be pairwise relatively prime polynomials which divide a polynomial $P$. We argue by induction on $r$, the case $r = 1$ being trivial. By the induction hypothesis,

$$P = P_1 \ldots P_{r-1} Q$$

for some polynomial $Q$. Since $P_r$ divides $P$, Lemma 5.9 shows that $P_r$ divides $Q$, so that $P_1 \ldots P_r$ divides $P$. $\qquad\square$

## 5.4 Roots

As in the preceding sections, $F$ denotes a field. For any polynomial

$$P = a_0 + a_1 X + \cdots + a_n X^n \in F[X]$$

we denote by $P(\cdot)$ the associated polynomial function

$$P(\cdot)\colon\ F \to F$$

which maps any $x \in F$ to $P(x) = a_0 + a_1 x + \ldots + a_n x^n$. It is readily verified that for any two polynomials $P,\ Q \in F[X]$ and any $x \in F$,

$$(P + Q)(x) = P(x) + Q(x) \qquad \text{and} \qquad (P \cdot Q)(x) = P(x) \cdot Q(x).$$

Therefore, the map $P \mapsto P(\cdot)$ is a homomorphism from the ring $F[X]$ to the ring of functions from $F$ to $F$.

**Definition 5.11.** An element $a \in F$ is a *root* of a polynomial $P \in F[X]$ if $P(a) = 0$.

**Theorem 5.12.** *An element $a \in F$ is a root of a polynomial $P \in F[X]$ if and only if $X - a$ divides $P$.*

**Proof.** Since $\deg(X - a) = 1$, the remainder $R$ of the division of $P$ by $X - a$ is a constant polynomial. Evaluating at $a$ the polynomial functions associated with each side of the equation

$$P = (X - a)Q + R$$

we get

$$P(a) = (a - a)Q(a) + R,$$

hence $P(a) = R$. This shows that $P(a) = 0$ if and only if the remainder of the division of $P$ by $X - a$ is 0. The theorem follows, since the last condition means that $X - a$ divides $P$. $\qquad\qquad\Box$

**Corollary 5.13.** *Let $P \in F[X]$ be a polynomial of degree 2 or 3. Then $P$ is irreducible over $F$ if and only if it has no roots in $F$.*

**Proof.** This readily follows from Theorem 5.12, since the hypothesis on $\deg P$ implies that if $P$ is not irreducible, then it has a factor of degree 1, hence of the form $X - a$. $\qquad\qquad\Box$

**Definitions 5.14.** The *multiplicity* of a root $a$ of a nonzero polynomial $P$ is the exponent of the highest power of $X - a$ which divides $P$. Thus, the multiplicity is $m$ if $(X - a)^m$ divides $P$ but $(X - a)^{m+1}$ does not divide $P$. A root is called *simple* when its multiplicity is 1; otherwise it is called *multiple*.

When the multiplicity of $a$ as a root of $P$ is considered as a function of $P$, it is also called the *valuation* of $P$ at $a$, and denoted $v_a(P)$; we then set $v_a(P) = 0$ if $a$ is not a root of $P$. By convention, we also set $v_a(0) = \infty$, so that the following relations hold for any $P, Q \in F[X]$ and any $a \in F$:

$$v_a(P + Q) \geq \min(v_a(P), v_a(Q)), \qquad v_a(PQ) = v_a(P) + v_a(Q).$$

These properties are exactly the same as those of the function

$$- \deg \colon F[X] \to \mathbb{Z} \cup \{\infty\}$$

which maps any polynomial to the *opposite* of its degree, see p. 42. Accordingly, $(- \deg)$ is sometimes considered as the valuation "at infinity."

**Theorem 5.15.** *Every non-zero polynomial $P \in F[X]$ has a finite number of roots. If $a_1, \ldots, a_r$ are the various roots of $P$ in $F$, with respective multiplicities $m_1, \ldots, m_r$, then $\deg P \geq m_1 + \ldots + m_r$ and*

$$P = (X - a_1)^{m_1} \cdots (X - a_r)^{m_r} Q$$

*for some polynomial $Q \in F[X]$ which has no roots in $F$.*

**Proof.** If $a_1, \ldots, a_r$ are distinct roots of $P$ in $F$, with respective multiplicities $m_1, \ldots, m_r$, then the polynomials $(X - a_1)^{m_1}, \ldots, (X - a_r)^{m_r}$ are relatively prime and divide $P$, hence Proposition 5.10 shows that

$$P = (X - a_1)^{m_1} \cdots (X - a_r)^{m_r} Q,$$

for some polynomial $Q$. It readily follows that

$$\deg P \geq m_1 + \cdots + m_r;$$

hence $P$ cannot have infinitely many roots. The rest follows from the observation that if $a_1, \ldots, a_r$ are *all* the roots of $P$ in $F$, then $Q$ has no root. $\square$

**Corollary 5.16.** *Let $P, Q \in F[X]$. If $F$ is infinite, then $P = Q$ if and only if the associated polynomial functions $P(\cdot)$ and $Q(\cdot)$ are equal.*

**Proof.** If $P(\cdot) = Q(\cdot)$, then all the elements in $F$ are roots of $P - Q$, hence $P - Q = 0$, since $F$ is infinite. The converse is clear. $\square$

## 5.5 Multiple roots and derivatives

The aim of this section is to devise a method to determine whether a polynomial has multiple roots without actually finding the roots, and to reduce to 1 the multiplicity of the roots. (More precisely, the method yields, for any given polynomial $P$, a polynomial $P_s$ which has the same roots as $P$, each with multiplicity 1).

This method is due to Johann Hudde (1633–1704). It uses the derivative of polynomials, which was introduced purely algebraically by Hudde in his letter "De Reductione Aequationum" (On the reduction of equations) (1657) [38] and subsequently applied to find maxima and minima of polynomials and rational functions in his letter "De Maximis et Minimis" (1658) [39] (see also Stedall [68, p. 52]).

**Definition 5.17.** The *derivative* $\partial P$ of a polynomial $P = a_0 + a_1 X + \cdots + a_n X^n$ with coefficients in a field $F$ is the polynomial

$$\partial P = a_1 + 2a_2 X + 3a_3 X^2 + \cdots + na_n X^{n-1}.$$

Straightforward calculations show that the following (familiar) relations hold:

$$\partial(P + Q) = \partial P + \partial Q, \qquad \partial(PQ) = (\partial P)Q + P(\partial Q).$$

**Remark.** The integers that appear in the coefficients of $\partial P$ are regarded as elements in $F$; thus, $n$ stands for $1 + 1 + \cdots + 1$ ($n$ terms), where 1 is the unit element of $F$. This requires some caution, since it could happen that $n = 0$ in $F$ even if $n \neq 0$ (as an integer). For instance, if $F$ is the field with two elements $\{0, 1\}$, then $2 = 1 + 1 = 0$ in $F$, hence over $F$ the derivative of non-constant polynomials like $X^2 + 1$ is the zero polynomial.

**Lemma 5.18.** *Let $a \in F$ be a root of a polynomial $P \in F[X]$. Then $a$ is a multiple root of $P$ (i.e., $v_a(P) \geq 2$) if and only if $a$ is a root of $\partial P$.*

**Proof.** Since $a$ is a root of $P$, one has $P = (X - a)Q$ for some $Q \in F[X]$, hence

$$\partial P = Q + (X - a)\partial Q.$$

This equality shows that $X - a$ divides $\partial P$ if and only if it divides $Q$. Since this last condition amounts to $(X - a)^2$ divides $P$, i.e., $v_a(P) \geq 2$, the lemma follows. $\qquad\square$

**Proposition 5.19.** *Let $P \in F[X]$ be a polynomial which splits into a product of linear factors[2] over some field $K$ containing $F$. A necessary and sufficient condition for the roots of $P$ in $K$ to be all simple is that $P$ and $\partial P$ be relatively prime.*

**Proof.** If some root $a$ of $P$ in $K$ is not simple, then, by the preceding lemma, $P$ and $\partial P$ have the common factor $X - a$ in $K[X]$; therefore, they are not relatively prime in $K[X]$, hence also not relatively prime in $F[X]$ (see Remark 5.5(b)).

Conversely, if $P$ and $\partial P$ are not relatively prime, they have in $K[X]$ a common irreducible factor, which has degree 1 since all the irreducible factors of $P$ in $K[X]$ are linear. We can thus find a polynomial $X - a \in K[X]$ that divides both $P$ and $\partial P$, and it follows from the preceding lemma that $a$ is a multiple root of $P$ in $K$. $\qquad\square$

To improve Lemma 5.18, we now assume that the characteristic of $F$ is 0, which means that every non-zero integer is non-zero in $F$. (The characteristic of a field $F$ is either 0 or the least integer $n > 0$ such that $n = 0$ in $F$).

**Proposition 5.20.** *Assume that* char $F = 0$ *and let* $a \in F$ *be a root of a non-zero polynomial* $P \in F[X]$. *Then*

$$v_a(\partial P) = v_a(P) - 1.$$

**Proof.** Let $m = v_a(P)$ and let $P = (X - a)^m Q$ where $Q \in F[X]$ is not divisible by $X - a$. Then

$$\partial P = (X - a)^{m-1}\big(mQ + (X - a)\partial Q\big),$$

and the hypothesis on the characteristic of $F$ implies that $mQ \neq 0$. Then, since $X - a$ does not divide $Q$, it does not divide $mQ + (X - a)\partial Q$ either, hence $v_a(\partial P) = m - 1$. $\qquad\square$

As an application of this result, we now derive Hudde's method to reduce to 1 the multiplicity of the roots of a non-zero polynomial $P$ over a field $F$ of characteristic zero. Let $D$ be a GCD of $P$ and $\partial P$, and let $P_s = P/D$.

**Theorem 5.21.** *Assume* char $F = 0$ *and let $K$ be an arbitrary field containing $F$. The roots of $P_s$ in $K$ are the same as those of $P$, and every root of $P_s$ in $K$ is simple, i.e., has multiplicity 1.*

---

[2]In §9.1, it is proved that this condition holds for every (non-constant) polynomial.

**Proof.** Since $P_s$ is a quotient of $P$, it is clear that every root of $P_s$ is a root of $P$. Conversely, if $a \in K$ is a root of $P$ of multiplicity $m$, then Proposition 5.20 shows that $v_a(\partial P) = m - 1$, and it follows that $(X - a)^{m-1}$ is the highest power of $X - a$ that divides both $P$ and $\partial P$. It is therefore the highest power of $X - a$ that divides $D$. Consequently, $P_s$ is divisible by $X - a$ but not by $(X - a)^2$, which means that $a$ is a simple root of $P_s$. $\square$

**Corollary 5.22.** *Assume* char $F = 0$. *If* $P \in F[X]$ *is irreducible, then its roots in every field* $K$ *containing* $F$ *are simple.*

**Proof.** Since $P$ is irreducible and does not divide $\partial P$ (because $\deg \partial P = (\deg P) - 1$), the constant polynomial 1 is a GCD of $P$ and $\partial P$. Therefore, $P_s = P$ and Theorem 5.21 shows that every root of $P$ in any field $K$ containing $F$ is simple. $\square$

Corollary 5.22 does not hold if char $F \neq 0$. For instance, if char $F = 2$ and $a \in F$ is not a square in $F$, then $X^2 - a$ is irreducible in $F[X]$, but it has $\sqrt{a}$ as a double root in $F(\sqrt{a})$. (Observe that $\sqrt{a} = -\sqrt{a}$ since the characteristic is 2.)

**Remarks 5.23.** (a) Since a GCD of $P$ and $\partial P$ can be calculated by Euclid's algorithm (see p. 45), it is not necessary to find the roots of $P$ in order to construct the polynomial $P_s$. Thus, there is no serious restriction if we henceforth assume, when trying to solve an equation $P = 0$, that all the roots of $P$ in $F$ or in any field containing $F$ are simple.

(b) In his work, Hudde does not explicitly introduce the derivative polynomial $\partial P$, but indirectly he uses it. His formulation [38, Reg. 10, pp. 433 ff] is as follows : to reduce to 1 the multiplicity of the roots of a polynomial

$$P = a_0 + a_1 X + a_2 X^2 + \cdots + a_n X^n,$$

form a new polynomial $P_1$ by multiplying the coefficients of $P$ by the terms of an arbitrary arithmetical progression $m, m + r, m + 2r, \ldots, m + nr$, so

$$P_1 = a_0 m + a_1(m + r)X + a_2(m + 2r)X^2 + \cdots + a_n(m + nr)X^n.$$

Then, the quotient of $P$ by a greatest common divisor $D_1$ of $P$ and $P_1$ is the required polynomial.

The relation between this rule and its modern translation is easy to see, since $P_1 = mP + rX\partial P$. Therefore, $D_1 = D$ up to a non-zero constant factor, and possibly to a factor $X$ in $D_1$, if 0 is a root of $P$. So, Hudde's method yields the same equation with simple roots as its modern equivalent, except that Hudde's equation lacks the root 0 whenever 0 is a root of the initial equation.

## 5.6   Common roots of two polynomials

As the preceding discussion shows, it is sometimes useful to determine whether two polynomials $P$, $Q$ are relatively prime or not. The most straightforward method is of course to calculate a GCD of $P$ and $Q$, but there is an alternative method, based on a construction of considerable theoretical interest: a certain algebraic expression involving the coefficients of $P$ and $Q$, called the *resultant* of $P$ and $Q$, vanishes if and only if $P$ and $Q$ have a (non-constant) common factor. This construction lies at the heart of elimination theory, which was initially developed during the eighteenth century by Newton, Euler, Bézout, and Cramer (see Stedall [68, Ch. 7]). We will put it to use in §6.4 and §10.1.

The basic idea is simple. Consider two non-constant polynomials $P$, $Q$ over a field $F$:

$$P = a_n X^n + a_{n-1} X^{n-1} + \cdots + a_1 X + a_0 \quad (n \geq 1, \ a_n \neq 0),$$
$$Q = b_m X^m + b_{m-1} X^{m-1} + \cdots + b_1 X + b_0 \quad (m \geq 1, \ b_m \neq 0).$$

(5.4)

Assume[3] $P$ and $Q$ split into products of linear factors over some field $K$ containing $F$:

$$P = a_n(X - u_1)\ldots(X - u_n), \quad Q = b_m(X - v_1)\ldots(X - v_m) \quad (5.5)$$

for some $u_1, \ldots, u_n, v_1, \ldots, v_m \in K$. The GCD of $P$ and $Q$ then also splits in $K[X]$; its irreducible factors are the $(X - u_i)$'s that also appear in the factorization of $Q$, i.e., such that $Q(u_i) = 0$. Therefore, $P$ and $Q$ are relatively prime if and only if

$$Q(u_1)\ldots Q(u_n) \neq 0.$$

The resultant of $P$ and $Q$ is (up to a factor, as we will see in (5.6) and Lemma 5.25 below), the product $Q(u_1)\ldots Q(u_n)$. The challenge is to express this product in terms of the coefficients of $P$ (proving that it lies in $F$, while its definition only shows that it lies in $K$). That such an expression exists was not in doubt to Euler, because the product is symmetric in $u_1$, $\ldots$, $u_n$ (see Chapter 8 for the discussion of symmetric polynomials). However, writing down an explicit formula leads to daunting computations even for small values of $m$ and $n$. For $m = n = 3$, Euler found an expression involving 34 terms (see Stedall [68, p. 136]). The development of matrix theory around the middle of the nineteenth century illuminated these computations. The following discussion is based on an approach to resultants due to James Joseph Sylvester (1814–1897).

---

[3]Theorem 9.3 (p. 110) will show that this hypothesis always holds.

Continuing with the notation in (5.4) and (5.5), we let

$$S(P,Q) = \prod_{i=1}^{n} \prod_{j=1}^{m} (u_i - v_j).$$

In view of the factorization of $Q$, we have

$$Q(u_1) \ldots Q(u_n) = b_m^n\, S(P,Q). \tag{5.6}$$

Likewise, because $P(v_j) = a_n \prod_{i=1}^{n}(v_j - u_i) = (-1)^n a_n \prod_{i=1}^{n}(u_i - v_j)$, we also have

$$P(v_1) \ldots P(v_m) = (-1)^{mn} a_n^m\, S(P,Q). \tag{5.7}$$

Our goal is to relate $S(P,Q)$ to the determinant of the following square matrix of order $m + n$:

$$M(P,Q) = \left. \begin{pmatrix} a_0 & a_1 & \cdots & a_{n-1} & a_n & 0 & \cdots & 0 \\ 0 & a_0 & a_1 & \cdots & a_{n-1} & a_n & & \vdots \\ \vdots & & \ddots & \ddots & & & \ddots & 0 \\ 0 & \cdots & 0 & a_0 & a_1 & \cdots & a_{n-1} & a_n \\ b_0 & b_1 & \cdots & b_{m-1} & b_m & 0 & \cdots & 0 \\ 0 & b_0 & b_1 & \cdots & b_{m-1} & b_m & & \vdots \\ \vdots & & \ddots & \ddots & & & \ddots & 0 \\ 0 & \cdots & 0 & b_0 & b_1 & \cdots & b_{m-1} & b_m \end{pmatrix} \right\} \begin{matrix} m \text{ rows} \\ \\ n \text{ rows} \end{matrix}$$

**Definition 5.24.** The determinant $\mathrm{Res}(P,Q) = \det M(P,Q)$ is called the *resultant* of $P$ and $Q$. From its definition, it is clear that $\mathrm{Res}(P,Q)$ has a polynomial expression with integral coefficients in terms of the coefficients of $P$ and $Q$.

**Lemma 5.25.** $\mathrm{Res}(P,Q) = (-1)^{mn} a_n^m b_m^n\, S(P,Q).$

**Proof.** We argue by induction on $m + n$. The first case to consider is $m = n = 1$. The computation is easy in this case: we have $u_1 = -\frac{a_0}{a_1}$, $v_1 = -\frac{b_0}{b_1}$, hence

$$S(P,Q) = -\frac{a_0}{a_1} + \frac{b_0}{b_1} = -\frac{a_0 b_1 - a_1 b_0}{a_1 b_1}$$

and

$$\mathrm{Res}(P,Q) = \det \begin{pmatrix} a_0 & a_1 \\ b_0 & b_1 \end{pmatrix} = a_0 b_1 - a_1 b_0.$$

For the rest of the proof, we assume $m + n > 2$. Note that each side of the equation we want to prove has the same behavior when $P$ and $Q$ are interchanged: from the definition it is clear that $S(Q, P) = (-1)^{mn} S(P, Q)$, and since $M(Q, P)$ is obtained from $M(P, Q)$ by cyclically permuting the rows of $M(P, Q)$ $m$ times from the bottom to the top (or $n$ times in the opposite direction) we also have $\mathrm{Res}(Q, P) = (-1)^{mn} \mathrm{Res}(P, Q)$. Therefore, up to interchanging $P$ and $Q$, we may also assume $n \geq m$. Reproducing the first step in the Euclidean division of $P$ by $Q$, we define a polynomial $P'$ by

$$P' = P - \frac{a_n}{b_m} X^{n-m} Q.$$

Let $n' = \deg P'$. Clearly, we have $n' < n$ and $P'(v_j) = P(v_j)$ for $j = 1$, $\ldots$, $m$. Therefore, letting $a'_{n'}$ denote the leading coefficient of $P'$ if $n' \geq 1$, we readily derive from (5.7)

$$a_n^m S(P, Q) = \begin{cases} (-1)^{m(n-n')} (a'_{n'})^m \, S(P', Q) & \text{if } n' \geq 1, \\ (-1)^{mn} P'^m & \text{if } P' \text{ is a constant.} \end{cases} \tag{5.8}$$

To prove the lemma, it suffices to establish the following:

$$\mathrm{Res}(P, Q) = \begin{cases} b_m^{n-n'} \, \mathrm{Res}(P', Q) & \text{if } n' \geq 1, \\ b_m^n P'^m & \text{if } P' \text{ is a constant.} \end{cases} \tag{5.9}$$

Indeed, if $P'$ is a constant the lemma readily follows by comparing (5.8) and (5.9). If $n' \geq 1$, the induction hypothesis yields $\mathrm{Res}(P', Q) = (-1)^{mn'} (a'_{n'})^m b_m^n \, S(P', Q)$, and again the lemma follows from (5.8) and (5.9).

To prove (5.9), observe that the first row of $M(P, Q)$, which is

$$(a_0, a_1, \ldots, a_n, \underbrace{0, \ldots, 0}_{m-1}),$$

consists of the coordinates of $P$ in the base $1, X, \ldots, X^{m+n-1}$ of the vector space $F[X]_{\leq m+n-1}$ of polynomials of degree at most $m + n - 1$. Similarly, we may view the second row as the $(m + n)$-tuple of coordinates of $XP$, the third as the coordinates of $X^2 P$, and so on. Now, the $(m + 1)$-st row, which is

$$(b_0, b_1, \ldots, b_m, \underbrace{0, \ldots, 0}_{n-1}),$$

is the $(m + n)$-tuple of coordinates of $Q$, and the $(n + 1)$-st row holds the coordinates of $X^{n-m} Q$. Therefore, as $P' = P - \frac{a_n}{b_m} X^{n-m} Q$, subtracting from the first row of $M(P,Q)$ the $(n{+}1)$-st row multiplied by $\frac{a_n}{b_m}$ amounts to substituting the coordinates of $P'$ for the coordinates of $P$ in the first row; this operation does not change the determinant. By the same argument, $\mathrm{Res}(P,Q)$ does not change when we substitute the coefficients of $XP'$, $X^2P'$, ..., $X^{m-1}P'$ for those of $XP$, $X^2P$, ..., $X^{m-1}P$ in the $m$ first rows of $M(P,Q)$. Note that the matrix $M'$ thus obtained is *not* $M(P',Q)$ because $M(P',Q)$ has order $m+n'$ whereas $M'$ has order $m+n$. In fact, if $P'$ is a constant we obtain

$$M' = \left. \begin{pmatrix} P' & 0 & \cdots & 0 & & & & \\ 0 & P' & & \vdots & & \mathbf{0} & & \\ \vdots & & \ddots & 0 & & & & \\ 0 & \cdots & 0 & P' & & & & \\ b_0 & b_1 & \cdots & b_{m-1} & b_m & 0 & \cdots & 0 \\ 0 & b_0 & b_1 & \cdots & b_{m-1} & b_m & & \vdots \\ \vdots & & \ddots & \ddots & & \ddots & \ddots & 0 \\ 0 & \cdots & 0 & b_0 & b_1 & \cdots & b_{m-1} & b_m \end{pmatrix} \right\} \begin{matrix} m \text{ rows} \\[1.2em] \\ n \text{ rows} \end{matrix}$$

hence $\mathrm{Res}(P,Q) = \det M' = P'^m b_m^n$. If $n' \geq 1$ the matrix $M'$ has the following form (where $*$ stands for entries whose precise value is irrelevant):

$$M' = \left( \begin{array}{c|c} M(P',Q) & \mathbf{0} \\ \hline & \begin{matrix} b_m & & 0 \\ & \ddots & \\ * & & b_m \end{matrix} \\ * & \end{array} \right) \begin{matrix} \} \ m+n' \text{ rows} \\[2em] \} \ n-n' \text{ rows} \end{matrix}$$

Therefore, $\mathrm{Res}(P,Q) = \det M' = b_m^{n-n'}\,\mathrm{Res}(P',Q)$, and (5.9) is proved. $\square$

**Theorem 5.26.** *Let $P$, $Q \in F[X]$ be non-constant polynomials over a field $F$, which split into products of linear factors over some field $K$ containing $F$. The following conditions are equivalent:*

(a) *$P$ and $Q$ are not relatively prime;*
(b) *$P$ and $Q$ have a common root in $K$;*
(c) *$\mathrm{Res}(P,Q) = 0$.*

**Proof.** (a) $\Longleftrightarrow$ (b) Let $D$ be a GCD of $P$ and $Q$. If (a) holds, $D$ is not constant. Since $D$ divides $P$, which splits into linear factors over $K$, the irreducible factors of $D$ in $K[X]$ have degree 1. Therefore, $D$ has at least one root in $K$, which is also a common root of $P$ and $Q$ since $D$ divides $P$ and $Q$. Conversely, if $P$ and $Q$ have a common root $u \in K$, then $X - u$ divides $P$ and $Q$, hence also $D$. Therefore, $D$ is not constant.

(b) $\Longleftrightarrow$ (c) From the definition of $S(P, Q)$, it is clear that (b) holds if and only if $S(P, Q) = 0$. This last condition is equivalent to (c) by Lemma 5.25.
$\square$

## Appendix: Decomposition of rational functions into sums of partial fractions

To find a primitive of a rational function $\frac{P}{Q}$ (where $P$, $Q \in \mathbb{R}[X]$ and $Q \neq 0$), it is customary to decompose it as a sum of partial fractions in the following way: factor $Q$ into irreducible factors

$$Q = Q_1^{m_1} \ldots Q_r^{m_r}$$

where $Q_1, \ldots, Q_r$ are distinct irreducible polynomials. Then there exist polynomials $P_0, P_1, \ldots, P_r$ such that

$$\frac{P}{Q} = P_0 + \frac{P_1}{Q_1^{m_1}} + \cdots + \frac{P_r}{Q_r^{m_r}}$$

and $\deg P_i < \deg Q_i^{m_i}$ for $i = 1, \ldots, r$.

The existence of the polynomials $P_0, \ldots, P_r$ follows by induction on $r$ from the following proposition:

**Proposition 5.27.** *If $Q = S_1 S_2$ where $S_1$ and $S_2$ are relatively prime polynomials, then there exist polynomials $P_0$, $P_1$, $P_2$ such that*

$$\frac{P}{Q} = P_0 + \frac{P_1}{S_1} + \frac{P_2}{S_2}$$

*and $\deg P_i < \deg S_i$ for $i = 1, 2$.*

**Proof.** Since $S_1$ and $S_2$ are relatively prime, Corollary 5.4 yields

$$1 = S_1 T_1 + S_2 T_2$$

for some polynomials $T_1$, $T_2$. Multiplying each side by $\frac{P}{Q}$, we get

$$\frac{P}{Q} = \frac{P T_1}{S_2} + \frac{P T_2}{S_1}. \tag{5.10}$$

By Euclidean division of $P T_1$ by $S_2$ and of $P T_2$ by $S_1$, we have

$$P T_1 = S_2 U_2 + P_2 \quad \text{and} \quad P T_2 = S_1 U_1 + P_1$$

for some polynomials $U_1$, $U_2$, $P_1$, $P_2$ with $\deg P_i < \deg S_i$ for $i = 1, 2$. Substituting for $PT_1$ and $PT_2$ in (5.10), we obtain

$$\frac{P}{Q} = (U_1 + U_2) + \frac{P_2}{S_2} + \frac{P_1}{S_1}. \qquad \square$$

To facilitate integration, each partial fraction $\frac{P}{Q^m}$ (where $Q$ is irreducible and $\deg P < \deg Q^m$) can be decomposed further as

$$\frac{P}{Q^m} = \frac{P_1}{Q} + \frac{P_2}{Q^2} + \cdots + \frac{P_m}{Q^m}$$

with $\deg P_i < \deg Q$ for $i = 1, \ldots, m$.

To obtain this decomposition, let $P_1$ be the quotient of the Euclidean division of $P$ by $Q^{m-1}$,

$$P = P_1 Q^{m-1} + R_1 \qquad \text{with} \qquad \deg R_1 < \deg Q^{m-1}.$$

Since $\deg P < \deg Q^m$ it follows that $\deg P_1 < \deg Q$. Let then $P_2$ be the quotient of the Euclidean division of $R_1$ by $Q^{m-2}$, and so on. Then

$$P = P_1 Q^{m-1} + P_2 Q^{m-2} + \cdots + P_{m-1} Q + P_m, \qquad (5.11)$$

and the required decomposition follows after the division of each side by $Q^m$.

**Remark.** The right-hand side of (5.11) is the "$Q$-adic" expansion of $P$. When $P$ and $Q$ are replaced by integers, equation (5.11) means that the integer $P$ is written as $P_1 P_2 \ldots P_m$ in the base $Q$.

## Chapter 6

# Alternative Methods for Cubic and Quartic Equations

With improved notation, mathematicians of the seventeenth century devised new methods for solving cubic and quartic equations. In this chapter, we review some of these advances, in particular the important method proposed by Ehrenfried Walther von Tschirnhaus in 1683.

## 6.1 Viète on cubic equations

Viète's contribution to the theory of cubic equations is twofold: in "De Recognitione Aequationum" he gave a trigonometric solution for the irreducible case and in "De Emendatione Aequationum" a solution for the general case, which requires the extraction of only one cube root. These methods were both posthumously published in "De Aequationum Recognitione et Emendatione Tractatus Duo" (1615) ("Two Treatises on the Understanding and Amendment of Equations," see [83]).

### 6.1.1  *Trigonometric solution for the irreducible case*

The irreducible case of cubic equations

$$x^3 + px + q = 0$$

occurs when $\left(\frac{p}{3}\right)^3 + \left(\frac{q}{2}\right)^2 < 0$ (see §2.3(c)). This inequality of course implies $p < 0$, hence there is no loss of generality if the above equation is written as

$$x^3 - 3a^2x = a^2b, \qquad (6.1)$$

and the condition $\left(\frac{p}{3}\right)^3 + \left(\frac{q}{2}\right)^2 < 0$ becomes $a > |\frac{b}{2}|$. (Note that one can obviously assume $a > 0$, since only $a^2$ occurs in the given equation.)

From the formula for the cosine of a sum of arcs (or from the general formula (4.1) in Chapter 4, p. 34), it follows that for all $\alpha \in \mathbb{R}$

$$(2a\cos\alpha)^3 - 3a^2(2a\cos\alpha) = 2a^3\cos 3\alpha.$$

Comparing with (6.1), we see that if $\alpha$ is an arc such that

$$\cos 3\alpha = \frac{b}{2a},$$

then $2a\cos\alpha$ is a solution of (6.1). The other two solutions are easily derived from this one: for $\alpha_k = \alpha + k\frac{2\pi}{3}$, with $k = 1, 2$, we also have

$$\cos 3\alpha_k = \frac{b}{2a},$$

hence the solutions of (6.1) are

$$2a\cos\alpha,$$
$$2a\cos\left(\alpha + \tfrac{2\pi}{3}\right) = -a\cos\alpha - a\sqrt{3}\sin\alpha,$$
$$\text{and}\quad 2a\cos\left(\alpha + \tfrac{4\pi}{3}\right) = -a\cos\alpha + a\sqrt{3}\sin\alpha.$$

Since Viète systematically avoids negative numbers, he gives only the first solution, which is positive if $b > 0$ (see [83, p. 174]). However, he points out immediately afterwards that $a\cos\alpha + a\sqrt{3}\sin\alpha$ and $a\cos\alpha - a\sqrt{3}\sin\alpha$ are solutions of the equation

$$3a^2 y - y^3 = a^2 b.$$

This shows how clear was Viète's notion of the number of roots of cubic equations.

### 6.1.2 *Algebraic solution for the general case*

Viète suggested (in [83, p. 287]) an ingenious change of variable to solve the equation

$$x^3 + px + q = 0. \tag{6.2}$$

Setting $x = \frac{p}{3y} - y$ and substituting in (6.2), he gets for $y$ the equation

$$y^6 - qy^3 - \left(\frac{p}{3}\right)^3 = 0,$$

hence $y^3$ can be found by solving a quadratic equation. The solutions are

$$y^3 = \frac{q}{2} \pm \sqrt{\left(\frac{p}{3}\right)^3 + \left(\frac{q}{2}\right)^2}.$$

Therefore, a solution of (6.2) is given by

$$x = \frac{p}{3y} - y,$$

where

$$y^3 = \frac{q}{2} + \sqrt{\left(\frac{p}{3}\right)^3 + \left(\frac{q}{2}\right)^2}.$$

**Remarks.** (a) If the other determination of $y^3$ is chosen, namely

$$y'^3 = \frac{q}{2} - \sqrt{\left(\frac{p}{3}\right)^3 + \left(\frac{q}{2}\right)^2},$$

then the value of $x$ does not change. Indeed, since $(yy')^3 = -\left(\frac{p}{3}\right)^3$, we have

$$\frac{p}{3y} = -y' \quad \text{and} \quad \frac{p}{3y'} = -y,$$

hence

$$\frac{p}{3y} - y = \frac{p}{3y'} - y'.$$

Incidentally, this remark also shows that Viète's method yields the same result as Cardano's formula, since after substituting for $y$ and $y'$ in the formula $x = -y - y'$, we obtain

$$x = \sqrt[3]{-\frac{q}{2} + \sqrt{\left(\frac{p}{3}\right)^3 + \left(\frac{q}{2}\right)^2}} + \sqrt[3]{-\frac{q}{2} - \sqrt{\left(\frac{p}{3}\right)^3 + \left(\frac{q}{2}\right)^2}}.$$

**(b)** The case where $y = 0$ occurs only if $p = 0$; this case is therefore readily solved.

**(c)** Viète gives only one root, because in the original formulation the equation is[1]

$$A^3 + 3B^p A = 2Z^s,$$

which has only one real root if $B^p$ is positive, since the function $A^3 + 3B^p A$ is then monotonically increasing and therefore takes the value $2Z^s$ only once.

---

[1] The exponents $p$, $s$ are for *plano* and *solido*; unknowns are always designated by vowels.

## 6.2    Descartes on quartic equations

New insights into the solution of equations arose from the arithmetic of polynomials. In "La Geometrie," Descartes recommends the following method to tackle equations of any degree:

> First, try to put the given equation into the form of an equation of the same degree obtained by multiplying together two others, each of a lower degree [23, p. 192].

He himself shows how this method can be successfully applied to quartic equations [23, pp. 180ff]. After canceling out the cubic term, as in Ferrari's method (Chapter 3), the general quartic equation is set in the form

$$x^4 + px^2 + qx + r = 0, \tag{6.3}$$

and we may assume $q \neq 0$, otherwise the equation is quadratic in $x^2$ and is therefore easily solved. We then determine $a$, $b$, $c$, $d$ in such a way that

$$x^4 + px^2 + qx + r = (x^2 + ax + b)(x^2 + cx + d).$$

Equating the coefficients of like powers of $x$, we obtain from this equation

$$0 = a + c, \tag{6.4}$$

$$p = b + d + ac, \tag{6.5}$$

$$q = ad + bc, \tag{6.6}$$

$$r = bd. \tag{6.7}$$

From (6.4), (6.5), (6.6), the values of $b$, $c$ and $d$ are easily derived in terms of $a$:

$$c = -a,$$

$$b = \frac{a^2}{2} + \frac{p}{2} - \frac{q}{2a},$$

$$d = \frac{a^2}{2} + \frac{p}{2} + \frac{q}{2a}.$$

(Observe that $a \neq 0$ since $q = a(d - b)$ by (6.4) and (6.6) and $q$ is assumed to be non-zero.) Substituting for $b$ and $d$ in (6.7), we get the following equation for $a$:

$$a^6 + 2pa^4 + (p^2 - 4r)a^2 - q^2 = 0. \tag{6.8}$$

This is a cubic equation in $a^2$, which can therefore be solved. If $a$ is a solution of this equation, then the given equation (6.3) factors into two quadratic equations

$$x^2 + ax + \frac{a^2}{2} + \frac{p}{2} - \frac{q}{2a} = 0 \quad \text{and} \quad x^2 - ax + \frac{a^2}{2} + \frac{p}{2} + \frac{q}{2a} = 0,$$

hence the solutions are easily found.

## 6.3 Rational solutions for equations with rational coefficients

The rational solutions of equations with rational coefficients of arbitrary degree can be found by a finite trial and error process. This seems to have been first observed by Albert Girard [33, D.4 v°]; it also appears in "La Geometrie" [23, p. 176].

Let

$$a_n x^n + a_{n-1} x^{n-1} + \cdots + a_1 x + a_0 = 0 \tag{6.9}$$

be an equation with rational coefficients $a_i \in \mathbb{Q}$ for $i = 0, \ldots, n$. Multiplying each side by a common multiple of the denominators of the coefficients if necessary, we may assume $a_i \in \mathbb{Z}$ for $i = 0, \ldots, n$. Multiplying then each side by $a_n^{n-1}$, the equation becomes

$$(a_n x)^n + a_{n-1}(a_n x)^{n-1} + a_{n-2} a_n (a_n x)^{n-2} + \cdots$$
$$+ a_1 a_n^{n-2}(a_n x) + a_0 a_n^{n-1} = 0.$$

Letting $y = a_n x$, we are then reduced to a *monic* equation with integral coefficients

$$y^n + b_{n-1} y^{n-1} + b_{n-2} y^{n-2} + \cdots + b_1 y + b_0 = 0 \qquad (b_i \in \mathbb{Z}). \tag{6.10}$$

**Theorem 6.1.** *All the rational roots of a monic equation with integral coefficients are integers; they divide the constant term.*

**Proof.** Discarding the null roots and dividing the left-hand side of (6.10) by a suitable power of $y$, we may assume $b_0 \neq 0$. Let then $y \in \mathbb{Q}$ be a rational root of (6.10). Write $y = \frac{y_1}{y_2}$ where $y_1$, $y_2$ are relatively prime integers. From

$$\left(\frac{y_1}{y_2}\right)^n + b_{n-1}\left(\frac{y_1}{y_2}\right)^{n-1} + \cdots + b_1\left(\frac{y_1}{y_2}\right) + b_0 = 0$$

it follows, after multiplication by $y_2^n$ and rearrangement of the terms,

$$y_1^n = -y_2\left(b_{n-1} y_1^{n-1} + \cdots + b_1 y_1 y_2^{n-2} + b_0 y_2^{n-1}\right).$$

This equation shows that each prime factor of $y_2$ divides $y_1^n$, hence also $y_1$; since $y_1$ and $y_2$ are relatively prime, this is impossible unless $y_2$ has no prime factor. Therefore $y_2 = \pm 1$ and $y \in \mathbb{Z}$.

To prove that $y$ divides $b_0$, consider again equation (6.10) and separate off on one side the constant term; we obtain

$$b_0 = -y\left(y^{n-1} + b_{n-1} y^{n-2} + \cdots + b_1\right).$$

This equation shows that $y$ divides $b_0$, since the factor between brackets is an integer. $\square$

Tracing back through the transformations from (6.9) to (6.10), we obtain the following result.

**Corollary 6.2.** *Each rational solution of the equation with integral coefficients*

$$a_n x^n + a_{n-1} x^{n-1} + \cdots + a_1 x + a_0 = 0 \qquad (a_i \in \mathbb{Z})$$

*has the form* $y a_n^{-n}$, *where* $y \in \mathbb{Z}$ *divides* $a_0 a_n^{n-1}$.

This last condition if very useful, in that it gives a bound on the number of trials that are necessary to find a rational root of the proposed equation, provided that $a_0 \neq 0$. Of course, this can always be assumed, after dividing by a suitable power of $x$.

For example, the theorem (or its corollary) shows that an equation like

$$x^n + a_{n-1} x^{n-1} + \cdots + a_1 x \pm 1 = 0$$

with $a_i \in \mathbb{Z}$ for $i = 1, \ldots, n-1$, has no rational root, except possibly $+1$ or $-1$.

"Other example, once very difficult" [33, E r°]: the rational solutions of

$$x^3 = 7x - 6$$

are among $\pm 1$, $\pm 2$, $\pm 3$, $\pm 6$. Trying successively the various possibilities, one finds 1, 2 and $-3$ as solutions.

## 6.4   Tschirnhaus' method

An idea that occurred to several mathematicians through the second half of the seventeenth century (see Stedall [68, Ch. 3]) is to simplify equations by changing variables. Cardano had shown that the term $ax^2$ can be removed from a cubic equation by the simple change of variables $y = x + \frac{a}{3}$ (see §2.2); perhaps more elaborate changes could remove two or more terms from an arbitrary equation? If one could remove *all* the intermediate terms between the first and the last, then the simplified equation is readily solved by a root extraction.

Specifically, to solve an arbitrary equation

$$x^n + a_{n-1} x^{n-1} + \cdots + a_1 x + a_0 = 0,$$

the idea is to try a change of variables of the form

$$y = x^{n-1} + b_{n-2} x^{n-2} + \cdots + b_1 x + b_0 \qquad (6.11)$$

(for suitable coefficients $b_0$, $b_1$, ..., $b_{n-2}$). Compute the resulting equation in $y$:

$$y^n + c_{n-1}y^{n-1} + \cdots + c_1 y + c_0 = 0,$$

and determine $b_0$, ..., $b_{n-2}$ so that $c_1 = \cdots = c_{n-1} = 0$. (Note that the number $n-1$ of conditions is equal to the number of parameters $b_0$, ..., $b_{n-2}$.) Then the equation in $y$ is easily solved; plugging the solution $y = \sqrt[n]{-c_0}$ in (6.11), we then find $x$ by solving an equation of degree $n-1$.

The problem is of course to determine $b_0$, ..., $b_{n-2}$ suitably. The difficulty of this task is not immediately apparent because the computation of the coefficients $c_1$, ..., $c_{n-1}$ in terms of $b_0$, ..., $b_{n-2}$ is already a serious hurdle (not unlike the computation of resultants in §5.6, as we will see below). Yet, in a paper of 1683 Tschirnhaus [75] boldly claimed that this method could be applied to solve equations of any degree—although he demonstrated it only on cubic equations. Leibniz was (rightfully) skeptical; he suggested that in degree 5 complications would arise that would not be simply due to the length of computations (see Leibniz [51, p. 449]).

To discuss Tschirnhaus' method in some detail, we start by explaining from a modern perspective how the equation in $y$ can be found (which Tschirnhaus himself does not spell out). Consider the polynomial

$$P = X^n + a_{n-1}X^{n-1} + \cdots + a_1 X + a_0 \in F[X]$$

over some field $F$, and let $x_1$, ..., $x_n$ be the roots of $P$ in some field $K$ containing $F$. Fix some integer $m$ between 1 and $n-1$, and some polynomial $f = X^m + b_{m-1}X^{m-1} + \cdots + b_1 X + b_0 \in F[X]$ of degree $m$. For $i = 1, \ldots, n$, let

$$y_i = f(x_i) \in K.$$

Then $y_1$, ..., $y_n$ are the roots of the following polynomial of degree $n$:

$$R(Y) = \big(Y - f(x_1)\big) \ldots \big(Y - f(x_n)\big), \tag{6.12}$$

which we write as $R(Y) = Y^n + c_{n-1}Y^{n-1} + \cdots + c_1 Y + c_0$. To see that the coefficients $c_0$, ..., $c_{n-1}$ lie in $F$, we may compare the construction above to the resultant of two polynomials (see §5.6): consider the expression $Q = Y - f(X)$ as a polynomial in $X$ with coefficients in the field $F(Y)$ of rational functions in $Y$. By comparing (5.6) and Lemma 5.25 with (6.12) we see that $R(Y)$ is (up to sign) the resultant $\mathrm{Res}(P, Q)$. We may therefore compute $R(Y)$ as the determinant of some matrix of order $m + n$ in which the nonzero entries are the coefficients $a_0, \ldots a_{n-1}, Y - b_0, -b_1, \ldots, -b_{m-1}$

of $P$ and $Q$. Thus, $c_0, \ldots, c_{n-1}$ lie in $F$. We may also use (6.12) to obtain the following explicit expressions:

$$-c_{n-1} = \sum_{i=1}^{n} f(x_i), \qquad c_{n-2} = \sum_{i<j} f(x_i)f(x_j),$$

$$-c_{n-3} = \sum_{i<j<k} f(x_i)f(x_j)f(x_k), \quad \ldots, \quad (-1)^n c_0 = f(x_1)\ldots f(x_n).$$

From these equations, it follows that $c_{n-1}$ is a polynomial of degree 1 in $b_0, \ldots, b_{m-1}$, that $c_{n-2}$ is a polynomial of degree 2, etc. To cancel out $c_1$, $\ldots, c_{n-1}$ following Tschirnhaus' strategy, consider now $m = n - 1$. The preceding discussion shows that $c_{n-1} = c_{n-2} = \ldots = c_1 = 0$ is a system of $n - 1$ equations of degrees 1, 2, 3, $\ldots$, $n - 1$ in the variables $b_0, \ldots,$ $b_{n-2}$. Between these equations, $n - 2$ variables can be eliminated, and the resulting equation in a single variable has degree $1 \cdot 2 \cdot 3 \cdot \ldots \cdot (n-1) = (n-1)!$ (see for instance Weber [85, §53]). This follows from a theorem of Bézout in elimination theory, which was proved much later. But by considering some examples one soon realizes that the solution of the above system of equations is far from easy. It thus appears that this method does not work for $n > 3$, unless the resulting equation of degree $(n - 1)!$ has some particular features that make it reducible to equations of degree less than $n$. This turns out to be the case for $n = 4$: the resulting sextic can be seen to factor into a product of polynomials of degree 2 whose coefficients are solutions of cubic equations (see Lagrange [49, Art. 41–45]), but for $n \geq 5$ no such simplification is apparent. (Note that for composite $n$, Tschirnhaus' method can be applied differently, and possibly more easily. For instance if $n = 4$, then canceling out the coefficients of $Y$ and $Y^3$ reduces the equation in $Y$ to a quadratic equation in $Y^2$.)

As an example, consider the cubic $P = X^3 + pX + q$. By Girard's theorem (see (4.5), p. 36), its roots $x_1$, $x_2$, $x_3$ satisfy

$$x_1 + x_2 + x_3 = 0, \quad x_1x_2 + x_1x_3 + x_2x_3 = p, \quad x_1x_2x_3 = -q.$$

Note for use in the calculations below that these equations imply

$$x_1^2 + x_2^2 + x_3^2 = (x_1 + x_2 + x_3)^2 - 2(x_1x_2 + x_1x_3 + x_2x_3) = -2p,$$

$$x_1^2x_2 + x_1x_2^2 + x_1^2x_3 + x_1x_3^2 + x_2^2x_3 + x_2x_3^2$$
$$= (x_1 + x_2 + x_3)(x_1x_2 + x_1x_3 + x_2x_3) - 3x_1x_2x_3 = 3q,$$

$$x_1^2x_2^2 + x_1^2x_3^2 + x_2^2x_3^2$$
$$= (x_1x_2 + x_1x_3 + x_2x_3)^2 - 2(x_1 + x_2 + x_3)x_1x_2x_3 = p^2.$$

Let $f(X) = X^2 + b_1 X + b_0$, where the parameters $b_0$, $b_1$ are to be determined. Under the change of variables $Y = f(X)$, the resulting cubic in $Y$ is

$$R(Y) = (Y - x_1^2 - b_1 x_1 - b_0)(Y - x_2^2 - b_1 x_2 - b_0)(Y - x_3^2 - b_1 x_3 - b_0).$$

Multiplying out and rearranging terms, we find

$$R(Y) = Y^3 + c_2 Y^2 + c_1 Y + c_0$$

with

$$
\begin{aligned}
c_2 = {} & -\big((x_1^2 + x_2^2 + x_3^2) + b_1(x_1 + x_2 + x_3) + 3b_0\big) = 2p - 3b_0, \\
c_1 = {} & (x_1^2 x_2^2 + x_1^2 x_3^2 + x_2^2 x_3^2) + 2b_0(x_1^2 + x_2^2 + x_3^2) \\
& + b_1(x_1^2 x_2 + x_1 x_2^2 + x_1^2 x_3 + x_1 x_3^2 + x_2^2 x_3 + x_2 x_3^2) \\
& + b_1^2(x_1 x_2 + x_1 x_3 + x_2 x_3) + 2b_1 b_0(x_1 + x_2 + x_3) + 3b_0^2 \\
= {} & p^2 - 4pb_0 + 3qb_1 + pb_1^2 + 3b_0^2, \\
c_0 = {} & -\big(x_1^2 x_2^2 x_3^2 + b_1(x_1^2 x_2^2 x_3 + x_1^2 x_2 x_3^2 + x_1 x_2^2 x_3^2) \\
& + b_0(x_1^2 x_2^2 + x_1^2 x_3^2 + x_2^2 x_3^2) + b_1^2(x_1^2 x_2 x_3 + x_1 x_2^2 x_3 + x_1 x_2 x_3^2) \\
& + b_1 b_0(x_1^2 x_2 + x_1 x_2^2 + x_1^2 x_3 + x_1 x_3^2 + x_2^2 x_3 + x_2 x_3^2) \\
& + b_0^2(x_1^2 + x_2^2 + x_3^2) + b_1^3 x_1 x_2 x_3 + b_1^2 b_0(x_1 x_2 + x_1 x_3 + x_2 x_3) \\
& + b_1 b_0^2(x_1 + x_2 + x_3) + b_0^3\big) \\
= {} & -\big(q^2 - pqb_1 + p^2 b_0 + 3qb_1 b_0 - 2pb_0^2 - qb_1^3 + pb_1^2 b_0 + b_0^3\big).
\end{aligned}
$$

Thus, in order to cancel out $c_2$ and $c_1$, it suffices to let $b_0 = \frac{2p}{3}$ and to choose for $b_1$ a root of the quadratic equation

$$pb_1^2 + 3qb_1 - \frac{p^2}{3} = 0,$$

for instance (assuming $p \neq 0$)

$$b_1 = \frac{3}{p}\left(\sqrt{\left(\frac{p}{3}\right)^3 + \left(\frac{q}{2}\right)^2} - \frac{q}{2}\right).$$

With the above choice of $b_0$ and $b_1$, and letting $A = \sqrt{\left(\frac{p}{3}\right)^3 + \left(\frac{q}{2}\right)^2}$, we have

$$c_0 = -2^3 A^3 \left(\frac{3}{p}\right)^3 \left(A - \frac{q}{2}\right).$$

Therefore, a root of the resulting cubic $R(Y)$ is

$$y = \frac{6A}{p} \sqrt[3]{A - \frac{q}{2}}.$$

A root of the proposed cubic $P(X)$ is then found by solving the quadratic equation $f(X) = y$, which is now

$$X^2 + \frac{3}{p}\left(A - \frac{q}{2}\right)X + \frac{2p}{3} = \frac{6A}{p}\sqrt[3]{A - \frac{q}{2}}. \qquad (6.13)$$

However, in general only one of the roots of this quadratic equation is a root of $P(X)$. A better way to solve $P(X) = 0$ is to find the common roots of $P$ and (6.13), which are the roots of their greatest common divisor.

Letting $B = \sqrt[3]{A - \frac{q}{2}}$, one gets by Euclid's algorithm (p. 45) the following greatest common divisor, if $A \neq 0$:

$$2A\left(\frac{3}{p}\right)^2\left(B^2 + \frac{p}{3}\right)\left(BX + \frac{p}{3} - B^2\right).$$

(It is easy to see that $B^2 + \frac{p}{3} \neq 0$ if $A \neq 0$ and $p \neq 0$.) There is thus only one common root of $P$ and (6.13); namely

$$X = \frac{B^2 - \frac{p}{3}}{B}.$$

Since $B = \sqrt[3]{-\frac{q}{2} + \sqrt{\left(\frac{p}{3}\right)^3 + \left(\frac{q}{2}\right)^2}}$, it is easily verified that

$$-\frac{p}{3B} = \sqrt[3]{-\frac{q}{2} - \sqrt{\left(\frac{p}{3}\right)^3 + \left(\frac{q}{2}\right)^2}};$$

thus the above formula for $X$ is identical to Cardano's formula. (Compare also Viète's method in §6.1.)

If $A = 0$, then the left-hand side of (6.13) divides $P$, so both roots of (6.13) are roots of $P$.

# Chapter 7

# Roots of Unity

New branches of mathematics, such as analytic geometry and differential calculus, came into being during the seventeenth century, and it is therefore not surprising that investigations in the algebraic theory of equations were given lower priority at the end of that century. However, progress in other branches indirectly brought some new advances in algebra. A case in point is the well-known "de Moivre's formula": for every integer $n$ and every $\alpha \in \mathbb{R}$,

$$(\cos \alpha + i \sin \alpha)^n = \cos(n\alpha) + i \sin(n\alpha). \tag{7.1}$$

This formula is easily proved by induction on $n$, since from the addition formulas for sines and cosines it readily follows that

$$(\cos \alpha + i \sin \alpha)(\cos \beta + i \sin \beta) = \cos(\alpha + \beta) + i \sin(\alpha + \beta). \tag{7.2}$$

Formula (7.1), and its proof through (7.2), were first given by Euler in 1748 (see Smith [67, vol. 2, p. 450]) but it was already implicit in earlier works by Cotes and by de Moivre. Actually, the above proof, simple as it is, is deceitful, since it does not keep any record of the slow evolution that led to de Moivre's formula. It is the purpose of this chapter to sketch this evolution and to discuss the significance of de Moivre's formula for the algebraic theory of equations.

## 7.1 The origins of de Moivre's formula

While the differential calculus was being shaped by Leibniz and Newton, the integration (or primitivation) of rational functions was unavoidable. Very soon, formulas equivalent to

$$\int x^n dx = \frac{x^{n+1}}{n+1} \quad \text{for } n \neq -1 \quad \text{and} \quad \int \frac{dx}{x} = \log x$$

became familiar, and the integration of any rational function in which the denominator is a power of a linear polynomial easily follows by a change of variable. Moreover, around 1675, Leibniz had also obtained

$$\int \frac{dx}{x^2 + 1} = \tan^{-1} x,$$

from which the integration of other rational functions can be derived.

The integration of rational functions is the main theme of a 1702 paper by Leibniz in the Acta Eruditorum of Leipzig : "Specimen novum Analyseos pro Scientia infiniti circa Summas et Quadraturas" ("New specimen of the Analysis for the Science of the infinite about Sums and Quadratures" [52, n° 24]). In this paper, Leibniz points out the usefulness of the decomposition of rational functions into sums of partial fractions (see the appendix to Chapter 5) to reduce the integration of rational functions to the integration of $\frac{dx}{x}$ and $\frac{dx}{x^2+1}$ or, in his words, to the quadrature of the hyperbola or the circle. Since this decomposition requires that the denominator be factored in a product of irreducible polynomials, he is thus led to investigate the factorization of real polynomials, coming close to the "fundamental theorem of algebra" according to which every real polynomial of positive degree is a product of factors of degree 1 or 2 (see Chapter 9).

> Now, this leads us to a question of utmost importance: whether all the rational quadratures can be reduced to the quadrature of the hyperbola and of the circle, which by our analysis above amounts to the following: whether every algebraic equation or real integral formula in which the indeterminate is rational can be decomposed into simple or plane real factors [= real factors of degree 1 or 2] [52, p. 359].

Leibniz then proposes the following counterexample: since

$$x^4 + a^4 = (x^2 + a^2\sqrt{-1})(x^2 - a^2\sqrt{-1}),$$

it follows that

$$x^4 + a^4$$
$$= \left(x + a\sqrt{\sqrt{-1}}\right)\left(x - a\sqrt{\sqrt{-1}}\right)\left(x + a\sqrt{-\sqrt{-1}}\right)\left(x - a\sqrt{-\sqrt{-1}}\right).$$

Failing to observe that

$$\sqrt{\sqrt{-1}} = \frac{1 + \sqrt{-1}}{\sqrt{2}} \qquad \text{and} \qquad \sqrt{-\sqrt{-1}} = \frac{1 - \sqrt{-1}}{\sqrt{2}},$$

he draws the erroneous conclusion that no non-trivial combination of the four factors above yields a real divisor of $x^4 + a^4$.

Therefore, $\int \frac{dx}{x^4+a^4}$ cannot be reduced to the squaring of the circle or the hyperbola by our analysis above, but founds a new kind of its own [52, p. 360].

Even without deeper considerations about complex numbers, Leibniz could have avoided this mistake if he had observed that, by adding and subtracting $2a^2x^2$, one gets[1]

$$x^4 + a^4 = (x^2 + a^2)^2 - 2a^2x^2$$
$$= (x^2 + a^2 + \sqrt{2}ax)(x^2 + a^2 - \sqrt{2}ax).$$

As it appears from [58, pp. 205 ff], Newton had also tried his hand at the same questions as early as 1676, and he had obtained this factorization of $x^4+a^4$, as well as factorizations of $1\pm x^n$ for various values of the integer $n$, (see the appendix to this chapter), but in 1702 he presumably did not care enough about mathematics any more to point out the mistake in Leibniz's paper, had he been aware of it.

Leibniz's argument was definitively refuted by Roger Cotes (1682–1716), who thoroughly investigated the factorization of the binomials $a^n \pm x^n$, obtaining the following formulas:

$$a^{2m} + x^{2m} = \prod_{k=0}^{m-1} \left(a^2 - 2a\cos\left(\frac{(2k+1)\pi}{2m}\right)x + x^2\right), \tag{7.3}$$

$$a^{2m+1} + x^{2m+1} = (a+x) \prod_{k=0}^{m-1} \left(a^2 - 2a\cos\left(\frac{(2k+1)\pi}{2m+1}\right)x + x^2\right), \tag{7.4}$$

$$a^{2m} - x^{2m} = (a-x)(a+x) \prod_{k=1}^{m-1} \left(a^2 - 2a\cos\left(\frac{2k\pi}{2m}\right)x + x^2\right), \tag{7.5}$$

$$a^{2m+1} - x^{2m+1} = (a-x) \prod_{k=1}^{m} \left(a^2 - 2a\cos\left(\frac{2k\pi}{2m+1}\right)x + x^2\right). \tag{7.6}$$

These formulas appear in a compilation of Cotes' papers entitled "Theoremata tum Logometrica tum Trigonometrica Datarum Fluxionum Fluentes exhibentia, per Methodum Mensurarum Ulterius extensam" (1722) ("Theorems, some logometric, some trigonometric, which yield the fluents of given fluxions by the method of measures further developed" [19, pp. 113–114]), in a very elegant form: to find the factors of $a^\lambda \pm x^\lambda$, it is prescribed to divide a circle of radius $a$ into $2\lambda$ equal parts $AB, BC, CD, DE, EF$, etc.

---

[1]This was pointed out by N. Bernoulli in the Acta Eruditorum of 1719.

Let $O$ be the center of the circle and let $P$ be a point on the radius $OA$, at a distance $OP = x$ $(< a)$ from $O$. Then

$$a^\lambda - x^\lambda = OA^\lambda - OP^\lambda = AP \cdot CP \cdot EP \cdot \text{etc.}, \qquad \text{and}$$
$$a^\lambda + x^\lambda = OA^\lambda + OP^\lambda = BP \cdot DP \cdot FP \cdot \text{etc.}$$

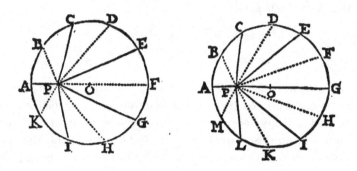

Exempli gratia fi λ fit 5, dividatur circumferentia in 10 partes æquales, eritque $AP \times CP \times EP \times GP \times IP = OA^5 - OP^5$ exiftente $P$ intra circulum: & $BP \times DP \times FP \times HP \times KP = OA^5 + OP^5$. Similiter fi λ fit 6, divifa circumferentia in 12 partes æquales: erit $AP \times CP \times EP \times GP \times IP \times LP = OA^6 - OP^6$, exiftente $P$ intra circulum; & $BP \times DP \times FP \times HP \times KP \times MP = OA^6 + OP^6$.

[19, p. 114] (Univ. Cath. Louvain, Centre général de Documentation)

To check that this formulation is equivalent to the previous one, consider the following figure:

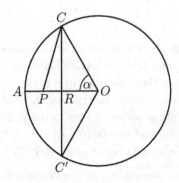

By Pythagoras' theorem, we have

$$CP^2 = PR^2 + RC^2.$$

Since

$$PR = OP - OR = x - a\cos\alpha \qquad \text{and} \qquad RC = a\sin\alpha,$$

it follows that

$$CP = \sqrt{x^2 - 2ax\cos\alpha + a^2}.$$

Therefore,

$$CP \cdot C'P = x^2 - 2ax\cos\alpha + a^2.$$

Cotes' formulas were given without justification, but a proof was eventually supplied in 1730 by Abraham de Moivre (1667–1754), who had already obtained some interesting results on the division of the circle. In a 1707 paper [20] (see also Smith [67, vol. 2, pp. 441 ff]), he had considered the equation

$$T_n(x) = a \qquad (n \text{ odd})$$

where $T_n$ is the polynomial by which $\cos n\alpha$ is expressed as a function of $\cos\alpha$. (The polynomials $T_n$ are the Chebyshev polynomials of the first kind; they are related to the polynomials $f_n$ defined in Chapter 4 (see (4.1), p. 34) by $2T_n(X) = f_n(2X)$.) De Moivre had observed that the equation $T_n(x) = a$ has the solution

$$x = \tfrac{1}{2}\sqrt[n]{a + \sqrt{a^2 - 1}} + \tfrac{1}{2}\sqrt[n]{a - \sqrt{a^2 - 1}},$$

for any value of $a$, whatsoever. In particular, if $a = \cos n\alpha$, it follows that

$$\cos\alpha = \tfrac{1}{2}\sqrt[n]{\cos n\alpha + \sqrt{-1}\sin n\alpha} + \tfrac{1}{2}\sqrt[n]{\cos n\alpha - \sqrt{-1}\sin n\alpha}, \qquad (7.7)$$

although this formula does not appear explicitly in de Moivre's paper of 1707.

De Moivre's result can be justified by observing that the equation $T_n(x) = a$ follows by elimination of $z$ between the two equations

$$1 - 2az^n + z^{2n} = 0, \qquad (7.8)$$

$$1 - 2xz + z^2 = 0. \qquad (7.9)$$

To see this, note that $T_1(x) = x$, $T_2(x) = 2x^2 - 1$, and $T_{n+1}(x) = 2xT_n(x) - T_{n-1}(x)$ for all $n \geq 2$; these equations readily follow from the formula for the cosine of a sum of arcs (see equation (4.3), p. 34). We may therefore rewrite (7.9) as $1 + z^2 = 2T_1(x)z$. By squaring each side, we obtain

$$1 + z^4 = 2(x^2 - 1)z^2 = 2T_2(x)z^2.$$

Next, multiply the left-hand side of the last equation by $1 + z^2$ and the right-hand side by $2xz$ to get

$$1 + z^2(1 + z^2) + z^6 = 4xT_2(x)z^3.$$

Because $1 + z^2 = 2T_1(x)z$, the last equation yields

$$1 + z^6 = \big(4xT_2(x) - 2T_1(x)\big)z^3 = 2T_3(x)z^3.$$

By induction on $n$, we see that (7.9) yields

$$1 + z^{2n} = 2T_n(x)z^n \qquad \text{for all } n \geq 1. \tag{7.10}$$

Comparing (7.10) with (7.8), we obtain $T_n(x) = a$. Now, by dividing (7.9) by $2z$, it follows that $x = \frac{1}{2}(z + z^{-1})$, while (7.8) yields

$$z^n = a \pm \sqrt{a^2 - 1}.$$

We thus obtain several equivalent expressions for $x$:

$$x = \frac{1}{2}\sqrt[n]{a + \sqrt{a^2 - 1}} + \frac{1}{2}\left(\sqrt[n]{a + \sqrt{a^2 - 1}}\right)^{-1},$$

or

$$x = \frac{1}{2}\sqrt[n]{a + \sqrt{a^2 - 1}} + \frac{1}{2}\sqrt[n]{a - \sqrt{a^2 - 1}},$$

or

$$x = \frac{1}{2}\left(\sqrt[n]{a - \sqrt{a^2 - 1}}\right)^{-1} + \frac{1}{2}\sqrt[n]{a - \sqrt{a^2 - 1}},$$

or

$$x = \frac{1}{2}\left(\sqrt[n]{a - \sqrt{a^2 - 1}}\right)^{-1} + \frac{1}{2}\left(\sqrt[n]{a + \sqrt{a^2 - 1}}\right)^{-1}.$$

The equivalence of these expressions is easily derived from the equation

$$(a + \sqrt{a^2 - 1})(a - \sqrt{a^2 - 1}) = 1.$$

De Moivre repeatedly returned to these questions in the sequel, displaying formula (7.7) quite explicitly on p. 1 of his book "Miscellanea Analytica" (1730) (see Smith [67, vol. 2, p. 446]). It is noteworthy that for $x = \cos\alpha$ and $a = \cos n\alpha$, the values of $z$ obtained by solving equations (7.8) and (7.9) are respectively

$$\sqrt[n]{a \pm \sqrt{a^2 - 1}} = \sqrt[n]{\cos n\alpha \pm \sqrt{-1}\sin n\alpha}$$

and

$$x \pm \sqrt{x^2 - 1} = \cos\alpha \pm \sqrt{-1}\sin\alpha,$$

so that

$$\sqrt[n]{\cos n\alpha \pm \sqrt{-1}\sin n\alpha} = \cos\alpha \pm \sqrt{-1}\sin\alpha, \qquad (7.11)$$

but this was never written out explicitly by de Moivre. Nevertheless, de Moivre's approach turned out to be quite fruitful, since Cotes' formulas can be easily proved by pushing the preceding calculations a little further (see Exercise 7.1).

In 1739, de Moivre used the trigonometric representation of complex numbers and presumably also his formula, which certainly was thoroughly familiar to him by then, to extract the $n$-th root of the "impossible binomial $a + \sqrt{-b}$" (see Smith [67, vol. 2, p. 449]). He states the procedure as follows: let $\varphi$ be an angle such that

$$\cos\varphi = \frac{a}{\sqrt{a^2 + b}};$$

then the $n$-th roots of $a + \sqrt{-b}$ are

$$\sqrt[2n]{a^2 + b}\left(\cos\psi + \sqrt{\cos^2\psi - 1}\right),$$

where $\psi$ ranges over $\frac{\varphi}{n}$, $\frac{2\pi - \varphi}{n}$, $\frac{2\pi + \varphi}{n}$, $\frac{4\pi - \varphi}{n}$, $\frac{4\pi + \varphi}{n}$, etc. until the number of them is equal to $n$. (This result is correct up to the sign of the imaginary part: see Proposition 7.1 below.)

As a result of this work, the credibility of the "fundamental theorem of algebra" was significantly enhanced, since the objection that Leibniz had raised was definitely answered: it was clear that extraction of roots of complex numbers does not produce imaginary numbers of a new kind. Moreover, since equations of degree at most 4 can be solved by radicals, it follows from de Moivre's result that polynomials of degree at most 4 split into products of linear factors over the field of complex numbers. It was not long afterwards that the first attempts to prove the fundamental theorem were made (without formulas for the solution of higher degree equations by radicals), and we will come back to this topic in Chapter 9.

Another consequence with far-reaching implications is that the $n$-th root of any (non-zero) number is ambiguous: it has $n$ different determinations. Therefore, every formula that involves the extraction of a root needs some clarification as to *which* root should be chosen. This observation, which was conspicuously used as a starting-point in Vandermonde's subsequent investigations, sheds a completely new light on the problem of solving equations by radicals, and even on known solutions. Indeed, Cardano's formula, as it appears in §2.2 (see equation (2.4), p. 16), involves the extraction of two cube roots; if we consider various determinations of these cube roots, we obtain the three solutions of the cubic equation: this solves the puzzle of §2.3(a).[2]

Moreover, even de Moivre's formula as it appears in (7.11) above is ambiguous. To express it properly, one has to raise $\cos\alpha + \sqrt{-1}\sin\alpha$ to the $n$-th power instead of extracting the $n$-th root of $\cos n\alpha + \sqrt{-1}\sin n\alpha$. This viewpoint was adopted by Euler in his "Introductio in Analysin Infinitorum" (1748) (see Smith [67, vol. 2, p. 450]), in which he proves de Moivre's formula (7.1) using the addition formula (7.2). Later in the same book, comparing the power series expansions of the exponential and of the sine and cosine functions, Euler also states

$$e^{\alpha\sqrt{-1}} = \cos\alpha + \sqrt{-1}\sin\alpha,$$

a relation from which de Moivre's formula readily follows. Of course, once de Moivre's formula is established, the other major results of this section, and in particular Cotes' formulas, can be seen as easy applications. We devote the next section to a streamlined exposition of de Moivre's results on the roots of complex numbers along these lines.

## 7.2   The roots of unity

Let $a$ and $b$ be real numbers, not both zero, and denote by $\sqrt{a^2 + b^2}$ the (real, positive) square root of $a^2 + b^2$. Since

$$\left(\frac{a}{\sqrt{a^2+b^2}}\right)^2 + \left(\frac{b}{\sqrt{a^2+b^2}}\right)^2 = 1,$$

there is a unique angle $\varphi$ such that $0 \le \varphi < 2\pi$,

$$\cos\varphi = \frac{a}{\sqrt{a^2+b^2}} \qquad \text{and} \qquad \sin\varphi = \frac{b}{\sqrt{a^2+b^2}}.$$

We thus obtain the trigonometric expression of the complex number $a+bi \ne 0$, namely

$$a + bi = \sqrt{a^2 + b^2}(\cos\varphi + i\sin\varphi).$$

**Proposition 7.1.** *For any positive integer $n$, the $n$ distinct $n$-th roots of* $a + bi$ *are*

$$\sqrt[2n]{a^2 + b^2}\left(\cos\frac{\varphi + 2k\pi}{n} + i\sin\frac{\varphi + 2k\pi}{n}\right) \tag{7.12}$$

*for $k = 0, \ldots, n - 1$.*

---

[2]With the notations of §2.2, the cube roots should be determined in such a way that their product be $-\frac{p}{3}$, as it appears from the proof of Cardano's formula in §2.2 or alternatively from Viète's method in §6.1.2.

In this formula, $\sqrt[2n]{a^2 + b^2}$ is the unique real positive $2n$-th root of $a^2 + b^2$.

**Proof.** De Moivre's formula (7.1) yields

$$\left( \sqrt[2n]{a^2 + b^2} \left( \cos \frac{\varphi + 2k\pi}{n} + i \sin \frac{\varphi + 2k\pi}{n} \right) \right)^n = \sqrt{a^2 + b^2} (\cos \varphi + i \sin \varphi),$$

so that each of the expressions (7.12) is an $n$-th root of $a + bi$. Moreover, these expressions are easily seen to be pairwise distinct for $k = 0, \ldots, n-1$, since for these values of $k$ it is impossible that two among the angles $\frac{\varphi + 2k\pi}{n}$ differ by a multiple of $2\pi$. $\qquad\square$

**Definition 7.2.** A complex number $\zeta$ is called an *n-th root of unity*, for some integer $n$, if $\zeta^n = 1$.

The set of all $n$-th roots of unity is denoted by $\mu_n$. Thus,

$$X^n - 1 = \prod_{\zeta \in \mu_n} (X - \zeta).$$

By the preceding proposition we have

$$\mu_n = \left\{ \cos \frac{2k\pi}{n} + i \sin \frac{2k\pi}{n} \mid k = 0, \ldots, n-1 \right\}, \qquad (7.13)$$

hence

$$X^n - 1 = \prod_{k=0}^{n-1} \left( X - \cos \frac{2k\pi}{n} - i \sin \frac{2k\pi}{n} \right). \qquad (7.14)$$

This formula can be used to obtain a factorization of $X^n - 1$ into real factors: if $k + \ell = n$, then

$$\cos \frac{2k\pi}{n} = \cos \frac{2\ell\pi}{n} \quad \text{and} \quad \sin \frac{2k\pi}{n} = -\sin \frac{2\ell\pi}{n},$$

hence

$$\left( X - \cos \frac{2k\pi}{n} - i \sin \frac{2k\pi}{n} \right) \left( X - \cos \frac{2\ell\pi}{n} - i \sin \frac{2\ell\pi}{n} \right)$$

$$= X^2 - 2\cos\left( \frac{2k\pi}{n} \right) X + 1.$$

Therefore, multiplying the corresponding pairs of factors in the right-hand side of (7.14), we obtain

$$X^n - 1 = (X - 1) \prod_{k=1}^{\frac{n-1}{2}} \left( X^2 - 2\cos\left( \frac{2k\pi}{n} \right) X + 1 \right) \qquad \text{if } n \text{ is odd,}$$

$$X^n - 1 = (X - 1)(X + 1) \prod_{k=1}^{\frac{n}{2}-1} \left( X^2 - 2\cos\left( \frac{2k\pi}{n} \right) X + 1 \right) \quad \text{if } n \text{ is even.}$$

Substituting $a/x$ for $X$ in these formulas and multiplying each side by $x^n$ to clear denominators, we recover Cotes' formulas (7.5) and (7.6). Formulas (7.3) and (7.4) can be similarly derived by considering $n$-th roots of $-1$ instead of $n$-th roots of 1.

It should be observed that, in a rectangular coordinate system, the points $\left(\cos\frac{2k\pi}{n}, \sin\frac{2k\pi}{n}\right)$ for $k = 0, 1, \ldots, n - 1$, which represent the $n$-th roots unity in the planar representation of $\mathbb{C}$, are the vertices of a regular polygon with $n$ sides: they divide the unit circle into $n$ equal parts. For this reason, the theory which is concerned with $n$-th roots of unity or with the values of the cosine and sine functions at $\frac{2k\pi}{n}$ for integers $k$, $n$, is called *cyclotomy*, meaning literally "division of the circle" [into equal parts].

Likewise, the $n$-th roots of any non-zero complex number are represented in the plane of complex numbers by the vertices of a regular $n$-gon, as Proposition 7.1 shows.

That the roots of 1 deserve special interest comes from the fact that, if an $n$-th root $u$ of some complex number $v$ has been found, then the various determinations of $\sqrt[n]{v}$ are the products $\omega u$, where $\omega$ runs over the set of $n$-th roots of unity. This readily follows from

$$(\omega u)^n = \omega^n u^n = 1 \cdot u^n = v$$

or, equivalently, from Proposition 7.1.

While the $n$-th roots of unity have been determined above by a trigonometric expression, yet the problem of deciding whether these roots of unity have an expression by radicals has been untouched. We now turn to this problem, and prove, after some ideas of de Moivre:

**Theorem 7.3.** *Let $n$ be a positive integer. If, for each prime factor $p$ of $n$, the $p$-th roots of unity can be expressed by radicals, then the $n$-th roots of unity can be expressed by radicals.*

This theorem follows by induction from the following result:

**Lemma 7.4.** *Let $r$ and $s$ be positive integers. If $\xi_1, \ldots, \xi_r$ (resp. $\eta_1, \ldots, \eta_s$) are the $r$-th roots of unity (resp. the $s$-th roots of unity), then the $rs$-th roots of unity are of the form $\xi_i \sqrt[r]{\eta_j}$ for $i = 1, \ldots, r$ and $j = 1, \ldots, s$.*

**Proof.** From the factorization of $Y^s - 1$, it follows by letting $Y = X^r$ that

$$X^{rs} - 1 = \prod_{j=1}^{s}(X^r - \eta_j).$$

Therefore, the $rs$-th roots of unity are the $r$-th roots of the various $\eta_j$, for $j = 1, \ldots, s$. $\qquad\qquad\square$

**Proof of Theorem 7.3.** We argue by induction on the number of factors of $n$. If $n$ is a prime number, then there is nothing to prove. Assume then that $n = rs$ for some positive integers $r$, $s \neq 1$. Then the number of factors of $r$ (resp. $s$) is strictly less than the number of factors of $n$, hence, by induction, the $r$-th roots of unity $\xi_1, \ldots, \xi_r$ and the $s$-th roots of unity $\eta_1,$ $\ldots, \eta_s$ can be expressed by radicals. Since the $n$-th roots of unity are of the form $\xi_i \sqrt[r]{\eta_j}$, these can also be expressed by radicals. $\qquad\square$

**Remark 7.5.** Of course, the expressions obtained by the induction method of the proof above are not necessarily the simplest ones or the most suitable for the actual calculation of $rs$-th roots of unity. For instance, since the 4-th roots of unity are $\pm 1$ and $\pm\sqrt{-1}$, the 8-th roots of unity are obtained as

$$\pm 1, \ \pm\sqrt{-1}, \ \pm\sqrt{\sqrt{-1}} \ \text{and} \ \pm\sqrt{-\sqrt{-1}},$$

while they can also be expressed as

$$\pm 1, \ \pm\sqrt{-1}, \ \frac{1 \pm \sqrt{-1}}{\sqrt{2}} \ \text{and} \ -\frac{1 \pm \sqrt{-1}}{\sqrt{2}},$$

since $\cos\frac{\pi}{4} = \sin\frac{\pi}{4} = \frac{1}{\sqrt{2}}$.

Moreover, the result of Lemma 7.4 can be improved when $r$ and $s$ are relatively prime: in this case, one of the determinations of $\sqrt[r]{\eta_j}$ is an $s$-th root of unity $\eta_k$, so that the $rs$-th roots of unity are the products of the form $\xi_i \eta_k$ (for $i = 1, \ldots, r$ and $k = 1, \ldots, s$), see Remark 7.11 and Exercise 7.3.

Theorem 7.3 reduces the problem to find expressions by radicals for the $n$-th roots of unity, for any integer $n$, to the case where $n$ is prime. Since the equation $x^n - 1 = 0$ has the obvious root $x = 1$, we may divide $x^n - 1$ by $x - 1$, and the question reduces to the following problem: solve by radicals the equation

$$x^{n-1} + x^{n-2} + \cdots + x + 1 = 0, \qquad (7.15)$$

for $n$ prime. This equation is readily solved for $n = 2$ and 3: the roots are $-1$ for $n = 2$, and $\frac{-1 \pm \sqrt{-3}}{2}$ for $n = 3$.

For $n \geq 5$, the following trick (due to de Moivre) is useful: after division by $x^{\frac{n-1}{2}}$, the change of variable $y = x + x^{-1}$ transforms equation (7.15) into an equation of degree $\frac{n-1}{2}$ in $y$. (This trick succeeds because in the polynomial (7.15) the coefficients of the terms that are symmetric with respect to the middle term are equal.) Thus, for $n = 5$, we first divide each side of

$$x^4 + x^3 + x^2 + x + 1 = 0$$

by $x^2$, and the change of variable $y = x + x^{-1}$ transforms the resulting equation $x^2 + x + 1 + x^{-1} + x^{-2} = 0$ into

$$y^2 + y - 1 = 0.$$

We thus find

$$y = \frac{-1 \pm \sqrt{5}}{2},$$

and the values of $x$ are obtained by solving $x + x^{-1} = y$ for the various values of $y$. Thus, the 5-th roots of unity (other than 1) are the roots of the equations

$$x^2 - \frac{-1 + \sqrt{5}}{2}x + 1 = 0 \quad \text{and} \quad x^2 - \frac{-1 - \sqrt{5}}{2}x + 1 = 0,$$

which are

$$\frac{\sqrt{5} - 1 \pm \sqrt{-10 - 2\sqrt{5}}}{4} \quad \text{and} \quad \frac{-\sqrt{5} - 1 \pm \sqrt{-10 + 2\sqrt{5}}}{4}.$$

Similarly, for $n = 7$, de Moivre's trick yields for $y$ ($= x + x^{-1}$) the cubic equation

$$y^3 + y^2 - 2y - 1 = 0$$

which can be solved by radicals; the 7-th roots of unity can therefore be expressed by radicals.

However, for the next prime number, which is 11, de Moivre's trick yields an equation of degree 5, for which no general formula by radicals is known. Solving this equation was one of the greatest achievements of Vandermonde (see Chapter 11).

**Remarks 7.6.** (a) Since the roots of equation (7.15) are $e^{2k\pi i/n}$ for $k = 1$, ..., $n - 1$, the roots of the equation in $y$ ($= x + x^{-1}$) are

$$e^{2k\pi i/n} + e^{-2k\pi i/n} = 2\cos\frac{2k\pi}{n} \quad \text{for } k = 1, \ldots, \frac{n-1}{2}.$$

Therefore, the calculations above yield expressions by radicals for $2\cos\frac{2\pi}{5}$ and $2\cos\frac{4\pi}{5}$ as the positive and the negative roots of $y^2 + y - 1 = 0$,

$$2\cos\frac{2\pi}{5} = \frac{\sqrt{5} - 1}{2} \quad \text{and} \quad 2\cos\frac{4\pi}{5} = -\frac{\sqrt{5} + 1}{2}.$$

(b) From Theorem 7.3 and the results above, it follows that the $2 \cdot 3^3 \cdot 5^2$-th roots of unity can be expressed by radicals. Hence, $\sin\frac{\pi}{3^3 \cdot 5^2}$ can also be expressed by radicals, since

$$\sin\frac{\pi}{3^3 \cdot 5^2} = \frac{1}{2i}\left(e^{2\pi i/2 \cdot 3^3 \cdot 5^2} - e^{-2\pi i/2 \cdot 3^3 \cdot 5^2}\right).$$

Thus, the problem raised by Van Roomen has a solution by radicals, see §4.2.

## 7.3 Primitive roots and cyclotomic polynomials

In this section, we complete our discussion of the elementary aspects of the theory of roots of unity by listing several results that are more or less straightforward consequences of de Moivre's formula.[3] They are a natural outgrowth of the theory developed so far, and became known during the second half of the eighteenth century. The central notion is the following:

**Definitions 7.7.** The *exponent* of a root of unity $\zeta$ is the least integer $e > 0$ such that $\zeta^e = 1$. For instance, the exponent of 1 is 1 and the exponent of $-1$ is 2, although 1 is an $n$-th root of unity for every $n$ and $-1$ is an $n$-th root of unity for every even $n$.

The $n$-th roots of unity of exponent $n$ are also called *primitive $n$-th roots of unity*.

Our goal is to give a complete description of the primitive $n$-th roots of unity and to show that these roots are indeed primitive, in the sense that the other $n$-th roots of unity can be obtained as powers of any such root.

**Lemma 7.8.** *Let $e$ be the exponent of a root of unity $\zeta$. The complex numbers $1, \zeta, \zeta^2, \ldots, \zeta^{e-1}$ are pairwise distinct. Moreover, for any integer $m$ we have $\zeta^m = 1$ if and only if $e$ divides $m$.*

**Proof.** If $\zeta^i = \zeta^j$ for some integers $i$, $j$ with $0 \le i < j \le e - 1$, then $\zeta^{j-i} = 1$ with $0 < j - i < e$, hence $e$ is not the least positive exponent for which $\zeta^e = 1$. This contradiction shows that $1, \zeta, \ldots, \zeta^{e-1}$ are distinct.

If $m$ is a multiple of $e$, say $m = ef$ for some integer $f$, then $\zeta^m = (\zeta^e)^f = 1$. Conversely, let $m$ be an arbitrary integer such that $\zeta^m = 1$, and let $m = eq + r$ with $q$, $r \in \mathbb{Z}$ and $0 \le r < e$ be the Euclidean division of $m$ by $e$. Then $\zeta^m = (\zeta^e)^f \zeta^r$, hence $\zeta^r = 1$ because $\zeta^m = \zeta^e = 1$. The minimality condition on $e$ implies $r = 0$, hence $e$ divides $m$. $\qquad\square$

The following characterization of primitive roots of unity easily follows:

**Proposition 7.9.** *Let $n$ be an arbitrary positive integer. A complex number $\zeta$ is a primitive $n$-th root of unity if and only if*

$$\mu_n = \{1, \zeta, \zeta^2, \ldots, \zeta^{n-1}\}.$$

**Proof.** If $\zeta$ is a primitive $n$-th root of unity, then $\zeta^k \in \mu_n$ for all $k \in \mathbb{Z}$ because $(\zeta^k)^n = (\zeta^n)^k = 1$. Therefore, $\{1, \zeta, \ldots, \zeta^{n-1}\} \subset \mu_n$. These two

---

[3]More precisely, these results follow from the fact that the set $\mu_n$ of $n$-th roots of unity is a finite subgroup of the multiplicative group of complex numbers.

sets have the same number of elements because Lemma 7.8 shows that $1$, $\zeta, \ldots, \zeta^{n-1}$ are pairwise distinct, hence the two sets coincide.

Conversely, if $\mu_n = \{1, \zeta, \ldots, \zeta^{n-1}\}$, then $\zeta$ is an $n$-th root of unity and $\zeta^k \neq 1$ for $k \in \{1, \ldots, n-1\}$, so the exponent of $\zeta$ is $n$.                           $\square$

Proposition 7.9 shows that the multiplicative group $\mu_n$ is generated by a single element; one says that $\mu_n$ is a *cyclic group*. More precisely, the proposition shows that the generators of $\mu_n$ are exactly the primitive $n$-th roots of unity. From the description of $\mu_n$ in (7.13), it already follows that $\cos \frac{2\pi}{n} + i \sin \frac{2\pi}{n}$ is a primitive $n$-th root of unity. A complete description of the primitive $n$-th roots of unity will be derived from the following result:

**Proposition 7.10.** *Let $\zeta$ be a primitive $n$-th root of unity and let $k$ be an integer. Then $\zeta^k$ is a primitive $n$-th root of unity if and only if $k$ is relatively prime to $n$.*

**Proof.** Assume first that $k$ and $n$ have a common factor $d \neq 1$. Then

$$(\zeta^k)^{n/d} = (\zeta^n)^{k/d},$$

hence $(\zeta^k)^{n/d} = 1$ since $\zeta^n = 1$. Therefore, the exponent of $\zeta^k$ divides $n/d$, hence $\zeta^k$ is not a primitive $n$-th root of unity.

Conversely, assume that $\zeta^k$ is not primitive; then there is a positive integer $m < n$ such that $(\zeta^k)^m = 1$. Since the exponent of $\zeta$ is $n$, it follows from Lemma 7.8 that $n$ divides $km$. If $n$ is relatively prime to $k$, then it divides $m$; but this is impossible because $m < n$. Therefore, $n$ and $k$ are not relatively prime.                                                                      $\square$

**Remark 7.11.** Let $r$ and $s$ be relatively prime integers and let $\eta$ be a primitive $s$-th root of unity. Proposition 7.10 shows that $\eta^r$ is a primitive $s$-th root of unity, hence by Proposition 7.9 the $s$-th roots of unity are

$$\mu_s = \{\eta^{ri} \mid i = 0, \ldots, s-1\}.$$

Therefore, every $s$-th root of unity can be written as $\eta^{ri}$ for some (unique) integer $i$ between $0$ and $s-1$, and it follows that every $s$-th root of unity has an $r$-th root in $\mu_s$.

**Corollary 7.12.** *The primitive $n$-th roots of unity are*

$$e^{2k\pi i/n} = \cos \frac{2k\pi}{n} + i \sin \frac{2k\pi}{n}$$

*where $k$ runs over the positive integers that are less than $n$ and relatively prime to $n$. In particular, if $n$ is prime, then every $n$-th root of unity except $1$ is primitive.*

**Proof.** Let $\zeta = \cos \frac{2\pi}{n} + i \sin \frac{2\pi}{n}$. We saw in (7.13) that

$$\mu_n = \left\{ \cos \frac{2k\pi}{n} + i \sin \frac{2k\pi}{n} \mid k = 0, \ldots, n-1 \right\}.$$

Therefore, by de Moivre's formula we have

$$\mu_n = \{ \zeta^k \mid k = 0, \ldots, n-1 \}.$$

Proposition 7.9 then shows that $\zeta$ is a primitive $n$-th root of unity, and it follows from Proposition 7.10 that every primitive $n$-th root of unity is of the form $\zeta^k$ where $k$ is a positive integer relatively prime to $n$ between $0$ and $n-1$. $\qquad\square$

For later use (in Chapter 12), we point out how roots of composite exponent can be obtained by multiplying roots of lower exponent:

**Proposition 7.13.** *Let $\zeta$ and $\eta$ be roots of unity of exponents $e$ and $f$ respectively. If $e$ and $f$ are relatively prime, then $\zeta\eta$ is a root of unity of exponent $ef$.*

**Proof.** Let $k$ be the exponent of $\zeta\eta$. Since $\zeta^e = 1$ and $\eta^f = 1$, we have $(\zeta\eta)^{ef} = 1$, hence $k$ divides $ef$ by Lemma 7.8. On the other hand, from $(\zeta\eta)^k = 1$, it follows that

$$\zeta^k = \eta^{-k},$$

and raising each side to the power $f$ yields

$$\zeta^{kf} = 1.$$

Lemma 7.8 then shows that $e$ divides $kf$. Since $e$ is relatively prime to $f$, by hypothesis, it follows that $e$ divides $k$. Likewise, interchanging $\zeta$ and $\eta$, we see that $f$ divides $k$. Since $e$ and $f$ are relatively prime and divide $k$, their product $ef$ divides $k$; but we have already observed that $k$ divides $ef$, hence $k = ef$. $\qquad\square$

We now introduce the polynomials that have as roots the primitive $n$-th roots of unity. Because of their relation with the division of the circle, these polynomials are called *cyclotomic polynomials*.

**Definition 7.14.** The cyclotomic polynomials $\Phi_n$ ($n = 1, 2, 3, \ldots$) are defined inductively by $\Phi_1(X) = X - 1$ and, for $n \geq 2$,

$$\Phi_n(X) = \frac{X^n - 1}{\displaystyle\prod_{\substack{d \mid n \\ d \neq n}} \Phi_d(X)}$$

where $d$ runs over the set of divisors of $n$, with $d \neq n$.

In particular, if $p$ is a prime number, we have

$$\Phi_p(X) = \frac{X^p - 1}{X - 1} = X^{p-1} + X^{p-2} + \cdots + X + 1.$$

However, if $n$ is not prime, it is not clear *a priori* that $\Phi_n$ is a polynomial. We prove this and the fact that the roots of $\Phi_n$ are the primitive $n$-th roots of unity simultaneously:

**Proposition 7.15.** *For every integer $n \geq 1$, the rational function $\Phi_n$ is a monic polynomial with integral coefficients, and*

$$\Phi_n(X) = \prod_\zeta (X - \zeta),$$

*where $\zeta$ runs over the set of primitive $n$-th roots of unity.*

**Proof.** We argue by induction on $n$. Since the proposition is trivial for $n = 1$, we assume that it holds for all integers up to $n - 1$. Then, for every divisor $d$ of $n$, with $d \neq n$, we have

$$\Phi_d(X) = \prod_\zeta (X - \zeta) \in \mathbb{Z}[X],$$

where $\zeta$ runs over the roots of unity of exponent $d$. Since the roots of exponent $d$ are $n$-th roots of unity, it follows that $\Phi_d$ divides $X^n - 1$. Moreover, if $d_1$ and $d_2$ are distinct divisors of $n$, then $\Phi_{d_1}$ and $\Phi_{d_2}$ are relatively prime since their roots are pairwise distinct. Therefore, by Proposition 5.10, the product $\prod_d \Phi_d(X) \in \mathbb{Z}[X]$ divides $X^n - 1$, and it follows that $\Phi_n(X)$ is a polynomial in $\mathbb{Q}[X]$, which is monic since $X^n - 1$ and $\Phi_d$ (for every proper divisor $d$ of $n$) are monic. Moreover, the Euclidean division property (Theorem 5.1, p. 43) shows that the quotient of a polynomial in $\mathbb{Z}[X]$ by a monic polynomial in $\mathbb{Z}[X]$ is in $\mathbb{Z}[X]$. Therefore, $\Phi_n(X) \in \mathbb{Z}[X]$.

Now, the effect of dividing $X^n - 1$ by $\prod_{\substack{d \mid n \\ d \neq n}} \Phi_d(X)$ is to remove from

$$X^n - 1 = \prod_{\zeta \in \mu_n} (X - \zeta)$$

all the factors $X - \zeta$ where $\zeta$ has exponent $d \neq n$. Therefore, in $\Phi_n(X)$ remain all the factors $X - \zeta$ where $\zeta$ has exponent $n$, and only those factors. Thus $\Phi_n(X) = \prod_\zeta (X - \zeta)$ where $\zeta$ runs over the set of primitive $n$-th roots of unity. $\qquad \square$

**Remark.** It will be shown in Theorem 12.31 (p. 183) that $\Phi_n$ is irreducible in $\mathbb{Q}[X]$, for every $n$. The proof is easier when $n$ is prime, see Theorem 12.10 (p. 162).

## Appendix: Leibniz and Newton on the summation of series

Around 1675, Leibniz obtained the following result, of which he was justifiably proud:

$$1 - \frac{1}{3} + \frac{1}{5} - \frac{1}{7} + \cdots = \frac{\pi}{4}.$$

His method was to use the formula

$$\int \frac{dx}{x^2 + 1} = \tan^{-1} x, \qquad (7.16)$$

which yields

$$\int_0^1 \frac{dx}{x^2 + 1} = \frac{\pi}{4},$$

together with the power series expansion of $(x^2 + 1)^{-1}$, namely

$$\frac{1}{x^2 + 1} = 1 - x^2 + x^4 - x^6 + x^8 - \cdots$$

from which it follows that

$$\int_0^1 \frac{dx}{x^2 + 1} = \int_0^1 dx - \int_0^1 x^2 dx + \int_0^1 x^4 dx - \cdots$$

$$= 1 - \frac{1}{3} + \frac{1}{5} - \cdots.$$

Of course, the fact that the integral of the power series expansion of $(x^2 + 1)^{-1}$ is equal to the sum of the series on the right-hand side needs some justification, which was supplied much later.

In 1676, Newton sent to Leibniz the following variant:

$$1 + \frac{1}{3} - \frac{1}{5} - \frac{1}{7} + \frac{1}{9} + \frac{1}{11} - \frac{1}{13} - \frac{1}{15} + \cdots = \frac{\pi}{2\sqrt{2}},$$

with the terse hint that it depended on the reduction of an integrand to partial fractions (see [58, p. 212]). It is very likely that Newton's result was obtained from the evaluation of $\int_0^1 \frac{(x^2+1)dx}{x^4+1}$. Indeed, the rational function $\frac{x^2+1}{x^4+1}$ has the following decomposition into partial fractions:

$$\frac{x^2 + 1}{x^4 + 1} = \frac{1}{2} \frac{1}{x^2 + \sqrt{2}x + 1} + \frac{1}{2} \frac{1}{x^2 - \sqrt{2}x + 1},$$

and, by the change of variable $y = \sqrt{2}x + 1$ (resp. $y = \sqrt{2}x - 1$), it easily follows from (7.16) that

$$\int \frac{dx}{x^2 + \sqrt{2}x + 1} = \sqrt{2}\tan^{-1}(\sqrt{2}x + 1),$$

resp. $\quad \int \frac{dx}{x^2 - \sqrt{2}x + 1} = \sqrt{2}\tan^{-1}(\sqrt{2}x - 1).$

Therefore,

$$\int_0^1 \frac{(x^2+1)dx}{x^4+1} = \frac{1}{\sqrt{2}}\left(\tan^{-1}(\sqrt{2}+1) + \tan^{-1}(\sqrt{2}-1)\right),$$

and since $(\sqrt{2}+1)(\sqrt{2}-1) = 1$, we have

$$\tan^{-1}(\sqrt{2}+1) + \tan^{-1}(\sqrt{2}-1) = \frac{\pi}{2},$$

hence

$$\int_0^1 \frac{(x^2+1)dx}{x^4+1} = \frac{\pi}{2\sqrt{2}}.$$

On the other hand, from the power series expansion

$$\frac{1}{x^4+1} = 1 - x^4 + x^8 - x^{12} + x^{16} - \cdots$$

it follows that

$$\frac{x^2+1}{x^4+1} = 1 + x^2 - x^4 - x^6 + x^8 + x^{10} - \cdots$$

hence

$$\int_0^1 \frac{(x^2+1)dx}{x^4+1} = \int_0^1 dx + \int_0^1 x^2 dx - \int_0^1 x^4 dx - \int_0^1 x^6 dx + \cdots$$

$$= 1 + \frac{1}{3} - \frac{1}{5} - \frac{1}{7} + \cdots.$$

This proves Newton's result.

## Exercises

*Exercise 7.1. The aim of this first exercise is to prove Cotes' formulas along the lines of de Moivre's calculations with polynomials (§7.1).* Denote by $T_n(X) \in \mathbb{Q}[X]$ the polynomial of degree $n$ such that $T_n(\cos\alpha) = \cos n\alpha$ (i.e., $T_n$ is the degree $n$ Chebyshev polynomial of the first kind) and let, for $n \geq 1$,

$$P_n(x,z) = 1 - 2T_n(x)z^n + z^{2n} \in \mathbb{Q}[x,z].$$

(a) Prove: $T_1(x) = x$, $T_2(x) = 2x^2 - 1$, and, for $n \geq 2$,

$$T_{n+1}(x) = 2xT_n(x) - T_{n-1}(x).$$

(b) Use (a) to show that for $n \geq 2$

$$P_{n+1}(x,z) = 2xzP_n(x,z) - z^2 P_{n-1}(x,z) + (1 + z^{2n})P_1(x,z).$$

(c) Use (b) to show that $P_1(x, z)$ divides $P_n(x, z)$ in $\mathbb{Q}[x, z]$ for every $n$.

(d) Prove:

$$P_n(\cos\alpha, z) = \prod_{k=0}^{n-1} P_1\left(\cos\left(\alpha + \tfrac{2k\pi}{n}\right), z\right) \qquad \text{for } \alpha \in \mathbb{R}.$$

[Hint: Prove it first for $\alpha \notin \frac{\pi}{n}\mathbb{Z}$, using (c) and observing that for $\alpha \notin \frac{\pi}{n}\mathbb{Z}$ the polynomials on the right-hand side are pairwise relatively prime. Since the coefficients of each side are continuous functions of $\alpha$ that are equal outside a discrete subset of $\mathbb{R}$, they are equal for every $\alpha \in \mathbb{R}$.]

(e) Show that for $\alpha = 0$ both sides of the equation in (d) are squares. Extracting the square root of each side, show that for all integers $m \geq 1$

$$1 - z^{2m} = (1 - z)(1 + z) \prod_{k=1}^{m-1} \left(1 - 2\cos\left(\tfrac{2k\pi}{2m}\right)z + z^2\right)$$

and

$$1 - z^{2m+1} = (1 - z) \prod_{k=1}^{m} \left(1 - 2\cos\left(\tfrac{2k\pi}{2m+1}\right)z + z^2\right).$$

For $\alpha = \frac{\pi}{n}$, derive similarly the formulas

$$1 + z^{2m} = \prod_{k=0}^{m-1} \left(1 - 2\cos\left(\tfrac{(2k+1)\pi}{2m}\right)z + z^2\right)$$

and

$$1 + z^{2m+1} = (1 + z) \prod_{k=0}^{m-1} \left(1 - 2\cos\left(\tfrac{(2k+1)\pi}{2m+1}\right)z + z^2\right).$$

Cotes' formulas (7.3)–(7.6) (p. 75) follow by substituting $\frac{x}{a}$ for $z$ and multiplying each side by an appropriate power of $a$ to clear denominators.

*Exercise 7.2. The following exercise aims to highlight some of the remarkable properties of the Chebyshev polynomials $T_n$ of Exercise 7.1.*

(a) Prove: $T_n(x) - \cos n\alpha = \prod_{k=0}^{n-1}\left(x - \cos\left(\alpha + \tfrac{2k\pi}{n}\right)\right)$ for $\alpha \in \mathbb{R}$. [Hint: Argue as in Exercise 7.1(d).]

(b) Prove: $T_m\left(T_n(x)\right) = T_{mn}(x)$ for all integers $m, n$.

(c) Prove: $2T_m(x) \cdot T_n(x) = T_{m+n}(x) + T_{|m-n|}(x)$ for all integers $m, n$. (To allow $m = n$, set $T_0(x) = 1$.) [Hint: Use induction on $n$. For $n = 1$, use the recurrence relation in Exercise 7.1(a).]

*Exercise 7.3.* Let $\mu_r = \{\xi_1, \ldots, \xi_r\}$ and $\mu_s = \{\eta_1, \ldots, \eta_s\}$. Assuming that $r$ and $s$ are relatively prime, show that

$$\mu_{rs} = \{\xi_i \eta_j \mid i = 1, \ldots, r \text{ and } j = 1, \ldots, s\}.$$

*Exercise 7.4.* Let $\zeta$ be a root of unity of exponent $e$ and let $k$ be an integer. Find the exponent of $\zeta^k$. (Compare Proposition 7.10.)

*Exercise 7.5.* Show that an $n$-th root of unity $\zeta$ is primitive if and only if $\zeta^d \neq 1$ for every proper factor $d$ of $n$.

*Exercise 7.6.* Let $p$ be a prime number. Prove that

$$\sum_{\omega \in \mu_p} \omega^i = \begin{cases} 0 & \text{if } i \in \mathbb{Z} \text{ is not divisible by } p, \\ p & \text{if } i \in \mathbb{Z} \text{ is divisible by } p. \end{cases}$$

[Hint: Use Newton's formulas (4.6) (p. 38).]

*Exercise 7.7. This last exercise is an introduction to Euler's "totient function"* $\varphi$. For any integer $n \geq 2$, let $\varphi(n)$ be the number of integers that are relatively prime to $n$ between 0 and $n - 1$.

(a) Show that $\varphi(n) = \deg \Phi_n$ and that $\varphi(n)$ is equal to the number of primitive $n$-th roots of unity.
(b) Show that $\varphi(mn) = \varphi(m)\varphi(n)$ if $m$ and $n$ are relatively prime. [Hint: Compare Exercise 7.3.]
(c) Show that $\varphi(p^k) = p^{k-1}(p - 1)$ for any prime number $p$.
(d) Derive from (b) and (c) the following formula: if $n = p_1^{k_1} \cdots p_r^{k_r}$ (for $p_1, \ldots, p_r$ distinct prime numbers), then

$$\varphi(n) = \prod_{i=1}^{r} p_i^{k_i - 1}(p_i - 1) = n \cdot \prod_{i=1}^{r}(1 - p_i^{-1}).$$

(e) Show that $n = \sum_{d \mid n} \varphi(d)$ where $d$ runs over the factors of $n$ (including $n$). [Hint: Compare Definition 7.14.]

# Chapter 8

# Symmetric Functions

During the first half of the eighteenth century, the structure of equations, as formerly investigated by Viète (§4.2), became clearer and clearer. Calculating formally with roots of equations, mathematicians became aware of the kind of information that can be gathered from the coefficients without solving equations. As Girard had shown (§4.2), for any polynomial

$$X^n - s_1 X^{n-1} + s_2 X^{n-2} - \cdots + (-1)^n s_n$$
$$= (X - x_1)(X - x_2) \ldots (X - x_n) \quad (8.1)$$

we have

$$s_1 = x_1 + \cdots + x_n,$$
$$s_2 = x_1 x_2 + \cdots + x_{n-1} x_n,$$
$$s_3 = x_1 x_2 x_3 + \cdots + x_{n-2} x_{n-1} x_n, \quad (8.2)$$
$$\vdots$$
$$s_n = x_1 x_2 \ldots x_n.$$

The following question naturally arises: what kind of function of the roots $x_1, \ldots, x_n$ can be calculated from $s_1, \ldots, s_n$?

To translate properly the results of this period, laying on firm ground the formal calculations with roots of polynomials, the roots $x_1, \ldots, x_n$ should be considered as independent indeterminates over some base field $F$ (usually, $F = \mathbb{Q}$, the field of rational numbers). Indeed, every calculation with indeterminates which does not involve divisions by non-constant polynomials can be done as well with arbitrary elements in a field $K$ containing the base field $F$. This is a loose translation of the fact that any map from $\{x_1, \ldots, x_n\}$ to $K$ can be (uniquely) extended to a ring homomorphism from the ring of polynomials $F[x_1, \ldots, x_n]$ to $K$, mapping a polynomial

$P(x_1, \ldots, x_n)$ to the element of $K$ obtained by substituting for $x_1, \ldots, x_n$ their assigned values in $K$. (If we wish to allow divisions by non-constant polynomials, caution is necessary since some denominators may vanish in $K$.) Therefore, we introduce the following definition:

**Definition 8.1.** If $x_1, \ldots, x_n$ are considered as independent indeterminates over some base field $F$, the polynomial (8.1) above is called the *general* (or *generic*) monic polynomial of degree $n$ over $F$. Thus, this general polynomial is a polynomial in one indeterminate $X$ with coefficients in $F[x_1, \ldots, x_n]$; it may thus be viewed as an element in $F[x_1, \ldots, x_n, X]$.

This polynomial is general (generic) in the following sense: if

$$P = X^n + a_{n-1}X^{n-1} + a_{n-2}X^{n-2} + \cdots + a_1 X + a_0$$

is an arbitrary monic polynomial of degree $n$ over some field $K$ containing $F$, which splits[1] into a product of linear factors in some field $L$ containing $K$, so that

$$P = (X - u_1) \ldots (X - u_n)$$

with $u_i \in L$ for $i = 1, \ldots, n$, then there is a homomorphism $F[x_1, \ldots, x_n] \to L$ mapping $x_i$ to $u_i$ for $i = 1, \ldots, n$. This ring homomorphism translates any calculation with $x_1, \ldots, x_n$ into a calculation with $u_1, \ldots, u_n$. For instance, it is easily seen that this ring homomorphism maps $s_1, s_2, \ldots, s_n$ onto $-a_{n-1}, a_{n-2}, \ldots, (-1)^n a_0 \in K$, which means that

$$s_1(u_1, \ldots, u_n) = -a_{n-1}, \quad \text{i.e.,} \quad u_1 + \cdots + u_n = -a_{n-1},$$
$$s_2(u_1, \ldots, u_n) = a_{n-2}, \quad \text{i.e.,} \quad u_1 u_2 + \cdots + u_{n-1} u_n = a_{n-2},$$
$$\vdots \qquad\qquad\qquad\qquad\qquad\qquad \vdots$$
$$s_n(u_1, \ldots, u_n) = (-1)^n a_0, \quad \text{i.e.,} \quad u_1 \ldots u_n = (-1)^n a_0.$$

After this slight change of viewpoint, from arbitrary elements to indeterminates, the question becomes: which are the rational functions in $n$ indeterminates $x_1, \ldots, x_n$ that can be expressed as rational functions in $s_1, \ldots, s_n$ (where $s_1, \ldots, s_n$ are defined by the equalities (8.2) above)?

The crucial condition turns out to be the following:

**Definition 8.2.** A polynomial $P(x_1, \ldots, x_n)$ in $n$ indeterminates is *symmetric* if it is not altered when the indeterminates are arbitrarily permuted

---

[1] We will see later that the condition that $P$ splits into a product of linear factors in some field containing $K$ is always fulfilled, see §9.1. At this point this provision cannot be disposed of, however. (Compare Remark 8.9(a) below.)

among themselves; i.e., for every permutation $\sigma$ of $1, \ldots, n$,

$$P(x_{\sigma(1)}, \ldots, x_{\sigma(n)}) = P(x_1, \ldots, x_n).$$

Similarly, a rational function $P/Q$ in $n$ indeterminates is *symmetric* if it is not altered when the indeterminates are permuted; i.e., for every permutation $\sigma$ of $1, \ldots, n$,

$$\frac{P(x_{\sigma(1)}, \ldots, x_{\sigma(n)})}{Q(x_{\sigma(1)}, \ldots, x_{\sigma(n)})} = \frac{P(x_1, \ldots, x_n)}{Q(x_1, \ldots, x_n)}.$$

Note that this does *not* imply that $P$ and $Q$ are each symmetric, since $P$ and $Q$ can be both multiplied by an arbitrary non-zero polynomial without changing the fraction $P/Q$, but we will see below (p. 99) that every symmetric rational function can be represented as a quotient of symmetric polynomials.

Since the polynomials $s_1, \ldots, s_n$ are symmetric, it is clear that every rational function in $s_1, \ldots, s_n$ is a symmetric rational function in $x_1, \ldots, x_n$. The converse turns out to be also true, so that the following result holds:

**Theorem 8.3.** *A rational function in $n$ indeterminates $x_1, \ldots, x_n$ over a field $F$ can be expressed as a rational function in $s_1, \ldots, s_n$ if and only if it is symmetric.*

This theorem is in fact a consequence of the analogous result for polynomials:

**Theorem 8.4.** *A polynomial in $n$ indeterminates $x_1, \ldots, x_n$ over a field $F$ can be expressed as a polynomial in $s_1, \ldots, s_n$ if and only if it is symmetric.*

These theorems are known as the *fundamental theorems* of symmetric functions or symmetric polynomials respectively. The polynomials $s_1, \ldots, s_n$ are sometimes called the *elementary* symmetric polynomials, since the others can be expressed in terms of these ones.

Because of their progressive emergence through the calculations of eighteenth century mathematicians, these theorems can hardly be credited to any specific author. (There is not much credit to give anyway, since the proofs are not difficult.) It seems that they first appeared in print around 1770, in "Meditationes Algebraicae" of Edward Waring (1736–1798) and in Vandermonde's "Mémoire sur la résolution des équations," and in presumably other works. It is noteworthy that Lagrange in 1770 qualifies the fundamental theorem of symmetric functions as "self-evident" [49, Art. 98, p. 372]. Therefore, the most interesting feature that one may expect of a proof is its effectiveness: it has to provide a method to express any symmetric polynomial as a polynomial in $s_1, \ldots, s_n$. In the next section, we explain the particularly simple method suggested by Waring [84, Chapter I, Problem III, Case 3], and thereafter we will discuss some applications, but first we point out a convenient notation that allows to designate symmetric polynomials without writing out all the terms.

**Notation 8.5.** We let $\sum x_1^{i_1} x_2^{i_2} \ldots x_n^{i_n}$ be the symmetric polynomial whose terms are the various *distinct* monomials obtained from $x_1^{i_1} x_2^{i_2} \ldots x_n^{i_n}$ by permutation of the indeterminates. Observe that this notation is slightly ambiguous, since the total number of variables is not clear from the notation, as some of the exponents $i_1, \ldots, i_n$ may be zero. Therefore, the total number of variables should always be indicated, unless it is clear from the context. For example, as a symmetric polynomial in two variables,

$$\sum x_1^2 x_2 = x_1^2 x_2 + x_1 x_2^2$$

whereas, as a symmetric polynomial in three variables,

$$\sum x_1^2 x_2 = x_1^2 x_2 + x_1 x_2^2 + x_1^2 x_3 + x_1 x_3^2 + x_2^2 x_3 + x_2 x_3^2.$$

With this notation, the elementary symmetric polynomials can be written simply as

$$s_1 = \sum x_1, \quad s_2 = \sum x_1 x_2, \quad \ldots, \quad s_{n-1} = \sum x_1 \ldots x_{n-1},$$
$$\text{and} \quad s_n = \sum x_1 \ldots x_n \quad (= x_1 \ldots x_n).$$

## 8.1  Waring's method

In order to define the degree of a polynomial in $n$ indeterminates, we endow the set $\mathbb{N}^n$ of $n$-tuples of integers with the lexicographic ordering. Thus $(i_1, \ldots, i_n) \geq (j_1, \ldots, j_n)$ if the first non-zero difference (if any) in the sequence $i_1 - j_1, \ldots, i_n - j_n$ is positive. For any non-zero polynomial $P = P(x_1, \ldots, x_n)$ in $n$ indeterminates $x_1, \ldots, x_n$ over a field, the *degree* of $P$ is then defined as the largest $n$-tuple $(i_1, \ldots, i_n) \in \mathbb{N}^n$ for which the coefficient of $x_1^{i_1} \ldots x_n^{i_n}$ in $P$ is non-zero. The degree of $P$ is denoted by

$\deg P$. For instance, we have

$$\deg s_1 = (1,0,0,\ldots,0),$$
$$\deg s_2 = (1,1,0,\ldots,0),$$
$$\vdots \tag{8.3}$$
$$\deg s_{n-1} = (1,1,\ldots,1,0),$$
$$\deg s_n = (1,1,\ldots,1,1).$$

By convention, we also set $\deg 0 = -\infty$, and the same relations as for polynomials in one indeterminate hold, namely

$$\deg(P + Q) \le \max(\deg P, \deg Q), \tag{8.4}$$
$$\deg(PQ) = \deg P + \deg Q. \tag{8.5}$$

**Proof of Theorem 8.4.** Waring's method to express any symmetric polynomial $P(x_1,\ldots,x_n)$ as a polynomial in $s_1$, ..., $s_n$ is quite similar to Euclid's division algorithm (Theorem 5.1, p. 43). The idea is to match $P$ with a polynomial in $s_1$, ..., $s_n$ that has the same degree as $P$. Adjusting the leading coefficient, we can arrange that the degree of the difference be less than $\deg P$, and we finish by induction on the degree.

Let $P \in F[x_1,\ldots,x_n]$ be a non-zero symmetric polynomial, and let

$$\deg P = (i_1, i_2, \ldots, i_n) \in \mathbb{N}^n.$$

We first observe that $i_1 \ge i_2 \ge \cdots \ge i_n$. Indeed, if we can find among the terms of $P$ a term like $ax_1^{i_1} \ldots x_n^{i_n}$ (with $a \ne 0$), then we can also find all the terms obtained from this one by permutation of $x_1$, ..., $x_n$, since $P$ is symmetric. The degrees of these terms are the various $n$-tuples obtained from $(i_1,\ldots,i_n)$ by permutation of the entries, and the greatest among these $n$-tuples is the one in which the entries are in not-increasing order.

We can therefore set

$$f = s_1^{i_1-i_2} s_2^{i_2-i_3} \ldots s_{n-1}^{i_{n-1}-i_n} s_n^{i_n}.$$

By (8.3) and (8.5), we have

$$\deg f = (i_1 - i_2)\deg(s_1) + (i_2 - i_3)\deg(s_2) + \cdots + i_n \deg(s_n)$$
$$= (i_1 - i_2, 0, \ldots, 0) + (i_2 - i_3, i_2 - i_3, 0, \ldots, 0) + \cdots + (i_n, \ldots, i_n)$$
$$= (i_1, i_2, \ldots, i_n).$$

Moreover, it is readily verified that the leading coefficient of $f$, which is the coefficient of $x_1^{i_1} \ldots x_n^{i_n}$, is 1, so that

$$f = x_1^{i_1} \ldots x_n^{i_n} + (\text{terms of lower degree}).$$

Therefore, if $a \in F^{\times}$ is the leading coefficient of $P$, so that

$$P = ax_1^{i_1} \ldots x_n^{i_n} + (\text{terms of lower degree}),$$

then, letting $P_1 = P - af$, we see that $\deg P_1 < \deg P$. Moreover, $P_1$ is symmetric (possibly zero), since $P$ and $f$ are symmetric. We can therefore apply the same arguments to $P_1$, which has lower degree than $P$.

In Waring's words:

> The first step of the solution is to find $S^a \times R^{b-a} \times Q^{c-b} \times P^{d-c} \ldots$; among these terms one particular product will be the required sum, while the remaining terms must be identified by the same method and then discarded [84, p. 13].

To complete the proof, it remains to prove that the process above, by which the degree of the initial symmetric polynomial has been reduced, terminates in a finite number of steps. This readily follows from the following observation:

**Lemma 8.6.** $\mathbb{N}^n$ *satisfies the descending chain condition, i.e., it does not contain any infinite strictly decreasing sequence of elements.*

**Proof.** As the lemma is obvious if $n = 1$, we argue by induction on $n$. If

$$(i_{11}, i_{12}, \ldots, i_{1n}) > (i_{21}, i_{22}, \ldots, i_{2n}) > \ldots > (i_{m1}, i_{m2}, \ldots, i_{mn}) > \cdots$$
$$(8.6)$$

is an infinite strictly decreasing sequence in $\mathbb{N}^n$, then the sequence of first entries is not increasing, so that

$$i_{11} \geq i_{21} \geq \cdots \geq i_{m1} \geq \cdots$$

Therefore, this sequence is eventually constant: there is an index $M$ such that

$$i_{m1} = i_{M1} \qquad \text{for all } m \geq M.$$

We then delete the first $M - 1$ terms in the sequence (8.6), and consider the last $n - 1$ entries of the elements of the remaining (infinite) sequence:

$$(i_{M2}, i_{M3}, \ldots, i_{Mn}) > (i_{(M+1)2}, i_{(M+1)3}, \ldots, i_{(M+1)n}) > \cdots$$

We thus obtain a strictly decreasing sequence of elements in $\mathbb{N}^{n-1}$. The existence of such a sequence contradicts the induction hypothesis. $\square$

**Example 8.7.** Let us express the symmetric polynomial in three variables

$$S = \sum x_1^4 x_2 x_3 + \sum x_1^3 x_2^3$$

(i.e., $S = x_1^4 x_2 x_3 + x_1 x_2^4 x_3 + x_1 x_2 x_3^4 + x_1^3 x_2^3 + x_1^3 x_3^3 + x_2^3 x_3^3$, see the notation 8.5) as a polynomial in $s_1$, $s_2$, $s_3$. Since $(4, 1, 1) > (3, 3, 0)$, the degree of $S$ is $(4, 1, 1)$, so we first calculate $s_1^{4-1} s_2^{1-1} s_3^1$,

$$s_1^3 s_3 = \left(\sum x_1\right)^3 (x_1 x_2 x_3) = \sum x_1^4 x_2 x_3 + 3\sum x_1^3 x_2^2 x_3 + 6x_1^2 x_2^2 x_3^2,$$

hence

$$S = \sum x_1^3 x_2^3 - 3\sum x_1^3 x_2^2 x_3 - 6x_1^2 x_2^2 x_3^2 + s_1^3 s_3.$$

It remains to express in terms of $s_1$, $s_2$, $s_3$ a polynomial of degree $(3, 3, 0)$. Therefore, we calculate the cube of $s_2$,

$$s_2^3 = \left(\sum x_1 x_2\right)^3 = \sum x_1^3 x_2^3 + 3\sum x_1^3 x_2^2 x_3 + 6x_1^2 x_2^2 x_3^2.$$

Substituting for $\sum x_1^3 x_2^3$ in the preceding expression of $S$, we obtain

$$S = -6\sum x_1^3 x_2^2 x_3 - 12 x_1^2 x_2^2 x_3^2 + s_1^3 s_3 + s_2^3.$$

Next, in order to eliminate $\sum x_1^3 x_2^2 x_3$, we calculate $s_1 s_2 s_3$,

$$s_1 s_2 s_3 = \left(\sum x_1\right)\left(\sum x_1 x_2\right)(x_1 x_2 x_3) = \sum x_1^3 x_2^2 x_3 + 3 x_1^2 x_2^2 x_3^2,$$

hence

$$S = 6 x_1^2 x_2^2 x_3^2 + s_1^3 s_3 + s_2^3 - 6 s_1 s_2 s_3.$$

Since $x_1^2 x_2^2 x_3^2 = s_3^2$, we finally obtain the required result:

$$S = s_1^3 s_3 + s_2^3 - 6 s_1 s_2 s_3 + 6 s_3^2.$$

From this brief example, it is already clear that the only difficulty in carrying out Waring's method is to write out the various monomials in products like $s_1^{i_1} \ldots s_n^{i_n}$ with their proper coefficient.

**Rational functions: proof of Theorem 8.3.** Let $P$, $Q$ be polynomials in $n$ indeterminates $x_1$, ..., $x_n$ such that the rational function $P/Q$ is symmetric. In order to prove that $P/Q$ is a rational function in $s_1$, ..., $s_n$, we represent $P/Q$ as a quotient of symmetric polynomials in $x_1$, ..., $x_n$, as follows: if $Q$ is symmetric then $P$ is symmetric too, since $P/Q$ is symmetric, and there is nothing to do. Otherwise, let $Q_1$, ..., $Q_r$ be the various distinct polynomials (other than $Q$) obtained from $Q$ by permutation of the indeterminates. The product $QQ_1 \ldots Q_r$ is then symmetric since any permutation of the indeterminates merely permutes the factors. Since $P/Q$ is symmetric and $P/Q = (PQ_1 \ldots Q_r)/(QQ_1 \ldots Q_r)$, it follows that the polynomial $PQ_1 \ldots Q_r$ is symmetric too. We have thus obtained the required representation of $P/Q$. Now, by the fundamental theorem of symmetric polynomials (Theorem 8.4), there are polynomials $f$, $g$ such that

$$PQ_1 \ldots Q_r = f(s_1, \ldots, s_n) \qquad \text{and} \qquad QQ_1 \ldots Q_r = g(s_1, \ldots, s_n),$$

hence $P/Q = f(s_1, \ldots, s_n)/g(s_1, \ldots, s_n)$. □

The fundamental theorem of symmetric polynomials asserts that every symmetric polynomial $P(x_1, \ldots, x_n)$ is of the form

$$P(x_1, \ldots, x_n) = f(s_1, \ldots, s_n)$$

for some polynomial $f$ in $n$ indeterminates. In other words, there is a polynomial $f(y_1, \ldots, y_n)$ in $n$ indeterminates that yields $P(x_1, \ldots, x_n)$ when $s_1$, $\ldots$, $s_n$ are substituted for the indeterminates $y_1$, $\ldots$, $y_n$. However, it is not clear *a priori* that the expression of $P$ as a polynomial in $s_1, \ldots, s_n$ is unique, or in other words, that there is only one polynomial $f$ for which the above equality holds. Admittedly, the contrary would be surprising, but no one seems to have cared to prove the uniqueness of $f$ before Gauss, who needed it for his second proof (1815) of the fundamental theorem of algebra [32, §5] (see also Smith [67, vol. I, pp. 292–306]).

**Theorem 8.8.** *Let $f$ and $g$ be polynomials in $n$ indeterminates $y_1, \ldots, y_n$ over a field $F$. If $f$ and $g$ yield the same polynomial in $x_1, \ldots, x_n$ when $s_1, \ldots, s_n$ are substituted for $y_1, \ldots, y_n$, i.e., if*

$$f(s_1, \ldots, s_n) = g(s_1, \ldots, s_n) \qquad in \ F[x_1, \ldots, x_n],$$

*then*

$$f(y_1, \ldots, y_n) = g(y_1, \ldots, y_n) \qquad in \ F[y_1, \ldots, y_n].$$

**Proof.** We compare the degree $(i_1, \ldots, i_n)$ of a non-zero monomial

$$m(y_1, \ldots, y_n) = ay_1^{i_1} \ldots y_n^{i_n}$$

to the degree of the monomial

$$m(s_1, \ldots, s_n) = as_1^{i_1} \ldots s_n^{i_n} \in F[x_1, \ldots, x_n].$$

By (8.3) and (8.5), we have

$$\deg m(s_1, \ldots, s_n) = (i_1 + \cdots + i_n, \, i_2 + \cdots + i_n, \, \ldots, \, i_{n-1} + i_n, i_n).$$

Since the map $\mathbb{N}^n \to \mathbb{N}^n$

$$(i_1, \ldots, i_n) \mapsto (i_1 + \cdots + i_n, \, i_2 + \cdots + i_n, \, \ldots, \, i_n)$$

is injective, it follows that monomials of different degrees in $F[y_1, \ldots, y_n]$ cannot cancel out in $F[x_1, \ldots, x_n]$ when $s_1, \ldots, s_n$ are substituted for $y_1$, $\ldots$, $y_n$. Therefore, every non-zero polynomial $h \in F[y_1, \ldots, y_n]$ yields a non-zero polynomial $h(s_1, \ldots, s_n)$ in $F[x_1, \ldots, x_n]$. Applying this result to $h = f - g$, the theorem follows. □

**Remarks 8.9. (a)** Let $\varphi\colon F[y_1, \ldots, y_n] \to F[x_1, \ldots, x_n]$ be the ring homomorphism that maps every polynomial $h(y_1, \ldots, y_n)$ to $h(s_1, \ldots, s_n)$. Theorem 8.8 asserts that $\varphi$ is injective. Therefore, the image of $\varphi$, which is the subring $F[s_1, \ldots, s_n]$ of $F[x_1, \ldots, x_n]$ generated by $s_1, \ldots, s_n$, is isomorphic under $\varphi$ to a ring of polynomials in $n$ indeterminates. In other words, the polynomials $s_1, \ldots, s_n$ in $F[x_1, \ldots, x_n]$ can be considered as independent indeterminates. This fact is expressed by saying that $s_1, \ldots, s_n$ are *algebraically independent*.

The point of this remark is that the generic monic polynomial of degree $n$ over $F$

$$X^n - s_1 X^{n-1} + s_2 X^{n-2} - \cdots + (-1)^n s_n$$

is really generic for all monic polynomials of degree $n$ over *any* field $K$ containing $F$. Indeed, if

$$X^n + a_{n-1} X^{n-1} + a_{n-2} X^{n-2} + \cdots + a_0$$

is such a polynomial, then, since $s_1, \ldots, s_n$ can be considered as independent indeterminates over $F$, there is a (unique) ring homomorphism from $F[s_1, \ldots, s_n]$ to $K$ that maps $s_1, s_2, \ldots, s_n$ to $-a_{n-1}, a_{n-2}, \ldots, (-1)^n a_0$. This homomorphism translates calculations with the coefficients of the generic polynomial into calculations with the coefficients of arbitrary polynomials. By contrast with the discussion following Definition 8.1, we do not need to restrict here to polynomials that split into products of linear factors over some extension of the base field; this is precisely why Theorem 8.8 is significant in Gauss' paper [32], since the purpose of this work was to prove the fundamental theorem of algebra, asserting that polynomials split into products of linear factors over the field of complex numbers.

**(b)** Inspection shows that the hypothesis that the base ring $F$ is a field has not been used in our exposition of Waring's method nor in the proof of Theorem 8.8. Therefore, Theorem 8.4 and Theorem 8.8 are valid over any base ring. Theorem 8.3 also holds over any (commutative) domain, but to generalize it further, some caution is necessary in the very definition of a rational function.

## 8.2 The discriminant

Let

$$\Delta(x_1, \ldots, x_n) = \prod_{1 \le i < j \le n} (x_i - x_j) \in \mathbb{Z}[x_1, \ldots, x_n].$$

Every permutation of $x_1, \ldots, x_n$ permutes the factors $x_i - x_j$ among themselves, changing the sign of some of them. So, $\Delta$ is either left unchanged or changed into its opposite by a permutation, and $\Delta^2$ is a symmetric polynomial. It follows from Theorem 8.4 (and Theorem 8.8) that

$$\Delta(x_1, \ldots, x_n)^2 = D(s_1, \ldots, s_n)$$

for some (uniquely defined) polynomial $D$ with integral coefficients, called the *discriminant* of the generic polynomial of degree $n$.

The discriminant of an arbitrary polynomial over a field $F$

$$X^n + a_{n-1}X^{n-1} + a_{n-2}X^{n-2} + \cdots + a_0$$

is then defined as $D(-a_{n-1}, a_{n-2}, \ldots, (-1)^n a_0)$, the element in $F$ obtained from $D(s_1, \ldots, s_n)$ by substituting the coefficients of the polynomial for $s_1$, $\ldots, s_n$.

In degree 2, we readily have

$$\Delta(x_1, x_2)^2 = (x_1 - x_2)^2 = x_1^2 + x_2^2 - 2x_1 x_2,$$

and this symmetric polynomial can be expressed in terms of elementary symmetric polynomials as

$$\Delta(x_1, x_2)^2 = (x_1 + x_2)^2 - 4x_1 x_2 = s_1^2 - 4s_2.$$

Therefore, the discriminant of the generic polynomial of degree 2 is

$$D(s_1, s_2) = s_1^2 - 4s_2.$$

For degree 3, we use some artifices to simplify the calculations. Written out as a sum of monomials, $\Delta(x_1, x_2, x_3) = (x_1 - x_2)(x_1 - x_3)(x_2 - x_3)$ appears as

$$\Delta(x_1, x_2, x_3) = A - B$$

where

$$A = x_1^2 x_2 + x_2^2 x_3 + x_3^2 x_1 \quad \text{and} \quad B = x_1 x_2^2 + x_2 x_3^2 + x_3 x_1^2.$$

Therefore,

$$\Delta(x_1, x_2, x_3)^2 = A^2 + B^2 - 2AB = (A + B)^2 - 4AB. \tag{8.7}$$

Now, $A + B$ and $AB$ are symmetric polynomials, which are not difficult to express in terms of $s_1$, $s_2$, $s_3$. Straightforward calculations by Waring's method yield the following results (with Notation 8.5, see Example 8.7):

$$A + B = \sum x_1^2 x_2 = s_1 s_2 - 3s_3,$$
$$AB = \sum x_1^4 x_2 x_3 + \sum x_1^3 x_2^3 + 3x_1^2 x_2^2 x_3^2 = s_1^3 s_3 + s_2^3 - 6s_1 s_2 s_3 + 9s_3^2.$$

The discriminant $D(s_1, s_2, s_3)$, which is equal to $\Delta(x_1, x_2, x_3)^2$, is easily calculated from (8.7) above. One finds

$$D(s_1, s_2, s_3) = s_1^2 s_2^2 + 18 s_1 s_2 s_3 - 27 s_3^2 - 4 s_1^3 s_3 - 4 s_2^3.$$

In particular, it follows that the discriminant of $X^3 + pX + q$ is $-27q^2 - 4p^3$. Denoting this discriminant by $d$, we thus have

$$d = -2^2 3^3 \left( \left( \frac{p}{3} \right)^3 + \left( \frac{q}{2} \right)^2 \right). \tag{8.8}$$

We now show what kind of information on the roots of polynomials with real coefficients can be obtained from the discriminant. We will need the following easy result:

**Lemma 8.10.** *If $a \in \mathbb{C}$ is a root of a polynomial $P \in \mathbb{R}[X]$, then the conjugate complex number $\bar{a}$ also is a root of $P$.*

**Proof.** Conjugating each side of the equation $P(a) = 0$, we obtain $P(\bar{a}) = 0$, since the coefficients of $P$ are equal to their own conjugate. $\square$

**Theorem 8.11.** *Let $P \in \mathbb{R}[X]$ be a monic polynomial with real coefficients, which splits[2] into a product of linear factors over $\mathbb{C}$, so that*

$$P = (X - u_1) \ldots (X - u_n)$$

*for some $u_1, \ldots, u_n \in \mathbb{C}$. Let $d \in \mathbb{R}$ be the discriminant of $P$. The equation $d = 0$ holds if and only if $P$ has a root of multiplicity at least 2 in $\mathbb{C}$. If all the roots of $P$ are real, then $d \geq 0$. The converse is true if $n = 2, 3$.*

**Proof.** Since the calculations with the roots of the generic polynomial are also valid with the roots $u_1, \ldots, u_n$ of $P$ (see the discussion following Definition 8.1), we have

$$d = \Delta(u_1, \ldots, u_n)^2 = \prod_{1 \leq i < j \leq n} (u_i - u_j)^2.$$

This readily shows that $d \geq 0$ if all the roots $u_1, \ldots, u_n$ are real. If $P$ has a root of multiplicity at least 2, then $u_i - u_j = 0$ for some indices $i$, $j$ with $i \neq j$, hence $d = 0$ since the product above has a zero factor. If the roots of $P$ are all simple, then each factor is non-zero, hence $d \neq 0$.

For $n = 2$,

$$d = (u_1 - u_2)^2.$$

---

[2]The fundamental theorem of algebra (Theorem 9.1, p. 109) will show that this is no restriction on $P$.

If $u_1$ is not real, then Lemma 8.10 shows that $u_2 = \overline{u_1}$. It follows that $\overline{(u_1 - u_2)} = -(u_1 - u_2)$, hence

$$(u_1 - u_2)^2 = -(u_1 - u_2)\overline{(u_1 - u_2)} = -|u_1 - u_2|^2,$$

and $d < 0$.

Similarly, for $n = 3$,

$$d = (u_1 - u_2)^2(u_1 - u_3)^2(u_2 - u_3)^2.$$

If one of the roots, $u_1$ say, is not real, then its conjugate $\overline{u_1}$ is among $u_2$, $u_3$. Renumbering $u_2$, $u_3$, we may assume $\overline{u_1} = u_2$. Then $(X - u_1)(X - u_2) \in \mathbb{R}[X]$, hence

$$X - u_3 = \frac{P}{(X - u_1)(X - u_2)} \in \mathbb{R}[X].$$

This shows $u_3 \in \mathbb{R}$. Now

$$\overline{(u_1 - u_2)} = -(u_1 - u_2), \qquad \overline{(u_1 - u_3)} = (u_2 - u_3),$$

$$\text{and} \quad \overline{(u_2 - u_3)} = (u_1 - u_3).$$

Therefore,

$$\overline{(u_1 - u_2)(u_1 - u_3)(u_2 - u_3)} = -(u_1 - u_2)(u_1 - u_3)(u_2 - u_3),$$

hence

$$d = -|(u_1 - u_2)(u_1 - u_3)(u_2 - u_3)|^2 < 0. \qquad \square$$

**Remarks 8.12.** (a) For $n \geq 4$, the sign of $d$ determines the number of real roots of $P$ up to a multiple of 4, see Exercise 8.4.

(b) The first statement (about multiple roots) in the preceding theorem is valid over an arbitrary field. Thus, we now have two necessary and sufficient conditions for a (monic) polynomial to have at least one multiple root in some extension of the base field: its discriminant has to be zero or, equivalently, the polynomial has to have a non-constant common divisor with its derivative (see Proposition 5.19, p. 54). The equivalence of these two conditions can be verified directly by using Theorem 5.26 (p. 59). Indeed, the discriminant of a monic polynomial $P$ is equal (up to sign) to the resultant of $P$ and its derivative $\partial P$; see Exercise 8.5.

**Corollary 8.13.** *Let $p$, $q \in \mathbb{R}$. The equation*

$$X^3 + pX + q = 0$$

*has three distinct real solutions if and only if $\left(\frac{p}{3}\right)^3 + \left(\frac{q}{2}\right)^2 < 0$.*

**Proof.** This readily follows from the preceding theorem, since the discriminant $d$ of $X^3 + pX + q$ is

$$d = -2^2 3^3 \left( \left(\frac{p}{3}\right)^3 + \left(\frac{q}{2}\right)^2 \right).$$

(See equation (8.8), p. 103.) $\qquad\qquad\qquad\qquad\qquad\qquad\square$

Corollary 8.13 shows that the "casus irreducibilis" of cubic equations (see §2.3(c)) is precisely the case where the equation has three distinct real roots.

## Appendix: Euler's summation of the series of reciprocals of perfect squares

Around 1735, Euler succeeded in finding the sum of the series $\sum_{k=1}^{\infty} \frac{1}{k^2}$, thus achieving a result that had baffled Leibniz and Jacques Bernoulli (see Boyer [10, Ch. 21, n° 4] or Goldstine [34, §3.2]). His method was to apply to a certain power series the relations between roots and coefficients of polynomials.

For the generic polynomial

$$X^n - s_1 X^{n-1} + s_2 X^{n-2} - \cdots + (-1)^n s_n = (X - x_1) \ldots (X - x_n),$$

we have seen that

$$s_1 = \sum x_1, \quad s_2 = \sum x_1 x_2, \quad \ldots \quad s_{n-1} = \sum x_1 \ldots x_{n-1},$$
$$\text{and} \quad s_n = x_1 \ldots x_n.$$

Therefore, using Notation 8.5 (p. 96) for rational functions,

$$\sum x_1^{-1} = s_{n-1} s_n^{-1}, \quad \sum (x_1 x_2)^{-1} = s_{n-2} s_n^{-1}, \quad \ldots,$$
$$\text{and} \quad \sum (x_1 \ldots x_{n-1})^{-1} = s_1 s_n^{-1}.$$

By the same calculations as for Newton's formulas (4.6) (p. 38), we then obtain

$$\sum x_1^{-2} = \left(\sum x_1^{-1}\right)^2 - 2\sum (x_1 x_2)^{-1} = (s_{n-1}^2 - 2s_{n-2} s_n) s_n^{-2},$$
$$\sum x_1^{-3} = \left(\sum x_1^{-1}\right)\left(\sum x_1^{-2}\right) - \left(\sum (x_1 x_2)^{-1}\right)\left(\sum x_1^{-1}\right) + 3\sum (x_1 x_2 x_3)^{-1}$$
$$= (s_{n-1}^3 - 3s_{n-2} s_{n-1} s_n + 3s_{n-3} s_n^2) s_n^{-3},$$

etc.

Now, if $u_1, \ldots, u_n$ are the roots of an arbitrary (not necessarily monic) polynomial

$$a_n X^n + a_{n-1} X^{n-1} + \cdots + a_1 X + a_0, \qquad (a_n \neq 0)$$

then $u_1, \ldots, u_n$ are the roots of the monic polynomial

$$X^n + a_{n-1} a_n^{-1} X^{n-1} + \cdots + a_1 a_n^{-1} X + a_0 a_n^{-1}.$$

Substituting $-a_{n-1} a_n^{-1}, a_{n-2} a_n^{-1}, \ldots, (-1)^n a_0 a_n^{-1}$ for $s_1, s_2, \ldots, s_n$ in the calculations above, it follows (provided that $a_0 \neq 0$) that

$$u_1^{-1} + \cdots + u_n^{-1} = -a_1 a_0^{-1}, \tag{8.9}$$

$$u_1^{-2} + \cdots + u_n^{-2} = (a_1^2 - 2a_2 a_0) a_0^{-2}, \tag{8.10}$$

$$u_1^{-3} + \cdots + u_n^{-3} = (-a_1^3 + 3a_2 a_1 a_0 - 3a_3 a_0^2) a_0^{-3}. \tag{8.11}$$

As Euler pointed out, these calculations yield interesting results when applied to the sine function, considered as the "infinite polynomial"

$$\sin z = z - \frac{z^3}{3!} + \frac{z^5}{5!} - \frac{z^7}{7!} + \cdots$$

which has as roots $0, \pm\pi, \pm 2\pi, \pm 3\pi, \ldots$ Dividing the series by $z$ to get rid of the root $0$, and changing the variable to $x = z^2$, we obtain the series

$$1 - \frac{x}{3!} + \frac{x^2}{5!} - \frac{x^3}{7!} + \cdots$$

with roots $\pi^2, (2\pi)^2, (3\pi)^2, \ldots$ Then equations (8.9), (8.10), (8.11), etc. with $n = \infty$, $a_0 = 1$, $a_1 = -\frac{1}{3!}$, $a_2 = \frac{1}{5!}$, $a_3 = -\frac{1}{7!}$, etc. yield

$$\sum_{k=1}^{\infty} (k\pi)^{-2} = \frac{1}{6}, \qquad \sum_{k=1}^{\infty} (k\pi)^{-4} = \frac{1}{90}, \qquad \sum_{k=1}^{\infty} (k\pi)^{-6} = \frac{1}{945}, \qquad \text{etc.}$$

Multiplying each side by the appropriate power of $\pi$, we obtain

$$\sum_{k=1}^{\infty} \frac{1}{k^2} = \frac{\pi^2}{6}, \qquad \sum_{k=1}^{\infty} \frac{1}{k^4} = \frac{\pi^4}{90}, \qquad \sum_{k=1}^{\infty} \frac{1}{k^6} = \frac{\pi^6}{945}, \qquad \text{etc.}$$

That Euler's calculations are valid is of course not obvious, but they can be rigorously verified. The point is that the sine function can be expressed as an infinite product

$$\sin z = z \prod_{k=1}^{\infty} \left(1 - \frac{z^2}{k^2 \pi^2}\right) \qquad \text{for } z \in \mathbb{C}.$$

Note also that the above calculations do not yield any information on the values of Riemann's zeta function

$$\zeta(s) = \sum_{k=1}^{\infty} \frac{1}{k^s}$$

at *odd* integers $s$. Indeed, very little is known about these values. Until relatively recently, it was not even known whether $\zeta(3)$ is rational or not. This question was answered in the negative by Roger Apéry in 1978 (see van der Poorten [76]).

**Exercises**

*Exercise 8.1. The following exercise provides an alternative procedure for calculating the discriminant of a polynomial.* Let

$$P(X) = X^n - s_1 X^{n-1} + \cdots + (-1)^n s_n = (X - x_1) \ldots (X - x_n),$$

let $\sigma_i = \sum_{j=1}^n x_j^i$ for $i = 1, 2, \ldots$ and let $A$ and $B$ be the $n \times n$ matrices

$$A = \begin{pmatrix} 1 & 1 & \cdots & 1 \\ x_1 & x_2 & \cdots & x_n \\ x_1^2 & x_2^2 & \cdots & x_n^2 \\ \vdots & \vdots & & \vdots \\ x_1^{n-1} & x_2^{n-1} & \cdots & x_n^{n-1} \end{pmatrix},$$

$$B = \begin{pmatrix} n & \sigma_1 & \sigma_2 & \cdots & \sigma_{n-1} \\ \sigma_1 & \sigma_2 & \sigma_3 & \cdots & \sigma_n \\ \sigma_2 & \sigma_3 & \sigma_4 & \cdots & \sigma_{n+1} \\ \vdots & \vdots & \vdots & & \vdots \\ \sigma_{n-1} & \sigma_n & \sigma_{n+1} & \cdots & \sigma_{2n-2} \end{pmatrix}.$$

(a) Show that $\det A = \prod_{i>j}(x_i - x_j)$. (The matrix $A$ is called a *Vandermonde matrix*).

(b) Show that $B = AA^t$ (where $A^t$ denotes the transpose of $A$).

(c) Derive from (a) and (b) that the discriminant of $P(X)$ is

$$D(s_1, \ldots, s_n) = \det B.$$

(d) Use this result to prove that the discriminant of the polynomial $X^n + pX + q$ is

$$(-1)^{\frac{n(n-1)}{2}} \left( (-1)^{n-1}(n-1)^{n-1} p^n + n^n q^{n-1} \right).$$

*Exercise 8.2.* Let $x_1$, $x_2$, $x_3$ be the roots of the cubic equation $X^3 + pX + q = 0$. Prove that the equation whose roots are $(x_1 - x_2)^2$, $(x_1 - x_3)^2$, and $(x_2 - x_3)^2$ is

$$Y^3 + 6pY^2 + 9p^2Y + (4p^3 + 27q^2) = 0.$$

This equation is called the *equation of squared differences* of the given cubic equation.

*Exercise 8.3.* Consider the general polynomial of degree 4

$$P(X) = X^4 - s_1X^3 + s_2X^2 - s_3X + s_4$$
$$= (X - x_1)(X - x_2)(X - x_3)(X - x_4),$$

and let

$$u_1 = x_1x_2 + x_3x_4, \qquad v_1 = (x_1 + x_2)(x_3 + x_4),$$
$$u_2 = x_1x_3 + x_2x_4, \qquad v_2 = (x_1 + x_3)(x_2 + x_4),$$
$$u_3 = x_1x_4 + x_2x_3, \qquad v_3 = (x_1 + x_4)(x_2 + x_3).$$

(a) Show that the equation which has as roots $u_1$, $u_2$, and $u_3$ is

$$Q(Y) = Y^3 - s_2Y^2 + (s_1s_3 - 4s_4)Y - (s_1^2s_4 - 4s_2s_4 + s_3^2) = 0$$

and that the equation which has as roots $v_1$, $v_2$, and $v_3$ is

$$R(Z) = Z^3 - 2s_2Z^2 + (s_2^2 + s_1s_3 - 4s_4)Z - (s_1s_2s_3 - s_1^2s_4 - s_3^2) = 0.$$

(Compare equations (3.5) and (3.6) in Chapter 3, pp. 23 and 24.)

(b) Show that the discriminants of $P$, $Q$, and $R$ are all equal. [Hint: Prove that $\Delta(X_1, X_2, X_3, X_4) = -\Delta(u_1, u_2, u_3) = -\Delta(v_1, v_2, v_3)$.]

*Exercise 8.4.* Let $P \in \mathbb{R}[X]$ be a monic polynomial with real coefficients, and suppose $P = (X - u_1) \ldots (X - u_n)$ for some pairwise distinct $u_i \in \mathbb{C}$. Let $r$ be the number of real roots among $u_1, \ldots, u_n$ and let $d$ be the discriminant of $P$. Show that $n - r$ is an even integer, which is divisible by 4 if and only if $d > 0$.

*Exercise 8.5.* Let $P = (X - x_1) \ldots (X - x_n)$ and let $d$ denote the discriminant of $P$. Show that the resultant of $P$ and its derivative $\partial P$ satisfies

$$\text{Res}(P, \partial P) = \prod_{i \neq j}(x_i - x_j) = (-1)^{\frac{n(n-1)}{2}}d.$$

[Hint: Calculate $\partial P(x_i)$ and use Lemma 5.25.]

Chapter 9

# The Fundamental Theorem of Algebra

The title of this chapter refers to the following result:

**Theorem 9.1.** *The number of roots of a non-zero polynomial over the field $\mathbb{C}$ of complex numbers, each root being counted with its multiplicity, is equal to the degree of the polynomial.*

Equivalently, by Theorem 5.15, the fundamental theorem of algebra asserts that every non-constant polynomial splits into a product of linear factors in $\mathbb{C}[X]$. There is also an equivalent formulation in terms of real polynomials only:

**Theorem 9.2.** *Every (non-constant) real polynomial can be decomposed into a product of real polynomials of degree 1 or 2.*

The equivalence of these statements will be proved in Proposition 9.6 below.

Theorem 9.1 can be traced back to Girard, in a considerably looser form (see §4.2, p. 36). Indeed, Leibniz's proposed counter-example (in §7.1, p. 75) clearly shows how remote a proof still was at the beginning of the eighteenth century. Yet, during the first half of that century, de Moivre's work prompted a deeper understanding of the operations on complex numbers and opened the way to the first attempts of proof.

Since the ultimate structure of $\mathbb{R}$ (or $\mathbb{C}$) is analytical, it is not surprising to us that the first idea of a proof, published in 1746 by Jean le Rond d'Alembert (1717–1783), used analytic techniques. However, an analytic proof for what was perceived as an algebraic theorem was hardly satisfactory, and Euler in 1749 tried a more algebraic method. Euler's idea was to prove the equivalent Theorem 9.2 by an induction argument on the highest power of 2 that divides the degree. Omitting several key details,

Euler failed to carry out his program completely, so that his proof is only a sketch. Some simplifications in Euler's proof were subsequently suggested by Daviet François de Foncenex (1734–1799), and Lagrange eventually gave in 1772 a complete proof, elaborating on Euler's and de Foncenex's ideas and correcting all the flaws in their proofs.

All, but one. As Gauss noticed in 1799 in his inaugural dissertation [30, §12], a critical flaw remained. Misled by the custom inherited from seventeenth century mathematicians, Lagrange implicitly takes for granted the _existence_ of $n$ "imaginary" roots for any equation of degree $n$, and he thus only proves that the _form_ of these imaginary roots is $a + b\sqrt{-1}$ with $a, b \in \mathbb{R}$. Gauss also shows that the same criticism can be addressed to the earlier proofs of Euler, de Foncenex and d'Alembert [30, §§6ff] and he then proceeds to give the first essentially complete proof of the fundamental theorem of algebra, along the lines of d'Alembert's proof.

In 1815, Gauss also found a way to mend the Euler–de Foncenex–Lagrange proof [32], and he subsequently gave two other proofs of the fundamental theorem.

The proof we give is based upon the Euler–de Foncenex–Lagrange ideas. However, instead of following Gauss' correction, we use some ideas from the late nineteenth century to first prove Girard's "theorem" on the existence of imaginary roots (of which nothing is known, except that operations can be performed on these roots as if they were numbers). Having thus justified the _postulate_ on which Euler's proof implicitly relied, we are then able to use Euler's arguments in a more direct way, to prove that the imaginary roots are of the form $a + b\sqrt{-1}$ (with $a, b \in \mathbb{R}$). We thus obtain a streamlined and almost completely algebraic proof of the fundamental theorem of algebra, also found in Samuel [65, p. 53].

## 9.1  Girard's theorem

In this section, we let $F$ be an arbitrary field. The modern translation of Girard's intuition is as follows.

**Theorem 9.3.**  _For any non-constant polynomial $P \in F[X]$, there is a field $K$ containing $F$ such that $P$ splits over $K$ into a product of linear factors, i.e.,_

$$P = a(X - x_1) \ldots (X - x_n) \qquad in \ K[X].$$

The field $K$ is constructed abstractly by successive quotients of polynomial rings by ideals.

We first recall that an *ideal* in a commutative ring $A$ is a subgroup $I$ of the additive group of $A$ that is stable under multiplication by elements in $A$, i.e., such that $ax \in I$ for $a \in A$ and $x \in I$. In order to define the quotient ring of $A$ by $I$, we set, for $a \in A$,

$$a + I = \{a + x \mid x \in I\}.$$

This is a subset of $A$, and from the hypothesis that $I$ is a subgroup of the additive group of $A$, it easily follows that, for $a$, $b \in A$,

$$a + I = b + I \quad \text{if and only if} \quad a - b \in I. \tag{9.1}$$

We then set

$$A/I = \{a + I \mid a \in A\}.$$

The condition that $I$ is an ideal implies that the operations on $A$ induce a ring structure on $A/I$, by

$$(a + I) + (b + I) = (a + b) + I \qquad \text{and} \qquad (a + I)(b + I) = ab + I.$$

The ring $A/I$ is called the *quotient ring* of $A$ by the ideal $I$.

The zero element in this quotient ring is $0 + I$ ($= I$), also denoted simply by $0$; therefore, (9.1) shows that in $A/I$

$$a + I = 0 \quad \text{if and only if} \quad a \in I. \tag{9.2}$$

We will need the following instance of this construction: let $A = F[X]$ and let $I$ be the set $(P)$ of multiples of a polynomial $P$,

$$(P) = \{PQ \mid Q \in F[X]\}.$$

It is readily verified that $(P)$ is an ideal of $F[X]$, so that we can construct the quotient ring $F[X]/(P)$. This construction is essentially due to Kronecker (1823–1897), although a special case had been considered earlier by Cauchy. (In 1847, Cauchy represented $\mathbb{C}$ as the quotient ring $\mathbb{R}[X]/(X^2 + 1)$.)

The basic lemma required for the proof of Girard's theorem is the following:

**Lemma 9.4.** *If $P \in F[X]$ is irreducible, then $F[X]/(P)$ is a field containing $F$ and a root of $P$.*

**Proof.** In order to show that $F[X]/(P)$ is a field, it suffices to prove that every non-zero element $Q + (P)$ in $F[X]/(P)$ is invertible. Since $Q + (P) \neq 0$, it follows from (9.2) above that $Q \notin (P)$, i.e., that $Q$ is not divisible by

$P$. Since $P$ is irreducible, $P$ and $Q$ are then relatively prime, hence, by Corollary 5.4 (p. 47) there exist polynomials $P_1$, $Q_1 \in F[X]$ such that
$$PP_1 + QQ_1 = 1.$$
This equation shows that $QQ_1 - 1$ is divisible by $P$, so that, by (9.1) above,
$$QQ_1 + (P) = 1 + (P).$$
Therefore,
$$\big(Q + (P)\big)\big(Q_1 + (P)\big) = 1 + (P),$$
which means that $Q_1 + (P)$ is the inverse of $Q + (P)$ in $F[X]/(P)$.

The map $a \mapsto a + (P)$ from $F$ to $F[X]/(P)$ is injective since no non-zero element in $F$ is divisible by $P$. Therefore, this map is an embedding of $F$ in $F[X]/(P)$, and $F$ is thus identified with a subfield of $F[X]/(P)$. We can therefore consider $P$ as a polynomial over $F[X]/(P)$, and it only remains to prove that $F[X]/(P)$ contains a root of $P$.

From the definition of the operations in $F[X]/(P)$ it follows that
$$P\big(X + (P)\big) = P(X) + (P).$$
Therefore, by (9.2) above,
$$P\big(X + (P)\big) = 0,$$
which means that $X + (P)$ is a root of $P$ in $F[X]/(P)$. $\qquad\square$

**Proof of Theorem 9.3.** We decompose $P$ into a product of irreducible factors in $F[X]$:
$$P = P_1 \ldots P_r.$$
Let $s$ be the number of linear factors among $P_1, \ldots, P_r$. The integer $s$ is thus the number of roots of $P$ in $F$, each root being counted with its multiplicity. (Possibly $s = 0$, since $P$ may have no roots in $F$.)

We argue by induction on $(\deg P) - s$, noting that this number is equal to the degree of the product of the non-linear factors among $P_1, \ldots, P_r$.

If $(\deg P) - s = 0$, then each of the factors $P_1, \ldots, P_r$ is linear, and we can choose $K = F$.

If $(\deg P) - s > 0$, then at least one of the factors $P_1, \ldots, P_r$ has degree greater than or equal to 2. Assume for instance that $\deg P_1 \geq 2$, and let
$$F_1 = F[X]/(P_1).$$
Since $P_1$ has a root in $F_1$, the decomposition of $P_1$ over $F_1$ involves at least one linear factor, by Theorem 5.12 (p. 51), hence the number $s_1$ of linear factors in the decomposition of $P$ into irreducible factors over $F_1$ is at least $s + 1$. Therefore, $(\deg P) - s_1 < (\deg P) - s$, and the induction hypothesis implies that there is a field $K$ containing $F_1$ such that $P$ splits into linear factors over $K$. Since $F_1$ contains $F$, the field $K$ also contains $F$ and satisfies all the requirements. $\qquad\square$

## 9.2 Proof of the fundamental theorem

Instead of proving Theorem 9.1 directly, we prove an equivalent formulation in terms of real polynomials. We first record for later reference the following easy special case of the fundamental Theorem 9.1:

**Lemma 9.5.** *Every quadratic polynomial over $\mathbb{C}$ splits into a product of linear factors in $\mathbb{C}[X]$.*

**Proof.** It suffices to show that the roots of every quadratic equation with complex coefficients are complex numbers. This readily follows from the usual formula by radicals for the roots (p. 1), since, by Proposition 7.1 (p. 80), every complex number has a square root in $\mathbb{C}$. □

We now prove the equivalence of several formulations of the fundamental theorem.

**Proposition 9.6.** *The following statements are equivalent:*

(a) *The number of roots of any non-zero polynomial over $\mathbb{C}$ is equal to its degree (each root being counted with its multiplicity).*

(b) *Every non-constant polynomial over $\mathbb{R}$ has at least one root in $\mathbb{C}$.*

(c) *Every non-constant real polynomial can be decomposed into a product of real polynomials of degree 1 or 2.*

**Proof.** (a) $\Rightarrow$ (b) This is clear.

(b) $\Rightarrow$ (c) By Theorem 5.8 (p. 49), it suffices to show, assuming (b), that every irreducible polynomial in $\mathbb{R}[X]$ has degree 1 or 2. Let $P$ be an irreducible polynomial in $\mathbb{R}[X]$, and let $a \in \mathbb{C}$ be a root of $P$.

If $a \in \mathbb{R}$, then $X - a$ divides $P$ in $\mathbb{R}[X]$, hence $\deg P = 1$ since, by definition, an irreducible polynomial cannot be divided by a non-constant polynomial of strictly smaller degree.

If $a \notin \mathbb{R}$, then $\bar{a} \neq a$, and $\bar{a}$ is also a root of $P$, by Lemma 8.10, p. 103. Therefore, by Proposition 5.10 (p. 50), $P$ is divisible by $(X - a)(X - \bar{a})$ in $\mathbb{C}[X]$. But $(X - a)(X - \bar{a})$ lies in $\mathbb{R}[X]$ since

$$(X - a)(X - \bar{a}) = X^2 - (a + \bar{a})X + a\bar{a}.$$

Therefore, $P$ is also divisible by $(X - a)(X - \bar{a})$ in $\mathbb{R}[X]$ (see Remark 5.5(b), p. 47), hence the same argument as above implies $\deg P = 2$.

(c) $\Rightarrow$ (a) Let $P \in \mathbb{C}[X]$ be a non-constant polynomial. We extend to $\mathbb{C}[X]$ the complex conjugation map from $\mathbb{C}$ to $\mathbb{C}$ by setting $\overline{X} = X$; namely, we set for $a_0, \ldots, a_n \in \mathbb{C}$

$$\overline{a_0 + a_1 X + \cdots + a_n X^n} = \overline{a_0} + \overline{a_1} X + \cdots + \overline{a_n} X^n.$$

The invariant elements are readily seen to be the polynomials with real coefficients. Therefore, $P\overline{P} \in \mathbb{R}[X]$ and it follows from the hypothesis (c) that

$$P\overline{P} = P_1 \ldots P_r$$

for some polynomials $P_1, \ldots, P_r \in \mathbb{R}[X]$ of degree 1 or 2.

By Lemma 9.5, the real polynomials of degree 2 split into products of linear factors in $\mathbb{C}[X]$, hence $P\overline{P}$ is a product of linear factors in $\mathbb{C}[X]$. Therefore, every irreducible factor of $P$ in $\mathbb{C}[X]$ has degree 1, so that $P$ splits into a product of linear factors in $\mathbb{C}[X]$, which proves (a).          $\square$

As we noted in the introduction, every proof of the fundamental theorem of algebra uses at some point an analytical (or topological) argument, since $\mathbb{R}$ (or $\mathbb{C}$) cannot be completely defined without reference to some of its topological properties. The only analytical result we will need in our proof is the following:

**Lemma 9.7.** *Every real polynomial $P$ of odd degree has at least one root in $\mathbb{R}$.*

**Proof.** Since $\deg P$ is odd, the polynomial function $P(\cdot) \colon \mathbb{R} \to \mathbb{R}$ changes sign when the variable runs from $-\infty$ to $+\infty$, so, by continuity, it must take the value 0 at least once.          $\square$

The continuity argument, according to which every continuous function that changes sign on an interval must take the value 0 at least once, may seem (and was for a long time considered as) evident by itself. It was first proved by Bernhard Bolzano (1781–1848) in 1817 (see Dieudonné [24, p. 340] or Kline [45, p. 952]), in an attempt to provide "arithmetic" proofs to the intuitive geometric arguments that Gauss used in his 1799 proof of the fundamental theorem.

**Proof of the fundamental theorem.** We prove the equivalent formulation (b) in Proposition 9.6, that every real non-constant polynomial has at least one root in $\mathbb{C}$.

Let $P \in \mathbb{R}[X]$ be a non-constant polynomial. Dividing $P$ by its leading coefficient if necessary, we may assume that $P$ is monic. We write the degree of $P$ in the form

$$\deg P = n = 2^e m$$

where $e \geq 0$ and $m$ is odd. If $e = 0$, then the degree of $P$ is odd, and Lemma 9.7 shows that $P$ has a root in $\mathbb{R}$. We then argue by induction on $e$, assuming that $e \geq 1$ and that the property holds when the exponent of the highest power of 2 that divides the degree of the polynomial is at most $e - 1$.

Let $K$ be a field containing $\mathbb{C}$, over which $P$ splits into a product of linear factors:

$$P = (X - x_1) \ldots (X - x_n).$$

(The existence of such a field $K$ follows from Theorem 9.3.) For $c \in \mathbb{R}$ and for $i, j = 1, \ldots, n$ with $i < j$, let

$$y_{ij}(c) = (x_i + x_j) + cx_ix_j.$$

Let also

$$Q_c(Y) = \prod_{1 \leq i < j \leq n} (Y - y_{ij}(c)).$$

The coefficients of $Q_c$ are the values of the elementary symmetric polynomials in the roots $y_{ij}(c)$. These coefficients are therefore the values of symmetric polynomials in $x_1, \ldots, x_n$ with real coefficients, hence they can be expressed in terms of the values of the elementary symmetric polynomials in $x_1, \ldots, x_n$, by the fundamental theorem of symmetric polynomials (Theorem 8.4, p. 95). Since the values of the elementary symmetric polynomials in $x_1, \ldots, x_n$ are the coefficients of $P$, which are real numbers, it follows that the coefficients of $Q_c$ also are real numbers. Moreover, the degree of $Q_c$ is $\frac{n(n-1)}{2}$, hence

$$\deg Q_c = 2^{e-1}\big(m(2^e m - 1)\big),$$

and the integer between brackets is odd. We may therefore apply the induction hypothesis to conclude that $Q_c$ has at least one root in $\mathbb{C}$, i.e., $y_{r(c),s(c)}(c) \in \mathbb{C}$ for some indices $r(c)$, $s(c) \in \{1, \ldots, n\}$. If we let $c$ run over the set of all real numbers, the indices $r(c)$, $s(c)$ for which $y_{r(c),s(c)}(c) \in \mathbb{C}$ cannot be all distinct, since the set of indices is finite, while $\mathbb{R}$ is infinite. Therefore, we can find some distinct real numbers $c_1$, $c_2$ such that $r(c_1) = r(c_2)$ and $s(c_1) = s(c_2)$. Writing simply $r$ and $s$ for these common indices, this means that

$$(x_r + x_s) + c_1 x_r x_s \in \mathbb{C} \qquad \text{and} \qquad (x_r + x_s) + c_2 x_r x_s \in \mathbb{C},$$

with $c_1, c_2 \in \mathbb{R}$ and $c_1 \neq c_2$. By subtraction, these relations imply

$$(c_1 - c_2)x_r x_s \in \mathbb{C},$$

hence $x_r x_s \in \mathbb{C}$. Since $(x_r + x_s) + c_1 x_r x_s \in \mathbb{C}$, we also have $x_r + x_s \in \mathbb{C}$. Therefore, the coefficients of the polynomial

$$X^2 - (x_r + x_s)X + x_r x_s$$

are complex numbers, and it then follows from Lemma 9.5 that its roots $x_r$ and $x_s$ are complex numbers. We have thus shown that at least one of the roots $x_1, \ldots, x_n$ of $P$ in $K$ is a complex number, as was required.     $\square$

**Corollary 9.8.** *Over* $\mathbb{C}$, *the irreducible polynomials are the polynomials of degree 1. Over* $\mathbb{R}$, *the irreducible polynomials are the polynomials of degree 1, and the polynomials of degree 2 that have no real roots.*

**Proof.** This readily follows from the fundamental Theorem 9.1 or the equivalent Theorem 9.2, since, by Theorem 5.12 (p. 51), the irreducible polynomials that have a root in the base field have degree 1.     $\square$

# Chapter 10

# Lagrange

## 10.1 The theory of equations comes of age

In the second half of the eighteenth century, the algebraic theory of equations is ripe for new advances. All the more or less elementary facts on polynomials are well-known, and computational skills are very high, even by modern standards. Moreover, deeper insights on the ambiguity of roots of (complex) numbers become available through de Moivre's work. The relevance of these insights for the problem of solving equations by radicals is obvious (see the end of §7.1), and one may venture the hypothesis that de Moivre's work provided an important stimulus to new research in the algebraic theory of equations.

Whatever its origin, it is clear that the spirit of the most significant research in this period is completely different from that of Cardano and his contemporaries: no direct application to the solution of numerical equations is expected, and no reference to any practical problem is made, even allusively. The subject has become pure mathematics, and is pursued for its own interest. Within less than a century, in the hands of several mathematicians of genius, it will undergo a rapid development which will dramatically change the whole subject of algebra.

The earliest works in this line appear in the sixties of the eighteenth century. They set forth new methods to solve equations of degree at most 4, which can seemingly be extended to equations of higher degree. One of these methods, proposed by Bézout in 1765, is of particular interest because of its explicit use of roots of unity; it is in fact very close to a method of Euler, and is deeply similar in its principle to Tschirnhaus' method.

The idea[1] is to eliminate the indeterminate $y$ between the two equations

$$x = a_0 + a_1 y + a_2 y^2 + \cdots + a_{n-1} y^{n-1}, \tag{10.1}$$

$$y^n = 1, \tag{10.2}$$

producing an equation of degree $n$ in $x$,

$$R_n(x) = 0,$$

as in Tschirnhaus' method (see §6.4). Dividing the polynomial $R_n$ by its leading coefficient if necessary, we can assume that $R_n$ is monic. The properties of $R_n$ imply that if $x$, $y$ are related by equations (10.1) and (10.2), i.e., if

$$x = a_0 + a_1 \omega + a_2 \omega^2 + \cdots + a_{n-1} \omega^{n-1}$$

for some $n$-th root of unity $\omega$, then $x$ is a solution of $R_n(x) = 0$, hence $R_n$ is divisible by $x - (a_0 + a_1 \omega + \cdots + a_{n-1} \omega^{n-1})$. Regarding $a_0, a_1, \ldots, a_{n-1}$ as independent indeterminates, the values of $x$ corresponding to the various $n$-th roots of unity $\omega$ are all different, so that, by Proposition 5.10 (p. 50),

$$R_n(x) = \prod_\omega \big(x - (a_0 + a_1 \omega + \cdots + a_{n-1} \omega^{n-1})\big), \tag{10.3}$$

where the product runs over the $n$ different $n$-th roots of unity $\omega$. The solutions of $R_n(x) = 0$ are thus known.

Now, to solve an arbitrary monic equation $P(x) = 0$ of degree $n$, the idea is to determine the parameters $a_0, a_1, \ldots, a_{n-1}$ in such a way that the polynomial $R_n$ be identical to $P$. The solutions of $P(x) = 0$ are then readily obtained in the form $a_0 + a_1 \omega + \cdots + a_{n-1} \omega^{n-1}$.

Of course, whether it is possible to assign some value to $a_0, \ldots, a_{n-1}$ in such a way that $R_n$ becomes identical to $P$ is not clear at all, but this turns out to be the case for $n = 2$, 3 or 4, as we are about to see.

The method for constructing $R_n$ by elimination of $y$ between equations (10.1) and (10.2) has already been discussed in §6.4. For the small values of $n$, the following results are found:

$$R_2(x) = (x - a_0)^2 - a_1^2,$$

$$R_3(x) = (x - a_0)^3 - 3a_1 a_2 (x - a_0) - (a_1^3 + a_2^3),$$

$$R_4(x) = (x - a_0)^4 - 2(a_2^2 + 2a_1 a_3)(x - a_0)^2 - 4a_2(a_1^2 + a_3^2)(x - a_0)$$
$$\quad - (a_1^4 - a_2^4 + a_3^4 + 4a_1 a_2^2 a_3 - 2a_1^2 a_3^2).$$

---

[1] According to Bézout's presentation; in Euler's work, equation (10.2) is replaced by $y^n = b$ and in (10.1), one of the coefficients $a_i$ is chosen to be 1. Tschirnhaus' method can be presented in a similar way, replacing equation (10.2) by $y^n = b$ and (10.1) by $y = a_0 + a_1 x + \cdots + a_{n-2} x^{n-2} + x^{n-1}$.

These results can be obtained from equation (10.3) by expanding the right-hand side.

To obtain the solutions of the cubic equation

$$x^3 + px + q = 0 \tag{10.4}$$

(to which the general cubic equation can be reduced by a linear change of variable, see §2.2), it now suffices to assign values to $a_0$, $a_1$, and $a_2$ in such a way that $R_3(x)$ takes the form $x^3 + px + q$. Thus, we choose $a_0 = 0$ and determine $a_1$ and $a_2$ by

$$-3a_1a_2 = p, \tag{10.5}$$

$$-(a_1^3 + a_2^3) = q \tag{10.6}$$

(compare §2.2). The first equation gives the value of $a_2$ as a function of $a_1$; substituting this value in the second equation yields the following quadratic equation in $a_1^3$:

$$a_1^6 + qa_1^3 - \left(\frac{p}{3}\right)^3 = 0.$$

A root of this equation is easily found: one can choose for $a_1$ any cube root of $-\frac{q}{2} + \sqrt{\left(\frac{p}{3}\right)^3 + \left(\frac{q}{2}\right)^2}$ or $-\frac{q}{2} - \sqrt{\left(\frac{p}{3}\right)^3 + \left(\frac{q}{2}\right)^2}$. Then for $a_2 = -\frac{p}{3a_1}$ equations (10.5) and (10.6) both hold, hence $R_3(x) = x^3 + px + q$. From (10.3) it then follows that the solutions of equation (10.4) are of the form $\omega a_1 + \omega^2 a_2$, where $\omega$ runs over the set of cube roots of unity. If $\zeta$ denotes one of these cube roots other than 1, then the cube roots of unity are 1, $\zeta$ and $\zeta^2$, and therefore the solutions of (10.4) are

$$a_1 + a_2, \qquad \zeta a_1 + \zeta^2 a_2, \qquad \text{and} \qquad \zeta^2 a_1 + \zeta a_2.$$

(Note that $\zeta^4 = \zeta$.)

**Remark.** The fact that one can choose $a_0 = 0$ obviously follows from the particular form of the proposed cubic equation (10.4), which lacks the term in $x^2$. The general case is in no way more difficult and could be treated in the same way, but the calculations are less transparent.

Similarly, for equations of degree 4 such as

$$x^4 + px^2 + qx + r = 0, \tag{10.7}$$

we seek values of $a_0$, $a_1$, $a_2$, and $a_3$ for which $R_4(x) = x^4 + px^2 + qx + r$. As above, we choose $a_0 = 0$ and we are left with the equations

$$-2(a_2^2 + 2a_1a_3) = p, \tag{10.8}$$

$$-4a_2(a_1^2 + a_3^2) = q, \tag{10.9}$$

$$-(a_1^4 - a_2^4 + a_3^4 + 4a_1a_2^2a_3 - 2a_1^2a_3^2) = r. \tag{10.10}$$

Substituting in the third equation $(a_1^2 + a_3^2)^2 - 2a_1^2a_3^2$ for $a_1^4 + a_3^4$, we get

$$-r = (a_1^2 + a_3^2)^2 - 4a_1^2a_3^2 + 4(a_1a_3)a_2^2 - a_2^4.$$

Equations (10.8) and (10.9) can then be used to eliminate $a_1$ and $a_3$ from this equation. The resulting equation is a cubic equation in $a_2^2$, from which a value of $a_2$ can be determined. Values for $a_1$ and $a_3$ are then easily found from equations (10.8) and (10.9), and the solutions of the proposed quartic equation (10.7) are obtained in the form $a_1\omega + a_2\omega^2 + a_3\omega^3$, where $\omega$ runs over the set of 4-th roots of unity. Letting $i$ denote (as usual) a square root of $-1$, the 4-th roots of unity are $1$, $i$, $-1$, $-i$ and the solutions of the quartic equation (10.7) are

$$a_1 + a_2 + a_3, \quad ia_1 - a_2 - ia_3, \quad -a_1 + a_2 - a_3, \quad \text{and} \quad -ia_1 - a_2 + ia_3.$$

As noted above, the principle of this method, hence also its difficulty, is not very different from that of Tschirnhaus' method. To its credit, one can nevertheless observe that the method of Euler and Bézout leads to easier calculations, that it is somewhat more direct and, what is more significant for later researches, that Bézout's method stresses the importance of the roots of unity. Altogether, it does not represent a very substantial progress toward the solution of equations of degree 5 or higher. (On the other hand, it led Bézout to incisive remarks on the degrees of the resolvents and elimination theory; see Stedall [68, Ch. 8].)

The first really important burst of activity in the theory of equations takes place only a few years later, around 1770, with the almost simultaneous publication of Lagrange's "Réflexions sur la résolution algébrique des équations" and Vandermonde's "Mémoire sur la résolution des équations," and of comparatively less important works such as Waring's "Meditationes Algebraicae," which we already quoted in Chapter 8. Among all the works of this period, Lagrange's massive paper clearly is the most lucid and the most comprehensive. Therefore, it proved to be also the most influential. Moreover, Lagrange provides an almost unhoped-for link between the early stages of the theory of equations and the subsequent period, by first reviewing the various methods for equations of degree 3 and 4, and the attempts at equations of higher degree so far proposed, before making his own highly original observations.

We will therefore begin our study of the two critical works of this period with Lagrange's paper, and discuss Vandermonde's memoir in the next chapter.

## 10.2 Lagrange's observations on previously known methods

Lagrange's discussion of the previously known methods is not a mere summary, it is a vast unification and reassessment of these methods. His very explicit aim is to determine not only *how* these methods work, but *why*.

> I propose in this Memoir to examine the various methods found so far for the algebraic solution of equations, to reduce them to general principles, and to let see *a priori* why these methods succeed for the third and the fourth degree, and fail for higher degrees.
>
> This examination will have a double advantage: on one hand, it will shed a greater light on the known solutions of the third and the fourth degree; on the other hand, it will be useful to those who will want to deal with the solution of higher degrees, by providing them with various views to this end and above all by sparing them a large number of useless steps and attempts [49, Intro.].

The phrase *a priori* keeps recurring throughout Lagrange's work. It is the hallmark of his fruitful new methodology. Lagrange starts from the rather obvious observation that the various methods for solving equations have a common feature: they all reduce the problem by some clever transformations to the solution of a certain auxiliary equation of smaller degree. *A posteriori*, when these clever transformations have been found, one can only ascertain that the method provides the required solutions from those of the auxiliary equation, but this does not yield any valuable insight into the solution of equations of higher degree. Indeed, the only evidence to support the belief that Tschirnhaus', Euler's or Bézout's method could be applied to equations of higher degree is that the approach is the same in all cases, that the first calculations are similar, and that it works for equations of degree 2, 3 or 4. This is rather scant evidence.

To find out *a priori* why a method works, Lagrange's highly original idea is to reverse the steps, and determine the roots of the auxiliary equations as functions of the roots of the proposed equation. The properties of the roots of the auxiliary equation then become apparent, and they clearly show why these roots provide the solution of the proposed equation.

As a first example, we take Cardano's solution of cubic equations, which is the first method scrutinized by Lagrange. Lagrange begins with a careful description of the procedure. The cubic equation

$$x^3 + ax^2 + bx + c = 0$$

is first reduced to the form

$$x'^3 + px' + q = 0$$

by the change of variable $x' = x + \frac{a}{3}$. Next, by setting $x' = y + z$, the equation becomes

$$(y^3 + z^3 + q) + (y + z)(3yz + p) = 0.$$

The solutions of the cubic equation are then obtained from the solutions of the system

$$\begin{cases} y^3 + z^3 + q = 0, \\ 3yz + p = 0. \end{cases}$$

From the second equation comes

$$z = -\frac{p}{3y} \tag{10.11}$$

and, substituting in the first equation, one gets $y^3 - (\frac{p}{3y})^3 + q = 0$, hence

$$y^6 + qy^3 - \left(\frac{p}{3}\right)^3 = 0. \tag{10.12}$$

This is the auxiliary equation, which Lagrange terms the "reduced" equation, on which Cardano's method depends. From this equation, the values of $y$ are easily obtained, since it is really a quadratic equation in $y^3$. The corresponding values of $z$ are derived from (10.11), and the solutions of the initial equation are then

$$x = -\frac{a}{3} + y + z.$$

Since the reduced equation (10.12) has degree 6, it has six roots, so we end up with six roots for the initial equation of degree 3. In fact, these six values of $x$ are pairwise equal, so that each root of the cubic equation is obtained twice. Indeed, let $y_1, y_2, \ldots, y_6$ be the six roots of (10.12). Since this equation is quadratic in $y^3$, their cubes $y_1^3, \ldots, y_6^3$ take only two values $v_1, v_2$, whose product is $v_1 v_2 = -(\frac{p}{3})^3$, since $-(\frac{p}{3})^3$ is the constant term of (10.12). Changing the numbering if necessary, we may assume

$$y_1^3 = y_2^3 = y_3^3 = v_1 \qquad \text{and} \qquad y_4^3 = y_5^3 = y_6^3 = v_2.$$

So, $y_1, y_2, y_3$ (resp. $y_4, y_5, y_6$) are the various cube roots of $v_1$ (resp. $v_2$). Therefore, writing $\omega$ for a cube root of unity other than 1, we may assume

$$y_2 = \omega y_1, \quad y_3 = \omega^2 y_1 \quad \text{and} \quad y_5 = \omega y_4, \quad y_6 = \omega^2 y_4.$$

Since $v_1 v_2 = -(\frac{p}{3})^3$, there are some determinations of the cube roots of $v_1$ and $v_2$ that multiply up to $-\frac{p}{3}$. Assume for instance (renumbering $y_1, \ldots, y_6$ if necessary) that

$$y_1 y_4 = -\frac{p}{3}.$$

Then

$$y_2 y_6 = \omega^3 y_1 y_4 = -\frac{p}{3}$$

and similarly

$$y_3 y_5 = -\frac{p}{3}.$$

Therefore, if we let $z_i$ denote the value of $z$ that corresponds to $y_i$ by (10.11), we have

$$z_1 = y_4, \quad z_2 = y_6, \quad z_3 = y_5, \quad z_4 = y_1, \quad z_5 = y_3, \quad \text{and} \quad z_6 = y_2,$$

and it follows that $y_i + z_i$ takes only three different values, namely

$$y_1 + y_4, \quad y_2 + y_6 = \omega y_1 + \omega^2 y_4, \quad \text{and} \quad y_3 + y_5 = \omega^2 y_1 + \omega y_4,$$

which yield the three roots of the initial cubic equation in the form

$$
\begin{aligned}
x_1 &= -\frac{a}{3} + y_1 + y_4, \\
x_2 &= -\frac{a}{3} + \omega y_1 + \omega^2 y_4, \\
x_3 &= -\frac{a}{3} + \omega^2 y_1 + \omega y_4.
\end{aligned}
\tag{10.13}
$$

So, we see *a posteriori* how the reduced equation provides the solution of the initial cubic equation.

To understand *a priori* why it does, Lagrange determines $y_1, \ldots, y_6$ as functions of $x_1$, $x_2$, and $x_3$. This is fairly easy: it suffices to solve the system (10.13) for $y_1$ and $y_4$, and the other $y_i$'s are multiples of $y_1$ or $y_4$ by $\omega$ or $\omega^2$. It is even easier if one notices that, since $\omega$ is a cube root of unity other than 1, it is a root of

$$\frac{X^3 - 1}{X - 1} = X^2 + X + 1,$$

so that $\omega^2 + \omega + 1 = 0$. Therefore, multiplying the second equation of (10.13) by $\omega^2$, the third by $\omega$ and adding to the first, one obtains

$$x_1 + \omega^2 x_2 + \omega x_3 = -\frac{a}{3}(1 + \omega + \omega^2) + 3y_1 + (1 + \omega + \omega^2)y_4,$$

hence

$$y_1 = \frac{1}{3}(x_1 + \omega^2 x_2 + \omega x_3).$$

Likewise, one obtains

$$y_4 = \frac{1}{3}(x_1 + \omega x_2 + \omega^2 x_3)$$

and the other roots of the reduced equation are easily obtained by multiplying $y_1$ or $y_4$ by $\omega$ or $\omega^2$. One thus gets

$$y_2 = \frac{1}{3}(\omega x_1 + x_2 + \omega^2 x_3),$$

$$y_3 = \frac{1}{3}(\omega^2 x_1 + \omega x_2 + x_3),$$

$$y_5 = \frac{1}{3}(\omega x_1 + \omega^2 x_2 + x_3),$$

$$y_6 = \frac{1}{3}(\omega^2 x_1 + x_2 + \omega x_3).$$

Thus, the roots of the reduced equation are all the expressions obtained from $\frac{1}{3}(x_1 + \omega x_2 + \omega^2 x_3)$ by permutation of $x_1$, $x_2$, and $x_3$, and the purpose of solving the reduced equation is to determine some (hence all) of these expressions.

From this observation, Lagrange draws some clever conclusions. First, it explains why the reduced equation has degree 6. Indeed, since the coefficients of the reduced equation are functions of the coefficients of the proposed equation, which are the elementary symmetric polynomials in $x_1$, $x_2$, $x_3$, it follows that these coefficients are symmetric. Therefore, if some expression of $x_1$, $x_2$, $x_3$ is a root of the reduced equation, every other expression obtained from this one by permutation of $x_1$, $x_2$, $x_3$ also is a root of this equation. Since $y_4$ takes six different values by permutation of $x_1$, $x_2$, $x_3$, these six values are the roots of the reduced equation, which has therefore degree 6.

Moreover, it explains why the reduced equation is a quadratic equation in $y^3$: this follows because $y_4^3$ takes only two values under permutation of $x_1$, $x_2$, $x_3$. Indeed, since $\omega^3 = 1$, one has for instance

$$\omega x_1 + \omega^2 x_2 + x_3 = \omega(x_1 + \omega x_2 + \omega^2 x_3)$$

and it follows that

$$(\omega x_1 + \omega^2 x_2 + x_3)^3 = (x_1 + \omega x_2 + \omega^2 x_3)^3.$$

Likewise, we obtain

$$(x_1 + \omega x_2 + \omega^2 x_3)^3 = (\omega x_1 + \omega^2 x_2 + x_3)^3 = (\omega^2 x_1 + x_2 + \omega x_3)^3$$

and

$$(x_1 + \omega^2 x_2 + \omega x_3)^3 = (\omega^2 x_1 + \omega x_2 + x_3)^3 = (\omega x_1 + x_2 + \omega^2 x_3)^3.$$

Therefore, the two values of $y_4^3$ are

$$\left(\frac{1}{3}(x_1 + \omega x_2 + \omega^2 x_3)\right)^3 \quad \text{and} \quad \left(\frac{1}{3}(x_1 + \omega^2 x_2 + \omega x_3)\right)^3,$$

which are the roots of a quadratic equation.

The general result behind these arguments is the following:

**Proposition 10.1.** *Let $f$ be a rational function in $n$ indeterminates $x_1$, ..., $x_n$. If $f$ takes $m$ different values[2] when the indeterminates $x_1$, ..., $x_n$ are permuted in all possible ways, then $f$ is a root of a monic equation $\Theta = 0$ of degree $m$, whose coefficients are symmetric in $x_1$, ..., $x_n$, hence expressible as functions of the elementary symmetric polynomials (by the fundamental theorem of symmetric functions, Theorem 8.3, p. 95). Moreover, if $f$ is a root of another equation $\Phi = 0$ with coefficients symmetric in $x_1$, ..., $x_n$, then*

$$\deg \Phi \geq m.$$

**Proof.** Let $f_1$, $f_2$, ..., $f_m$ be the various values of $f$ obtained by permutation of $x_1$, ..., $x_n$ (with $f = f_1$, say), and let

$$\Theta(Y) = (Y - f_1) \dots (Y - f_m).$$

Since every permutation of $x_1$, ..., $x_n$ permutes the $f_i$'s among themselves, the coefficients of $\Theta$, which are the elementary symmetric polynomials in $f_1$, ..., $f_m$, are not altered when the indeterminates $x_1$, ..., $x_n$ are permuted. Therefore, these coefficients are symmetric in $x_1$, ..., $x_n$, and the equation $\Theta = 0$ satisfies the required properties.

If $\Phi$ is another polynomial with symmetric coefficients such that $\Phi(f) = 0$, then, for any $i = 1$, ..., $m$, the permutation of $x_1$, ..., $x_n$ that gives to $f$ the value $f_i$ transforms $\Phi(f)$ into $\Phi(f_i)$, since it does not change the coefficients of $\Phi$. Thus, $\Phi(f_i) = 0$ for $i = 1$, ..., $m$, hence $\Phi$ has $m$ different roots and it follows that $\deg \Phi \geq m$ (and, in fact, $\Phi$ is divisible by $\Theta$, by Proposition 5.10, p. 50). $\qquad\square$

For instance, the polynomial $x_1 + \omega x_2 + \omega^2 x_3$ takes six different values by permutation of $x_1$, $x_2$, $x_3$. It is therefore a root of an equation of degree 6 with symmetric coefficients, and of no equation of smaller degree. By contrast, $(x_1 + \omega x_2 + \omega^2 x_3)^3$ takes only two different values, hence it is a root of a quadratic equation.

After Cardano's method, Lagrange investigates Tschirnhaus' method. If the change of variable

$$y = b_0 + b_1 x + x^2$$

---

[2] Properly speaking, one should say "if the permutations of $x_1$, ..., $x_n$ in $f$ give rise to $m$ different rational functions." However, Lagrange's use of the term "values of a rational function" will be retained in the sequel since it is more suggestive and should not cause any confusion.

transforms a given cubic equation in $x$ with roots $x_1$, $x_2$, $x_3$ into an equation like

$$y^3 = c,$$

which has as roots $\sqrt[3]{c}$, $\omega \sqrt[3]{c}$ and $\omega^2 \sqrt[3]{c}$ (where, as above, $\omega$ denotes a cube root of unity other than 1), then we can assume

$$x_1^2 + b_1 x_1 + b_0 = \sqrt[3]{c},$$
$$x_2^2 + b_1 x_2 + b_0 = \omega \sqrt[3]{c}, \qquad\qquad (10.14)$$
$$x_3^2 + b_1 x_3 + b_0 = \omega^2 \sqrt[3]{c}.$$

Multiplying the second equation by $\omega$, the third by $\omega^2$ and adding to the first, we obtain

$$(x_1^2 + \omega x_2^2 + \omega^2 x_3^2) + b_1(x_1 + \omega x_2 + \omega^2 x_3) + b_0(1 + \omega + \omega^2)$$
$$= \sqrt[3]{c}(1 + \omega + \omega^2).$$

Since $1 + \omega + \omega^2 = 0$, it follows that

$$b_1 = -\frac{x_1^2 + \omega x_2^2 + \omega^2 x_3^2}{x_1 + \omega x_2 + \omega^2 x_3}. \qquad\qquad (10.15)$$

This rational function takes only two values under the permutations of $x_1$, $x_2$, $x_3$, namely

$$b_1 = -\frac{x_1^2 + \omega x_2^2 + \omega^2 x_3^2}{x_1 + \omega x_2 + \omega^2 x_3} \quad \text{and} \quad b_1' = -\frac{x_1^2 + \omega^2 x_2^2 + \omega x_3^2}{x_1 + \omega^2 x_2 + \omega x_3}.$$

Therefore, $b_1$ can be determined by solving the quadratic equation

$$y^2 - (b_1 + b_1')y + b_1 b_1' = 0.$$

The coefficients of this quadratic equation are symmetric in $x_1$, $x_2$, $x_3$, and can therefore be calculated from the coefficients of the proposed equation. On the other hand, adding the equations (10.14) and taking into account the fact that $1 + \omega + \omega^2 = 0$, we obtain

$$(x_1^2 + x_2^2 + x_3^2) + b_1(x_1 + x_2 + x_3) + 3b_0 = 0.$$

Since $x_1^2 + x_2^2 + x_3^2$ and $x_1 + x_2 + x_3$ are symmetric in $x_1$, $x_2$, $x_3$, they can be calculated from the coefficients of the proposed equation. This last equation then shows that $b_0$ can be rationally calculated from $b_1$ and the coefficients of the proposed equation. Similarly, multiplying the equations (10.14), we obtain

$$c = (x_1^2 + b_1 x_1 + b_0)(x_2^2 + b_1 x_2 + b_0)(x_3^2 + b_1 x_3 + b_0).$$

Since this expression is symmetric in $x_1$, $x_2$, $x_3$, it can be rationally calculated from $b_1$, $b_0$ and the coefficients of the proposed equation. Once $b_0$, $b_1$, and $c$ have been calculated, the roots $x_1$, $x_2$, and $x_3$ can be rationally calculated as explained in §6.4 (p. 72). Therefore, Tschirnhaus' method for the cubic equation requires only the solution of a quadratic equation, beyond rational calculations; ultimately, this follows because the rational function $b_1$ in (10.15) takes only two values.

Thereafter, Lagrange successively analyzes Euler's and Bézout's methods, and the various methods for equations of degree 4. Each time, he shows how the roots of the auxiliary equations can be expressed in terms of those of the proposed equation, and he observes that the number of values of these expressions is less than the degree of the proposed equation.

For degree 4, Ferrari's method reduces the quartic equation

$$x^4 + px^2 + qx + r = 0$$

to

$$\left(x^2 + \frac{p}{2} + u\right)^2 = \left(\sqrt{2u}x - \frac{q}{2\sqrt{2u}}\right)^2$$

by the choice of a suitable $u$ (see §3.2). This last equation splits into two quadratic equations

$$x^2 + \frac{p}{2} + u = \sqrt{2u}x - \frac{q}{2\sqrt{2u}} \quad \text{and} \quad x^2 + \frac{p}{2} + u = -\left(\sqrt{2u}x - \frac{q}{2\sqrt{2u}}\right),$$

which yield the roots $x_1$, $x_2$, $x_3$, $x_4$ of the proposed quartic equation. Renumbering the roots if necessary, we may assume that $x_1$ and $x_2$ are the roots of the first quadratic equation and that $x_3$ and $x_4$ are the roots of the second one. Then, since the constant term is the product of the roots, it follows that

$$x_1 x_2 = \frac{p}{2} + u + \frac{q}{2\sqrt{2u}} \quad \text{and} \quad x_3 x_4 = \frac{p}{2} + u - \frac{q}{2\sqrt{2u}},$$

hence

$$x_1 x_2 + x_3 x_4 = p + 2u. \tag{10.16}$$

Since $p$ is the coefficient of $x^2$ in the quartic equation, it is the second symmetric polynomial in $x_1$, $x_2$, $x_3$, $x_4$, i.e.,

$$p = x_1 x_2 + x_1 x_3 + x_1 x_4 + x_2 x_3 + x_2 x_4 + x_3 x_4.$$

Substituting for $p$ in (10.16), the following value of $u$ is found:

$$u = -\frac{1}{2}(x_1 + x_2)(x_3 + x_4).$$

This expression takes only three values when $x_1$, $x_2$, $x_3$, and $x_4$ are permuted. This explains why $u$ is a root of a cubic equation.

Lagrange concludes his review with the attempted applications of Tschirnhaus', Euler's and Bézout's methods to equations of higher degree. He then observes that, according to Bézout's method (see §10.1), the roots of the proposed equation of degree $n$ are obtained in the form $a_0 + a_1\omega + a_2\omega^2 + \cdots + a_{n-1}\omega^{n-1}$, where $\omega$ runs over the set of $n$-th roots of unity. If $\zeta$ is a primitive $n$-th root of unity, then by Proposition 7.9 (p. 85) the $n$-th roots of unity are 1, $\zeta$, $\zeta^2$, ..., $\zeta^{n-1}$. Substituting successively 1, $\zeta$, $\zeta^2$, ..., $\zeta^{n-1}$ for $\omega$, we obtain the following expressions of the roots:

$$x_1 = a_0 + a_1 + a_2 + \cdots + a_{n-1},$$
$$x_2 = a_0 + a_1\zeta + a_2\zeta^2 + \cdots + a_{n-1}\zeta^{n-1},$$
$$x_3 = a_0 + a_1\zeta^2 + a_2\zeta^4 + \cdots + a_{n-1}\zeta^{2(n-1)},$$
$$\vdots$$
$$x_n = a_0 + a_1\zeta^{n-1} + a_2\zeta^{2(n-1)} + \cdots + a_{n-1}\zeta^{(n-1)^2}.$$

Thus, in general,

$$x_i = \sum_{j=0}^{n-1} a_j \zeta^{(i-1)j} \qquad \text{for } i = 1, \ldots, n. \tag{10.17}$$

This system is easily solved for $a_0$, $a_1$, ..., $a_{n-1}$: to obtain the value of $a_k$, it suffices to multiply each of these equations by a suitable power of $\zeta$ so that the coefficient of $a_k$ be 1, and to add the equations thus produced. We obtain

$$\sum_{i=1}^{n} \zeta^{-(i-1)k} x_i = \sum_{j=0}^{n-1} a_j \left( \sum_{i=1}^{n} \zeta^{(j-k)(i-1)} \right). \tag{10.18}$$

If $j \neq k$, then $\zeta^{j-k}$ is an $n$-th root of unity different from 1. Therefore, $\zeta^{j-k}$ is a root of

$$\frac{X^n - 1}{X - 1} = X^{n-1} + X^{n-2} + \cdots + X + 1,$$

hence

$$\sum_{i=1}^{n} \zeta^{(j-k)(i-1)} = 0.$$

Therefore, in the right-hand side of (10.18), all the terms vanish except the term corresponding to the index $j = k$, which is $na_k$. Thus, (10.18) yields

$$a_k = \frac{1}{n} \left( \sum_{i=1}^{n} \zeta^{-(i-1)k} x_i \right). \tag{10.19}$$

It is easily seen that, if $x_1, \ldots, x_n$ are considered as independent indeterminates, all the values of $a_k$ obtained by the various permutations of $x_1, \ldots, x_n$ are distinct. Therefore, $a_k$ is a root of an equation of degree $n!$. However, Lagrange shows[3] that $a_k^n$ takes only $(n-1)!$ values. Moreover, if $n$ is prime, then $a_k^n$ is a root of an equation of degree $n-1$ whose coefficients can be determined from the solution of a single equation of degree $(n-2)!$. Thus, for $n = 5$, the determination of $a_k^5$ still requires the solution of an equation of degree $3! = 6$.

If $n$ is not prime, the result is more complicated. Using the same arguments as in the case where $n$ is prime, Lagrange shows that if $n = pq$ with $p$ prime, and if $k$ is divisible by $q$, then $a_k^p$ is a root of an equation of degree $p - 1$ whose coefficients depend on a single equation of degree $\frac{n!}{(p-1)p(q!)^p}$. For $n = 4$, it follows that $a_2^2$ can be found by solving an equation of degree $\frac{4!}{1.2.(2!)^2} = 3$, but for $n = 6$ the determination of $a_3^2$ requires the solution of an equation of degree $\frac{6!}{1.2.(3!)^2} = 10$. These results lead Lagrange to doubt the possibility of solving algebraically the general equations of degree 5 or higher, although he cautiously avoids rejecting this possibility too categorically. The conclusion of his investigations of previously known methods is given in [49, Art. 86], which is quoted here extensively to give an idea of Lagrange's leisurely style of writing.

> As should be clear from the analysis that we have just given of the main known methods for the solution of equations, all these methods reduce to the same general principle, namely to find functions of the roots of the proposed equation that are such: 1° that the equation or the equations by which they are given, i.e., of which they are the roots (equations that are usually called the *reduced* equations), happen to be of a degree smaller than that of the proposed equation, or at least decomposable in other equations of a degree smaller than this one; 2° that the values of the sought roots can be easily deduced from them.
>
> The art of solving equations thus consists of discovering functions of the roots that have the above-mentioned properties; but is it always possible to find such functions, for equations of any degree, i.e., for any number of roots? That is a question that seems very difficult to decide in general.
>
> As to equations that do not exceed the fourth degree, the simplest functions that yield their solution can be represented by the general formula
>
> $$x_1 + \omega x_2 + \omega^2 x_3 + \cdots + \omega^{n-1} x_n,$$

---

[3] Lagrange's arguments are elementary, but can be more easily explained when they are related to some results found in the sequel of Lagrange's paper and with the help of an appropriate notation. Therefore, we postpone the proof of this result to the next section (see Proposition 10.9, p. 141).

$x_1$, $x_2$, $x_3$, ..., $x_n$ being the roots of the proposed equation, which is assumed to be of degree $n$, and $\omega$ being an arbitrary root other than 1 of the equation

$$\omega^n - 1 = 0,$$

i.e., an arbitrary root of the equation

$$\omega^{n-1} + \omega^{n-2} + \omega^{n-3} + \cdots + 1 = 0,$$

as follows from what has been shown in the first two sections about the solution of equations of the third and the fourth degree. [...]

It thus seems that one could conclude from this by induction that every equation of any degree will also be solvable with the help of a reduced equation whose roots are represented by the same formula

$$x_1 + \omega x_2 + \omega^2 x_3 + \omega^3 x_4 + \cdots .$$

However, after what we proved about the methods of MM. Euler and Bezout, which readily lead to such reduced equations, one has, it seems, the occasion to convince oneself beforehand that this conclusion will be flawed from the fifth degree on; hence it follows that, if the algebraic solution of the equations of degree higher than four is not impossible, it must rely on some functions of the roots, other than the above [49, Art. 86].

The polynomials of the form

$$x_1 + \omega x_2 + \omega^2 x_3 + \cdots + \omega^{n-1} x_n,$$

where $\omega$ is an $n$-th root of unity, were subsequently christened *Lagrange resolvents*. As we have seen, they originate from the works of Euler and Bézout, and it will be clear later that they play a prominent role in Galois' theory of equations. For the convenience of later reference, we recapitulate the formula which yields the roots of equations of any degree $n$ in terms of Lagrange resolvents.

**Proposition 10.2.** *If $t(\omega)$ denotes the Lagrange resolvent*

$$t(\omega) = x_1 + \omega x_2 + \omega^2 x_3 + \cdots + \omega^{n-1} x_n,$$

*where $\omega$ is an $n$-th root of unity, then for $i = 1, \ldots, n$,*

$$x_i = \frac{1}{n}\Big(\sum_\omega \omega^{-(i-1)} t(\omega)\Big),$$

*where the sum runs over all the $n$-th roots of unity.*

This was shown above: see equations (10.17) and (10.19).

## 10.3 First results of group theory and Galois theory

In the final section of his paper, Lagrange draws from his investigations some general conclusions concerning the degree of the equations by which functions of the roots of a given equation can be determined. Proposition 10.1 above (p. 125) is a first instance of Lagrange's observations, but in his conclusions Lagrange goes much farther. In effect, he begins to calculate with permutations of the roots, obtaining first results in group theory and Galois theory.

Remarkably enough, these results were achieved without even devising a notation for permutations, which were indeed very new objects to calculate with. Unfortunately, this makes Lagrange's arguments sometimes hard to follow. To facilitate our exposition of Lagrange's results, we will not refrain from using modern notation. Thus, for any integer $n \geq 1$, we let $S_n$ denote the symmetric group on $\{1, \ldots, n\}$, i.e., the group of permutations of 1, ..., $n$. For $\sigma \in S_n$ and for any rational function $f$ in $n$ indeterminates $x_1$, ..., $x_n$, we set

$$\sigma\big(f(x_1, \ldots, x_n)\big) = f(x_{\sigma(1)}, \ldots, x_{\sigma(n)}).$$

So, $S_n$ can be considered as the group of permutations of $x_1$, ..., $x_n$, and $S_n$ acts on the rational functions in $x_1$, ..., $x_n$ by permuting the indeterminates. For any rational function $f$, we let $I(f)$ denote the subgroup of permutations $\sigma \in S_n$ that leave $f$ invariant, (sometimes called the *isotropy group* of $f$), i.e.,

$$I(f) = \big\{\sigma \in S_n \mid \sigma\big(f(x_1, \ldots, x_n)\big) = f(x_1, \ldots, x_n)\big\}.$$

We generally write $|E|$ for the number of elements in a finite set $E$, which is called the *order* of $E$, if $E$ is a group. Thus, for instance

$$|S_n| = n!.$$

In his Article 97, Lagrange proves the following theorem:

**Theorem 10.3.** *Let $f = f(x_1, \ldots, x_n)$ be a rational function in $n$ indeterminates. The number $m$ of different values that $f$ takes under the permutations of $x_1$, ..., $x_n$ is equal to the quotient of $n!$ by the number of permutations that leave $f$ invariant,*

$$m = \frac{n!}{|I(f)|}.$$

**Proof.** Let $f_1, \ldots, f_m$ be the various values of $f$ (with $f = f_1$, say). For $i = 1, \ldots, m$, let $I(f \mapsto f_i)$ be the set of permutations $\sigma \in S_n$ such that $\sigma(f) = f_i$; thus,

$$I(f \mapsto f_1) = I(f).$$

Now, fix some $i = 1, \ldots, m$ and some $\sigma \in I(f \mapsto f_i)$. If $\tau \in I(f)$, then

$$\sigma \circ \tau(f) = \sigma(f) = f_i,$$

hence $\sigma \circ \tau \in I(f \mapsto f_i)$. Conversely, if $\rho \in I(f \mapsto f_i)$, then $\sigma^{-1} \circ \rho \in I(f)$. Since

$$\rho = \sigma \circ (\sigma^{-1} \circ \rho),$$

it follows that every element in $I(f \mapsto f_i)$ is of the form $\sigma \circ \tau$ where $\tau \in I(f)$. Therefore, composition on the left with $\sigma$ defines a bijection from $I(f)$ onto $I(f \mapsto f_i)$, hence

$$|I(f)| = |I(f \mapsto f_i)|.$$

Since every permutation in $S_n$ maps $f$ onto one of its values $f_1, \ldots, f_m$, we have a decomposition of $S_n$ as a disjoint union

$$S_n = \bigcup_{i=1}^{m} I(f \mapsto f_i),$$

hence

$$|S_n| = \sum_{i=1}^{m} |I(f \mapsto f_i)|.$$

Since $|S_n| = n!$ and since each term in the right-hand side is equal to $|I(f)|$, this last equation yields

$$n! = m \cdot |I(f)|. \qquad \square$$

Here now are Lagrange's own words (in free translation, and with slight notational changes to fit the notation of this section). To understand the following quotation, one needs to know that $\Theta$ is the polynomial

$$\Theta(Y) = \prod_{\sigma \in S_n} \Big( Y - \sigma\big(f(x_1, \ldots, x_n)\big) \Big),$$

whose roots are the values of $f$ under the permutations of $x_1, \ldots, x_n$ (compare Proposition 10.1, p. 125).

Although the equation $\Theta = 0$ must in general be of degree $1 \cdot 2 \cdot 3 \cdot \ldots \cdot n = n!$, which is equal to the number of permutations of $x_1, \ldots, x_n$, yet if it happens that the function be such that it does not receive any change by some or several permutations, then the equation in question will necessarily reduce to a smaller degree.

Assume, for instance, that the function $f(x_1, x_2, x_3, x_4, \ldots)$ be such that it keeps the same value when $x_1$ is changed into $x_2$, $x_2$ into $x_3$ and $x_3$ into $x_1$, so that

$$f(x_1, x_2, x_3, x_4, \ldots) = f(x_2, x_3, x_1, x_4, \ldots),$$

it is clear that the equation $\Theta = 0$ will already have two equal roots; but I am going to prove that with this hypothesis all the other roots will be pairwise equal too. Indeed, let us consider an arbitrary root of the same equation, represented by the function $f(x_4, x_3, x_1, x_2, \ldots)$. As this one derives from the function $f(x_1, x_2, x_3, x_4, \ldots)$ by changing $x_1$ into $x_4$, $x_2$ into $x_3$, $x_3$ into $x_1$, $x_4$ into $x_2$, it follows that it will have to keep the same value when we change in it $x_4$ into $x_3$, $x_3$ into $x_1$ and $x_1$ into $x_4$; so that we will also have

$$f(x_4, x_3, x_1, x_2, \ldots) = f(x_3, x_1, x_4, x_2, \ldots).$$

Therefore, in this case, the quantity $\Theta$ will be equal to a square $\theta^2$ and consequently the equation $\Theta = 0$ will be reduced to this one $\theta = 0$, which will have dimension $\frac{n!}{2}$.

Likewise, one will prove that if the function $f$ is by its own nature such that it keeps the same value when two or three or a greater number of different permutations are made among the roots $x_1$, $x_2$, $x_3$, $x_4$, $\ldots$, $x_n$, the roots of the equation $\Theta = 0$ will be equal three by three or four by four or, etc.; so that the quantity $\Theta$ will be equal to a cube $\theta^3$ or to a square-square $\theta^4$ or, etc., and therefore the equation $\Theta = 0$ will reduce to this one $\theta = 0$, whose degree will be equal to $\frac{n!}{3}$, or equal to $\frac{n!}{4}$, or, etc. [49, Art. 97].

To see the link between Lagrange's argument in the quotation above and the preceding proof, let $f = f(x_1, x_2, x_3, x_4, \ldots)$ and $f_i = f(x_4, x_3, x_1, x_2, \ldots)$, and let $\sigma$ be the permutation $x_1 \mapsto x_4$, $x_2 \mapsto x_3$, $x_3 \mapsto x_1$, $x_4 \mapsto x_2$, $\ldots$ so that

$$\sigma(f) = f_i, \qquad \text{i.e., } \sigma \in I(f \mapsto f_i).$$

Lagrange's observation is that if $\tau\colon x_1 \mapsto x_2 \mapsto x_3 \mapsto x_1$ is in $I(f)$, then $\sigma \circ \tau\colon x_1 \mapsto x_3$, $x_2 \mapsto x_1$, $x_3 \mapsto x_4$, $x_4 \mapsto x_2$, $\ldots$ is such that

$$\sigma \circ \tau(f) = f_i, \qquad \text{i.e., } \sigma \circ \tau \in I(f \mapsto f_i).$$

This is indeed the crucial step in the proof.

The theorem often referred to as "Lagrange's theorem" nowadays deals with the order (i.e., the number of elements) of subgroups of a group. It is stated as follows:

**Theorem 10.4.** *Let $H$ be a subgroup of a finite group $G$. Then $|H|$ divides $|G|$.*

**Proof.** For $g \in G$, define the (left) coset $gH$ by

$$gH = \{gh \mid h \in H\}.$$

We readily have $|gH| = |H|$, since multiplication by $g$ defines a bijection from $H$ onto $gH$. Since $g = g1 \in gH$, it is clear that every element of $G$ is in some coset. Moreover, if two cosets have a common element, then they are equal. To see this, suppose there exists an element $x \in g_1 H \cap g_2 H$, and let $h_1$, $h_2 \in H$ be such that $x = g_1 h_1 = g_2 h_2$. Every element $g_1 h \in g_1 H$ can then be written as

$$g_1 h = g_2(h_2 h_1^{-1} h) \in g_2 H,$$

so that $g_1 H \subset g_2 H$. Interchanging the indices 1 and 2, we obtain $g_2 H \subset g_1 H$, hence $g_1 H = g_2 H$.

This shows that the group $G$ decomposes into a disjoint union of cosets. Since the number of elements of each of these cosets is equal to $|H|$, it follows that $|H|$ divides $|G|$ (and the quotient of $|G|$ by $|H|$ is the number of different cosets in a decomposition of $G$, which is called the *index* of $H$ in $G$). $\qquad\square$

Although the pattern of this proof is quite similar to that of Theorem 10.3 (observe that $I(f \mapsto f_i)$ is a left coset of $I(f)$), Lagrange did not reach this generality, nor did he need to. His primary concern was to obtain some information on the number of values of functions, hence, by Proposition 10.1 above (p. 125), on the degree of the equation by which a given function of the roots can be determined.

In this respect, his achievement is even more stunning: he proves a "relative version" of Proposition 10.1 above, which can be seen as a part of the fundamental theorem of Galois theory for the splitting field of the general polynomial:

Now, as soon as the value of a given function of the roots $x_1, \ldots, x_n$ has been found, either by the solution of the equation $\theta = 0$ or otherwise, I claim that the value of another arbitrary function of the same roots can be found, and that, generally speaking, simply by a linear equation, except for some particular cases that demand an equation of the second degree, or of the third, etc. This Problem seems to me to be one of the most important of the theory of equations, and the general Solution that we are going to give will shed a new light on this part of Algebra [49, Art. 100].

Lagrange's result can be stated more precisely as follows:

**Theorem 10.5.** *Let $f$ and $g$ be two rational functions in $n$ indeterminates $x_1$, $\ldots$, $x_n$. If $f$ takes $m$ different values by the permutations that leave $g$ invariant, then $f$ is a root of an equation of degree $m$ whose coefficients are rational functions in $g$ and in the elementary symmetric polynomials $s_1$, $\ldots$, $s_n$.*

In particular, if $f$ is invariant by the permutations that leave $g$ invariant, then $f$ is a rational function in $g$ and $s_1$, $\ldots$, $s_n$.

**Proof.** We begin with the special case above, i.e., we first assume $m = 1$. Then, let $g_1$, $\ldots$, $g_r$ be the different values of $g$ under the permutations of $x_1$, $\ldots$, $x_n$ (with $g_1 = g$, say). Let $f_1 = f$, $f_2$, $\ldots$, $f_r$ be the corresponding values of $f$, in the sense that if a permutation of $x_1$, $\ldots$, $x_n$ gives to $g$ the value $g_i$ (for some $i$), then it gives to $f$ the value $f_i$. The possibility of defining such a correspondence between the values of $g$ and the values of $f$ follows from the hypothesis that $f$ is invariant under the permutations that leave $g$ invariant. Indeed, this hypothesis means that $I(g) \subset I(f)$; thus, if $\sigma$ and $\rho$ both give to $g$ the value $g_i$, then the proof of Theorem 10.3 shows that $\rho = \sigma \circ \tau$ for some $\tau \in I(g)$, and the hypothesis implies that $\tau \in I(f)$, hence $\rho(f) = \sigma(f)$. We may therefore define $f_i = \sigma(f)$ for *any* $\sigma \in S_n$ such that $\sigma(g) = g_i$.

Consider then the following expressions, denoted by $a_0$, $\ldots$, $a_{r-1}$:

$$a_0 = f_1 + f_2 + \cdots + f_r$$
$$a_1 = f_1 g_1 + f_2 g_2 + \cdots + f_r g_r$$
$$a_2 = f_1 g_1^2 + f_2 g_2^2 + \cdots + f_r g_r^2 \qquad (10.20)$$
$$\vdots$$
$$a_{r-1} = f_1 g_1^{r-1} + f_2 g_2^{r-1} + \cdots + f_r g_r^{r-1}.$$

From the definition of $f_i$ and $g_i$, it follows that every permutation of $x_1$, $\ldots$, $x_n$ merely permutes the terms in $a_0$, $\ldots$, $a_{r-1}$, so that each of these expressions is symmetric in $x_1$, $\ldots$, $x_n$ and can therefore be calculated as a rational function in $s_1$, $\ldots$, $s_n$, by the fundamental theorem of symmetric functions (Theorem 8.3, p. 95).

Now, the idea is to solve system (10.20) for $f_1, \ldots, f_r$. However, the usual elimination method would yield $f_1$ in terms of $a_0, \ldots, a_{r-1}$ (thus, eventually, in terms of $s_1, \ldots, s_n$) but also of $g_1, \ldots, g_r$, while we need an expression of $f_1$ in terms of $s_1, \ldots, s_n$, and $g_1$ only.

Lagrange then uses the following trick: let

$$\theta(Y) = (Y - g_1)(Y - g_2) \ldots (Y - g_r) = Y^r + b_{r-1}Y^{r-1} + \cdots + b_0.$$

Dividing this polynomial by $Y - g_1$, we obtain

$$\psi(Y) = (Y - g_2) \ldots (Y - g_r) = Y^{r-1} + c_{r-2}Y^{r-2} + \cdots + c_0,$$

and the coefficients $c_0, c_1, \ldots, c_{r-2}$ are rational functions in $b_0, b_1, \ldots, b_{r-1}$, and $g_1$, as is easily seen by carrying out explicitly the division of $\theta(Y)$ by $Y - g_1$. Now, the coefficients $b_0, \ldots, b_{r-1}$ of $\theta$ are symmetric in $g_1, \ldots, g_r$ hence also in $x_1, \ldots, x_n$, and can therefore be calculated in terms of $s_1, \ldots, s_n$. Thus, $c_0, c_1, \ldots, c_{r-2}$ are rational functions in $g_1$ and $s_1, \ldots, s_n$.

Multiplying the first equation of (10.20) by $c_0$, the second by $c_1$, the third by $c_2$, etc. and the last by 1, and adding the equations thus obtained, we obtain an equation in which the coefficient of $f_i$ (for $i = 1, \ldots, r$) is the polynomial $\psi(Y)$ calculated at $Y = g_i$, namely

$$a_0c_0 + a_1c_1 + \cdots + a_{r-1} = f_1\psi(g_1) + f_2\psi(g_2) + \cdots + f_r\psi(g_r).$$

Since $\psi(g_2) = \cdots = \psi(g_r) = 0$, we thus end up with an expression of $f_1$ as a rational function in $g$ and $s_1, \ldots, s_n$,

$$f_1 = (a_0c_0 + a_1c_1 + \cdots + a_{r-1})\psi(g_1)^{-1},$$

as was required.

This proves Theorem 10.5 in the special case where $m = 1$, but the general case follows easily. To see this, assume now that $f$ takes $m$ values $f_1, \ldots, f_m$ under the permutations that leave $g$ invariant. Then $f$ is a root of the equation

$$(Y - f_1) \ldots (Y - f_m) = 0,$$

and this equation satisfies the required properties since its coefficients are symmetric in $f_1, \ldots, f_m$, hence invariant under the permutations that leave $g$ invariant, hence, by the special case above, rational functions in $g$ and $s_1, \ldots, s_n$.                                                                $\square$

Lagrange's result is even more general than the above, since he also considers the case where $x_1, \ldots, x_n$ are related by some algebraic relations (this occurs when $x_1, \ldots, x_n$ are the roots of some particular equation instead of the general equation of degree $n$), but the theorem above gives the gist of Lagrange's proof and covers the essential part of the applications that Lagrange had in mind, since his purpose was to investigate the solution of general equations.

As envisioned by Lagrange, his breakthrough (Theorem 10.5) sheds much light on the solution of general equations indeed. The strategy appears as follows: to solve the general equation of degree $n$, one has to find a (finite) sequence of rational functions $V_0$, $V_1$, ..., $V_r$ in $n$ indeterminates $x_1$, ..., $x_n$ such that the first function $V_0$ is symmetric in $x_1$, ..., $x_n$, the last function $V_r$ is one of the roots, say $V_r = x_1$, and for $i = 1$, ..., $r$, the function $V_i$ satisfies either

(1) $V_i^n = V_{i-1}$, or
(2) the number of values of $V_i$ under the permutations that leave $V_{i-1}$ invariant is strictly less than $n$.

In case (1), the function $V_i$ can be calculated from the preceding ones by extracting an $n$-th root, and in case (2) it can be found by solving an equation of degree less than $n$, by Theorem 10.5. Since the last function is a root of the proposed equation, it means that a root can be found by successive extractions of roots and solutions of equations of lower degree. The sequence $V_0$, $V_1$, ..., $V_r$ indicates in which order the calculations can be arranged. The other roots can be found likewise, substituting for $V_0$, $V_1$, ..., $V_r$ similar functions. (More precisely, for any $\sigma \in S_n$, the root $\sigma(V_r)$ can be found by the sequence $\sigma(V_0) = V_0$, $\sigma(V_1)$, ..., $\sigma(V_r)$.) For $n = 2$, one chooses

$$V_0 = (x_1 - x_2)^2,$$
$$V_1 = x_1 - x_2,$$
$$V_2 = x_1.$$

Thus, $V_1$ can be found by extracting the square root of $V_0$, and $V_2$ can then be found rationally, since $V_2 = \frac{1}{2}(V_1 + (x_1 + x_2))$.

For $n = 3$, one can choose for $V_0$ any symmetric function, next (writing $\omega$ for a cube root of unity other than 1)

$$V_1 = (x_1 + \omega x_2 + \omega^2 x_3)^3,$$
$$V_2 = x_1 + \omega x_2 + \omega^2 x_3,$$
$$V_3 = x_1.$$

Since $V_1$ takes only two values by all the permutations of $x_1$, $x_2$, $x_3$, it can be found by solving a quadratic equation. Next, $V_2$ is found by extracting a cube root of $V_1$ and finally, since $V_3$ is invariant under the permutations that leave $V_2$ invariant (as only the identity leaves $V_2$ invariant), it can be determined rationally from $V_2$, i.e., by solving an equation of degree 1. Likewise, $x_2$ and $x_3$ can be determined rationally from $V_2$.

For $n = 4$, one can choose for $V_0$ any symmetric function, next

$$V_1 = (x_1 + x_2)(x_3 + x_4),$$
$$V_2 = x_1 + x_2,$$
$$V_3 = x_1.$$

Indeed, $V_1$ can be determined by solving a cubic equation since it takes only three values by the permutations of $x_1$, $x_2$, $x_3$, $x_4$. Next, $V_2$ can be determined by a quadratic equation since it takes only two values under the permutations that leave $V_1$ invariant, and finally $V_3$ can be found by a quadratic equation since it takes only two values under the permutations that leave $V_2$ invariant. The root $x_2$ is then readily found, since it is the other root of the quadratic equation that yields the value of $V_3$ ($= x_1$), and the other roots $x_3$ and $x_4$ are found by similar calculations: $x_3 + x_4$ is the other root of the equation that yields the value of $V_2$, and $x_3$, $x_4$ are the roots of a quadratic equation. This is the pattern suggested by Ferrari's solution of quartic equations. Of course, other choices are possible; for instance, using Lagrange's resolvents, one could choose

$$V_1 = (x_1 - x_2 + x_3 - x_4)^2,$$
$$V_2 = x_1 - x_2 + x_3 - x_4,$$
$$V_3 = x_1.$$

The first function $V_1$ is the root of a cubic equation, and $V_2$, $V_3$ are obtained by solving successively two quadratic equations.

> These are, if I am not mistaken, the genuine principles of the solution of equations and the analysis that is most suitable to lead to it; everything is reduced, as is seen, to a kind of calculus of combinations, by which the results to which one is led are found *a priori*. It would be opportune to apply it to the equations of the fifth degree and higher degrees, whose solution is so far unknown; but this application requires a too large amount of researches and combinations, whose success is, for that matter, still very dubious, for us to tackle this problem now; we hope however to come back to it at another time, and we will be content to have here set the foundations of a theory which seems to us new and general [49, Art. 109].

To conclude this chapter, we now apply the results above to sketch a proof of the properties of "Lagrange resolvents," which we have pointed out in §10.2, at least for the case where $n$ is prime. We will need the following result on the existence of rational functions that are invariant under a prescribed group:

**Proposition 10.6.** *For any subgroup $G$ in $S_n$, there exists a rational function $f$ in $n$ indeterminates such that $I(f) = G$.*

**Proof.** Choose a monomial $m$ that is not invariant under any (non-trivial) permutation of the indeterminates, for instance $m = x_1 x_2^2 x_3^3 \ldots x_n^n$, and let

$$f = \sum_{\sigma \in G} \sigma(m).$$

Since for any $\tau \in G$ the set of products $\{\tau \circ \sigma \mid \sigma \in G\}$ is $G$, it follows that

$$\sum_{\sigma \in G} \tau \circ \sigma(m) = \sum_{\sigma \in G} \sigma(m),$$

hence

$$\tau(f) = f \qquad \text{for any } \tau \in G.$$

Therefore, $G \subset I(f)$.

On the other hand, if $\rho \notin G$, then the monomial $\rho(m)$ appears in $\rho(f)$ but not in $f$, so $\rho(f) \neq f$. Thus, $G = I(f)$. $\qquad\square$

Henceforth, to simplify notation we index the indeterminates from 0. We will thus consider $S_n$ as the group of permutations of $\{0, 1, \ldots, n-1\}$, and we now define certain permutations that have interesting properties in relation with Lagrange's resolvents.

For any integer $k$ relatively prime to $n$, we write

$$\sigma_k \colon \{0, 1, \ldots, n-1\} \to \{0, 1, \ldots, n-1\}$$

for the map defined as follows: for any $i \in \{0, 1, \ldots, n-1\}$, the image $\sigma_k(i)$ is the unique integer $j$ between 0 and $n-1$ such that $ik - j$ is divisible by $n$ (i.e., $ik \equiv j \bmod n$, using a notation that will be introduced in §12.1). In other words, $\sigma_k(i)$ is the remainder of the division of $ik$ by $n$.

**Proposition 10.7.** *For any integer $k$ relatively prime to $n$, the map $\sigma_k$ is a permutation of $\{0, 1, \ldots, n-1\}$.*

**Proof.** Since $\sigma_k$ is a map from the finite set $\{0, 1, \ldots, n-1\}$ to itself, it suffices to show that $\sigma_k(i)$ does not take twice the same value as $i$ ranges from $0$ to $n - 1$. Suppose $\sigma_k(i) = \sigma_k(i')$ for $i$, $i' \in \{0, 1, \ldots, n-1\}$. Then the divisions of $ik$ and of $i'k$ by $n$ yield the same remainder, hence $n$ divides $ik - i'k$. As $k$ is prime to $n$, it follows that $n$ divides $i - i'$, hence $i = i'$ since $i$ and $i'$ both lie between $0$ and $n - 1$. □

Proposition 10.7 shows in particular that there exists $\ell \in \{0, 1, \ldots, n-1\}$ such that $\sigma_k(\ell) = 1$, i.e., such that $n$ divides $k\ell - 1$. It then follows that $n$ divides $ik\ell - i$ for every integer $i$. But $n$ also divides $ik - \sigma_k(i)$, hence

$$n \text{ divides } \sigma_k(i)\ell - i \quad \text{for } i \in \{0, 1, \ldots, n-1\}. \tag{10.21}$$

From now on, we assume that $n$ *is a prime number*, so that $\sigma_k$ is defined for all $k = 1, \ldots, n - 1$. Let $\tau$ be the cyclic permutation $0 \mapsto 1 \mapsto 2 \mapsto \cdots \mapsto n - 1 \mapsto 0$ and let $\mathrm{GA}(n)$ be the subgroup of $S_n$ generated by $\sigma_1$, $\ldots, \sigma_{n-1}$ and $\tau$. It can be shown that

$$|\mathrm{GA}(n)| = n(n - 1)$$

(see Exercise 10.5 or §14.5.1). In fact, from a less elementary point of view, $\mathrm{GA}(n)$ can be identified with the group of affine transformations of the affine line over the field with $n$ elements: $\tau$ generates the group of translations while $\sigma_1, \ldots, \sigma_{n-1}$ are homotheties.

Let $V$ be a rational function in $x_0, x_1, \ldots, x_{n-1}$ such that $I(V) = \mathrm{GA}(n)$ (the existence of such a function is established in Proposition 10.6) and, for any $n$-th root of unity $\omega$, let $t(\omega)$ denote the following Lagrange resolvent:

$$t(\omega) = x_0 + \omega x_1 + \omega^2 x_2 + \cdots + \omega^{n-1} x_{n-1}.$$

**Theorem 10.8.** *Assume $n$ is prime. If $\omega \neq 1$, then $t(\omega)^n$ is a root of an equation of degree $n - 1$ whose coefficients are rational functions in $V$ and in the elementary symmetric polynomials in $x_0, \ldots, x_{n-1}$. Moreover, $V$ is a root of an equation of degree $(n - 2)!$ whose coefficients are rational functions in the elementary symmetric polynomials.*

**Proof.** The fact that $V$ is a root of an equation of degree $(n - 2)!$ readily follows from Proposition 10.1 and Theorem 10.3, since $I(V)$ has $n(n - 1)$ elements. To prove the rest, it suffices, by Theorem 10.5, to show that $t(\omega)^n$ takes $n - 1$ values under the permutations that leave $V$ invariant, i.e., under the permutations in $\mathrm{GA}(n)$.

First, we consider the action of $\sigma_k$:

$$\sigma_k\big(t(\omega)\big) = x_0 + \omega x_{\sigma_k(1)} + \omega^2 x_{\sigma_k(2)} + \cdots + \omega^{n-1} x_{\sigma_k(n-1)}.$$

Since $\omega^n = 1$, (10.21) yields
$$\omega^i = (\omega^\ell)^{\sigma_k(i)} \qquad \text{for } i = 0, 1, \ldots, n-1.$$
Therefore,
$$\sigma_k\big(t(\omega)\big)$$
$$= x_0 + (\omega^\ell)^{\sigma_k(1)}x_{\sigma_k(1)} + (\omega^\ell)^{\sigma_k(2)}x_{\sigma_k(2)} + \cdots + (\omega^\ell)^{\sigma_k(n-1)}x_{\sigma_k(n-1)},$$
which shows
$$\sigma_k\big(t(\omega)\big) = t(\omega^\ell). \tag{10.22}$$
Next, we consider the action of $\tau$. Since
$$\tau\big(t(\omega)\big) = x_1 + \omega x_2 + \omega^2 x_3 + \cdots + \omega^{n-1}x_0,$$
we have
$$\tau\big(t(\omega)\big) = \omega^{-1}t(\omega),$$
for any $n$-th root of unity $\omega$. Since $\omega^{-n} = 1$, this last equation yields
$$\tau\big(t(\omega)^n\big) = t(\omega)^n. \tag{10.23}$$
This result, together with (10.22), shows that under any product of the permutations $\sigma_1, \ldots, \sigma_{n-1}$, $\tau$ the function $t(\omega)^n$ takes one of the values $t(\omega)^n$, $t(\omega^2)^n$, $\ldots$, $t(\omega^{n-1})^n$, which are pairwise different if $\omega \neq 1$. Since $GA(n)$ is generated by $\sigma_1, \ldots, \sigma_{n-1}, \tau$, this means that $t(\omega)^n$ takes $n-1$ values under the permutations in $GA(n)$, and the proof is complete. $\square$

Simpler arguments yield the number of values of $t(\omega)^n$:

**Proposition 10.9.** *The function $t(\omega)^n$ takes $(n-1)!$ values under the permutations of $x_0, \ldots, x_{n-1}$.*

**Proof.** Let $k$ be the number of values of $t(\omega)^n$. Equation (10.23) shows that $t(\omega)^n$ is invariant under $\tau$, hence also under all the powers $\tau^2$, $\tau^3$, $\ldots$, $\tau^{n-1}$. Thus, $\big|I\big(t(\omega)^n\big)\big| \geq n$, and Theorem 10.3 yields
$$k \leq (n-1)!.$$
On the other hand, it follows from Proposition 10.1 (p. 125) that $t(\omega)^n$ is a root of an equation $\Theta(Y) = 0$ of degree $k$. Thus, $t(\omega)$ is a root of $\Theta(Y^n) = 0$, which has degree $kn$; but since $t(\omega)$ takes $n!$ different values under the permutations of the variables, it cannot be a root of an equation of degree less than $n!$, by Proposition 10.1. Therefore, $kn \geq n!$, hence
$$k \geq (n-1)!. \qquad \square$$

This proposition remains valid with the same proof when $n$ is not prime, since the permutations $\sigma_k$ were not used.

## Exercises

*Exercise 10.1.* As in Exercise 8.3 (p. 108), let

$$u_1 = x_1 x_2 + x_3 x_4, \qquad v_1 = (x_1 + x_2)(x_3 + x_4),$$
$$u_2 = x_1 x_3 + x_2 x_4, \qquad v_2 = (x_1 + x_3)(x_2 + x_4),$$
$$u_3 = x_1 x_4 + x_2 x_3, \qquad v_3 = (x_1 + x_4)(x_2 + x_3).$$

Show that $u_1$, $u_2$, and $u_3$ are rational functions in $v_1$, $v_2$, $v_3$ with symmetric coefficients. Use this result to show how the cubic equation which has as roots $u_1$, $u_2$, $u_3$ is related to the equation with roots $v_1$, $v_2$, $v_3$. (Compare Exercise 8.3). Same questions with $w_1 = (x_1 - x_2 + x_3 - x_4)^2$, $w_2 = (x_1 + x_2 - x_3 - x_4)^2$, $w_3 = (x_1 - x_2 - x_3 + x_4)^2$ instead of $u_1$, $u_2$, $u_3$.

*Exercise 10.2.* Use the arguments in the proof of Lagrange's theorem (Theorem 10.5, p. 135) to express $x_1 x_2$ as a rational function of $x_1 + x_2$ with coefficients symmetric in the three indeterminates $x_1$, $x_2$, $x_3$. Is this expression unique?

*Exercise 10.3.* Find all the polynomials $f = a x_1 + b x_2 + c x_3$ (with $a$, $b$, $c \in \mathbb{C}$) that have the property that $x_1$, $x_2$, $x_3$ can be rationally expressed from $f$ with symmetric coefficients and such that $f^3$ takes only two values by the permutations of $x_1$, $x_2$, $x_3$.

*Exercise 10.4.* Let $n$ be a prime number. For any $n$-th root of unity $\omega$, let

$$t(\omega) = x_0 + \omega x_1 + \cdots + \omega^{n-1} x_{n-1}.$$

Show that $t(\omega^k) t(\omega)^{-k}$ is a rational function in $t(\omega)^n$ with symmetric coefficients, for any integer $k$.

*Exercise 10.5.* Let $n$ be a prime number and use the notation of Proposition 10.7. Prove that $\tau \circ \sigma_i = \sigma_i \circ \tau^k$ for some $k$. Deduce that

$$\mathrm{GA}(n) = \{\sigma_i \circ \tau^j \mid i = 1, \ldots, n-1 \text{ and } j = 0, \ldots, n-1\}$$

and that $|\mathrm{GA}(n)| = n(n-1)$.

*Exercise 10.6.* Show that for any group $G$ and any subgroup $H$, the map $g \mapsto g^{-1}$ (which is an anti-automorphism of $G$) induces a bijection between the set of left cosets of $H$ in $G$ and the set of right cosets of $H$.

# Chapter 11

# Vandermonde

Alexandre-Théophile Vandermonde (1735–1796) is not a mathematician in the same league as Lagrange or Euler. His contributions to mathematics were scarce and hardly influential. Ironically enough, he is most often remembered nowadays for a determinant that bears his name but is not to be found in his papers: Vandermonde determinants (see Exercise 8.1, p. 107) may have been so christened because someone misread indices for exponents (see Lebesgue [50, pp. 206–207]).

Nevertheless, his work, remarkably described by Lebesgue [50], shows that brilliant ideas and deep insights come not only from first-class mathematicians. Several of Lagrange's ideas were indeed discovered simultaneously or perhaps even a little earlier by Vandermonde. Most notably, Vandermonde performed calculations with permutations and singled out the functions known as Lagrange resolvents, but his exposition is less clear, less authoritative than Lagrange's.

Moreover, the delay in publication was such that Vandermonde's "Mémoire sur la résolution des équations" [81] appeared two years after the first part of Lagrange's "Réflexions sur la résolution algébrique des équations." Lagrange was already famous at that time, and Vandermonde's self-effacing comment (in a footnote added in proof)

> One will notice some conformities between this [Lagrange's] work and mine, of which I cannot feel but flattered [81, p. 365].

did not help to secure notoriety for his paper. However, Vandermonde can be credited with a real breakthrough in the theory of equations: the solution of cyclotomic equations. This was definitely not obtained previously by Lagrange.

We will divide up our discussion of Vandermonde's memoir into two parts: the discussion of general equations, which is somewhat analogous to Lagrange's, and the solution of cyclotomic equations.

## 11.1   The solution of general equations

Vandermonde's starting point is that the formula that yields the solutions of an equation in terms of the coefficients is necessarily ambiguous, since it must take as values the various roots. He then separates the solution into three "heads" [81, p. 370]:

1° To find a function of the roots, of which it can be said, in some sense, that it equals such of the roots that one wants.

2° To put this function in such a form that it be indifferent to interchange the roots in it.

3° To substitute in it the values of the sum of the roots, the sum of pairwise products, etc.

Consider for instance the solution of quadratic equations

$$X^2 - s_1 X + s_2 = 0,$$

with roots $x_1$, $x_2$. The function

$$F_2(x_1, x_2) = \frac{1}{2}\big((x_1 + x_2) + \sqrt{(x_1 - x_2)^2}\big)$$

satisfies the condition in 1°, since its value is $x_1$ or $x_2$, depending on the choice of the square root of $(x_1 - x_2)^2$, namely

$$\sqrt{(x_1 - x_2)^2} = \pm(x_1 - x_2).$$

Since moreover $F_2(x_1, x_2)$ is not altered when the roots $x_1$ and $x_2$ are interchanged, it is already in the form which is called for by 2°. Finally, 3° requires the evaluation of $F_2(x_1, x_2)$ in terms of $s_1$ and $s_2$. This is quite easy:

$$x_1 + x_2 = s_1 \quad \text{and} \quad (x_1 - x_2)^2 = s_1^2 - 4s_2,$$

hence

$$F_2(x_1, x_2) = \frac{1}{2}\big(s_1 + \sqrt{s_1^2 - 4s_2}\big).$$

Vandermonde first solves in full generality problem 3°. He thus proves the fundamental theorem of symmetric functions, which states that every symmetric function can be evaluated in terms of the elementary symmetric polynomials (Theorem 8.3, p. 95).

He then solves problem 1°, displaying the following formula:

$$F_n(x_1, \ldots, x_n) = \frac{1}{n}\left((x_1 + \cdots + x_n) + \sum_{i=1}^{n-1} \sqrt[n]{V_i^n}\right), \tag{11.1}$$

where

$$V_i = \rho_1^i x_1 + \cdots + \rho_n^i x_n$$

and $\rho_1, \ldots, \rho_n$ denote the $n$-th roots of unity (including 1).

To see that this function indeed answers to head 1°, we have to prove that for any $k = 1, \ldots, n$, some determination of the $n$-th roots $\sqrt[n]{V_i^n}$ can be chosen in such a way that $F_n(x_1, \ldots, x_n) = x_k$. This can be done as follows: choose $\sqrt[n]{V_i^n} = \rho_k^{-i} V_i$, i.e.,

$$\sqrt[n]{V_i^n} = x_k + \sum_{j \neq k} (\rho_k^{-1} \rho_j)^i x_j.$$

Then

$$F_n(x_1, \ldots, x_n) = \frac{1}{n}\left(nx_k + \sum_{j \neq k}\left(\sum_{i=0}^{n-1}(\rho_k^{-1}\rho_j)^i\right)x_j\right). \tag{11.2}$$

Now, $\rho_k^{-1}\rho_j$ is an $n$-th root of unity, different from 1 if $k \neq j$, hence it is a root of

$$\frac{X^n - 1}{X - 1} = 1 + X + \cdots + X^{n-1}.$$

Therefore,

$$\sum_{i=0}^{n-1}(\rho_k^{-1}\rho_j)^i = 0 \qquad \text{for } k \neq j,$$

and equation (11.2) simplifies to

$$F_n(x_1, \ldots, x_n) = x_k.$$

Of course, if $n \geq 3$ the function $F_n(x_1, \ldots, x_n)$ also has other determinations besides $x_1, \ldots, x_n$, but this does not seem to matter to Vandermonde.

It is instructive to compare Vandermonde's formula (11.1) to Lagrange's formula (Proposition 10.2, p. 130). It turns out that the functions $V_i$ are none others than Lagrange resolvents. To see this, choose a primitive $n$-th root of unity $\omega$; the various $n$-th roots of unity are then powers of $\omega$, and we can set $\rho_k = \omega^{k-1}$ for $k = 1, \ldots, n$. Then

$$V_i = (\omega^0)^i x_1 + (\omega^1)^i x_2 + (\omega^2)^i x_3 + \cdots + (\omega^{n-1})^i x_n,$$

hence

$$V_i = x_1 + \omega^i x_2 + (\omega^i)^2 x_3 + \cdots + (\omega^i)^{n-1} x_n$$

and it follows that $V_i$ is the Lagrange resolvent that was denoted by $t(\omega^i)$ in the formula of Proposition 10.2 (p. 130).

Problems $1°$ and $3°$ are thus completely solved by Vandermonde; the real stumbling-block is of course problem $2°$. For $n = 3$, Vandermonde observes that, choosing $\rho_1 = 1$, $\rho_2 = \omega$ and $\rho_3 = \omega^2$, where $\omega$ is a cube root of unity other than 1, the functions involved in $F_3(x_1, x_2, x_3)$, which are

$$V_1^3 = (x_1 + \omega x_2 + \omega^2 x_3)^3 \quad \text{and} \quad V_2^3 = (x_1 + \omega^2 x_2 + \omega x_3)^3,$$

are not invariant under all the permutations of $x_1$, $x_2$, $x_3$, but every permutation either leaves $V_1^3$ and $V_2^3$ invariant or interchanges $V_1^3$ and $V_2^3$. Therefore, in order to make the function $F_3(x_1, x_2, x_3)$ invariant under all the permutations, it suffices to substitute for $V_1^3$ and $V_2^3$ an ambiguous function that takes the values $V_1^3$ and $V_2^3$. Such a function has been found previously in the solution of quadratic equations: it is

$$F_2(V_1^3, V_2^3) = \frac{1}{2}\left( (V_1^3 + V_2^3) + \sqrt{(V_1^3 - V_2^3)^2} \right).$$

So, problem $2°$ is solved for $n = 3$.

Vandermonde argues similarly for $n = 4$, using $F_4(x_1, x_2, x_3, x_4)$. He also points out that in this case, since $n$ is not a prime number, other functions can be chosen instead of $F_4(x_1, x_2, x_3, x_4)$, for instance

$$G_4(x_1, x_2, x_3, x_4) = \frac{1}{4}\left( (x_1 + x_2 + x_3 + x_4) + \sqrt{W_1^2} + \sqrt{W_2^2} + \sqrt{W_3^2} \right)$$

where

$$W_1 = x_1 + x_2 - x_3 - x_4,$$
$$W_2 = x_1 - x_2 + x_3 - x_4,$$
$$W_3 = x_1 - x_2 - x_3 + x_4.$$

It is easy to put $G_4$ in such a form that it is not altered when $x_1$, $x_2$, $x_3$, $x_4$ are permuted, since every permutation interchanges $W_1^2$, $W_2^2$ and $W_3^2$. It therefore suffices to replace them by $F_3(W_1^2, W_2^2, W_3^2)$, which takes the values $W_1^2$, $W_2^2$ and $W_3^2$ and can be put in symmetric form, as previously observed.

For $n \geq 5$, the problem is that the functions $V_i^n$ for $i = 1, \ldots, n-1$ are not interchanged among themselves when the indeterminates are permuted. Indeed, the function $V_1^n$ takes $(n-1)!$ values under the permutations of the indeterminates (see Proposition 10.9, p. 141). Nevertheless, for $n = 5$

Vandermonde succeeds in reducing the determination of $V_1^5$ to the solution of an equation of degree 6 (compare Theorem 10.8, p. 140). For $n = 6$, he shows that his method requires the solution of an equation of degree 10 or 15.

Inconclusive as it is, this section is not devoid of interest, since it prompts Vandermonde to initiate fairly explicit calculations with permutations. He decomposes the symmetric polynomials (which he calls "types") into sums of "partial types" that are, in fact, sums of the values that a monomial takes under a subgroup of the symmetric group (especially, but not exclusively, cyclic subgroups). For instance, for three variables $a$, $b$, $c$, he writes

$$[\alpha \ \beta \ \gamma] = a^\alpha b^\beta c^\gamma + a^\gamma b^\alpha c^\beta + a^\beta b^\gamma c^\alpha$$
$$\text{ii iii i}$$

(where $\alpha$, $\beta$, $\gamma$ are pairwise distinct integers). The Latin subscripts indicate that in the second term the exponents $\alpha$, $\beta$, $\gamma$ must be changed in such a way that $\gamma$ takes the first place, $\alpha$ the second and $\beta$ the third; the third term is obtained from the second as the second was obtained from the first, and so on for the next terms, as long as this process yields new monomials. The function thus produced is obviously invariant under the (cyclic) subgroup of $S_3$ generated by the permutation $a \mapsto b \mapsto c \mapsto a$. (Compare the proof of Proposition 10.6, p. 139.)

Sometimes, Vandermonde also uses a more general notation, which includes all the partial types that are invariant under the same group of permutations, but he stops short of devising a notation for permutations. For instance,

$$[a \ b \ c \ d \ e]$$
$$\text{v i iv ii iii}$$

(where $a$, $b$, $c$, $d$, $e$ are the indeterminates) is a generic notation for the various partial types that are invariant under the permutation that sets the letters $a$, $b$, $c$, $d$, $e$ in the order $b$, $d$, $e$, $c$, $a$ indicated by the Latin numerals, i.e., under the permutation $a \mapsto b \mapsto d \mapsto c \mapsto e \mapsto a$.

This notation allows Vandermonde to perform coherently some very complicated explicit calculations, but he cannot elude the conclusion that his method for equations of degree at least 5 leads to equations of ever higher degree, and that it may therefore not work eventually.

That is all that the calculations taught me on this object, and I do not have enough faith in conjectures in such a thorny matter to dare try one here. I will only add that I have not found any partial type involving five letters which depends on an equation of the fourth or the third degree, and I am convinced that such a type does not exist [81, p. 414].

However, that is not the end of the story. In the final two articles of his paper, Vandermonde briefly considers cyclotomic equations.

## 11.2  Cyclotomic equations

Recall from §7.2 (see Theorem 7.3, p. 82) that the problem of determining radical expressions for the roots of unity had been reduced to the solution by radicals of the cyclotomic equations

$$\Phi_p(X) = X^{p-1} + X^{p-2} + \cdots + X + 1 = 0$$

for $p$ prime. Moreover, for $p$ odd, de Moivre had shown that the change of variable $Y = X + X^{-1}$ converts $\Phi_p(X) = 0$ into an equation of degree $\frac{p-1}{2}$. Thus, for $p = 11$, the solution of the cyclotomic equation requires the solution of

$$Y^5 + Y^4 - 4Y^3 - 3Y^2 + 3Y + 1 = 0,$$

which de Moivre had been unable to solve by radicals. We also recall from Remark 7.6 (p. 84) that, since the roots of $\Phi_p$ are the complex numbers $e^{2k\pi i/11}$ for $k = 1, \ldots, 10$, it follows that the roots of the equation in $Y$ are the values of $2\cos\frac{2k\pi}{11}$ for $k = 1, \ldots, 5$.

Vandermonde in fact uses the (obviously equivalent) change of variable $Z = -(X + X^{-1})$, which yields the equation

$$Z^5 - Z^4 - 4Z^3 + 3Z^2 + 3Z - 1 = 0. \tag{11.3}$$

The roots of this equation, which are denoted by $a$, $b$, $c$, $d$, $e$, are chosen as

$$a = -2\cos\frac{2\pi}{11}, \qquad b = -2\cos\frac{4\pi}{11},$$

$$c = -2\cos\frac{6\pi}{11}, \qquad d = -2\cos\frac{8\pi}{11}, \qquad e = -2\cos\frac{10\pi}{11}.$$

It is useful to note, with Vandermonde, that the trigonometric formula

$$2\cos\alpha\cos\beta = \cos(\alpha + \beta) + \cos(\alpha - \beta) \tag{11.4}$$

yields relations between $a$, $b$, $c$, $d$, and $e$. For instance, substituting $\frac{2\pi}{11}$ for $\alpha$ and $\beta$, we obtain

$$2\cos^2\frac{2\pi}{11} = \cos\frac{4\pi}{11} + \cos 0,$$

hence

$$a^2 = -b + 2.$$

Likewise, substituting $\frac{2\pi}{11}$ for $\alpha$ and $\frac{4\pi}{11}$ for $\beta$, we find

$$ab = -c - a,$$

and so on. Thus, substituting for $\alpha$ and $\beta$ successively the various angles $\frac{2k\pi}{11}$ for $k = 1, \ldots, 5$, the trigonometric equation (11.4) yields linear expressions for the products of roots. Observing these expressions, Vandermonde draws an amazing conclusion, which enables him to find expressions by radicals for the eleventh roots of unity. Here are Vandermonde's own words, in the penultimate article of his paper [81, pp. 415–416]:

> In the particular cases where there are equations between the roots, the method just explained may be used to solve, without resorting to the general solution formulas. The equation $r^{11} - 1 = 0$ will provide us with an example: it leads (article VI) to this one
>
> $$X^5 - X^4 - 4X^3 + 3X^2 + 3X - 1 = 0,$$
>
> and denoting its roots by $a$, $b$, $c$, $d$, $e$, it will readily follow from article XI
>
> $$a^2 = -b + 2, \quad b^2 = -d + 2, \quad c^2 = -e + 2, \quad d^2 = -c + 2, \quad e^2 = -a + 2,$$
> $$ab = -a - c, \quad bc = -a - e, \quad cd = -a - d, \quad de = -a - b,$$
> $$ac = -b - d, \quad bd = -b - e, \quad ce = -b - c,$$
> $$ad = -c - e, \quad be = -c - d,$$
> $$ae = -d - e,$$
>
> and all the partial types of the form $[\,a\ b\ c\ d\ e\,]$ will have a purely
> $$\text{v i iv ii iii}$$
> rational value; thus, taking everywhere in article XXVIII $[\,\alpha\ \beta\ \varepsilon\ \delta\ \gamma\,]$
> $$\text{v i iv ii iii}$$
> instead of $[\,\alpha\ \beta\ \gamma\ \delta\ \varepsilon\,]$ we will find $[\ldots]$
> $$\text{v iii iv i ii}$$
>
> $$X = \frac{1}{5}[1 + \Delta' + \Delta'' + \Delta''' + \Delta^{iv}]$$

with[1]

$$\Delta' = \sqrt[5]{\frac{11}{4}\left(89 + 25\sqrt{5} - 5\sqrt{-5+2\sqrt{5}} + 45\sqrt{-5-2\sqrt{5}}\right)},$$

$$\Delta'' = \sqrt[5]{\frac{11}{4}\left(89 + 25\sqrt{5} + 5\sqrt{-5+2\sqrt{5}} - 45\sqrt{-5-2\sqrt{5}}\right)},$$

$$\Delta''' = \sqrt[5]{\frac{11}{4}\left(89 - 25\sqrt{5} - 5\sqrt{-5+2\sqrt{5}} - 45\sqrt{-5-2\sqrt{5}}\right)},$$

$$\Delta^{iv} = \sqrt[5]{\frac{11}{4}\left(89 - 25\sqrt{5} + 5\sqrt{-5+2\sqrt{5}} + 45\sqrt{-5-2\sqrt{5}}\right)}.$$

Vandermonde's brilliant (but not quite explicit) observation is that the permutation $a \mapsto b \mapsto d \mapsto c \mapsto e \mapsto a$ preserves the relations between the roots. For instance, applying this permutation to the relation

$$a^2 = -b + 2,$$

i.e., changing $a$ into $b$ and $b$ into $d$, we obtain

$$b^2 = -d + 2,$$

and this relation actually holds! (Compare Exercise 11.2.)

This is very significant since the relations between the roots can be used to lower the degree of any polynomial in $a$, $b$, $c$, $d$, $e$, eventually providing a linear expression. Thus, suppose $f$ is a polynomial in five variables (of any degree). Using the relations between the roots, we can eventually find

$$f(a,b,c,d,e) = Aa + Bb + Cc + Dd + Ee + F \qquad (11.5)$$

for some numbers $A$, $B$, ..., $F$ that can be explicitly determined from the coefficients of $f$. Now, since the permutation $a \mapsto b \mapsto d \mapsto c \mapsto e \mapsto a$ preserves the relations that have been used to simplify the expression of $f(a,b,c,d,e)$, we can perform the *same* simplification procedure changing $a$ into $b$, $b$ into $d$, $d$ into $c$, $c$ into $e$ and $e$ into $a$ at each step, and we end up with

$$f(b,d,e,c,a) = Ab + Bd + Ce + Dc + Ea + F.$$

---

[1]Oscar Luis Palacios-Vélez pointed out (private communication) that there are misprints in Vandermonde's formulas for $\Delta'$ and $\Delta''$. They should be

$$\Delta' = \sqrt[5]{\frac{11}{4}\left(89 + 25\sqrt{5} - 5\sqrt{-5-2\sqrt{5}} + 45\sqrt{-5+2\sqrt{5}}\right)},$$

$$\Delta'' = \sqrt[5]{\frac{11}{4}\left(89 + 25\sqrt{5} + 5\sqrt{-5-2\sqrt{5}} - 45\sqrt{-5+2\sqrt{5}}\right)}.$$

This point is somewhat delicate, since the expression (11.5) of $f(a,b,c,d,e)$ is *not* unique. Indeed, since the coefficient of $Z^4$ in (11.3) is the opposite of the sum of the roots, it follows that

$$a + b + c + d + e = 1. \tag{11.6}$$

Therefore, we have for instance

$$Aa + Bb + Cc + Dd + Ee + F$$
$$= (A + F)a + (B + F)b + (C + F)c + (D + F)d + (E + F)e.$$

This does not matter, however, as long as we use the same procedure to simplify $f(a,b,c,d,e)$ and (after the permutation $a \mapsto b \mapsto d \mapsto c \mapsto e \mapsto a$) $f(b,d,e,c,a)$.

If we apply this observation to the Vandermonde (–Lagrange) resolvents

$$V_i(a,b,c,d,e)^5 = (\rho_1^i a + \rho_2^i b + \rho_3^i c + \rho_4^i d + \rho_5^i e)^5$$

where $\rho_1, \ldots, \rho_5$ are the 5-th roots of unity, we find

$$V_i(a,b,c,d,e)^5 = Aa + Bb + Cc + Dd + Ee + F \tag{11.7}$$

where $A, B, \ldots, F$ are rational expressions in $\rho_1, \ldots, \rho_5$.

Applying the permutation four times, we obtain

$$V_i(b,d,e,c,a)^5 = Ab + Bd + Ce + Dc + Ea + F,$$
$$V_i(d,c,a,e,b)^5 = Ad + Bc + Ca + De + Eb + F,$$
$$V_i(c,e,b,a,d)^5 = Ac + Be + Cb + Da + Ed + F, \tag{11.8}$$
$$V_i(e,a,d,b,c)^5 = Ae + Ba + Cd + Db + Ec + F.$$

Now, if we choose $\rho_1 = 1$, $\rho_2 = \omega$, $\rho_3 = \omega^3$, $\rho_4 = \omega^2$ and $\rho_5 = \omega^4$, where $\omega$ is some primitive 5-th root of unity, then

$$V_i(a,b,c,d,e) = a + \omega^i b + \omega^{2i} d + \omega^{3i} c + \omega^{4i} e,$$

and the permutation $a \mapsto b \mapsto d \mapsto c \mapsto e \mapsto a$ then leaves $V_i(a,b,c,d,e)^5$ invariant. Indeed, this permutation changes $V_i(a,b,c,d,e)$ into

$$V_i(b,d,e,c,a) = b + \omega^i d + \omega^{2i} c + \omega^{3i} e + \omega^{4i} a$$

and since $\omega^5 = 1$, we have

$$V_i(b,d,e,c,a) = \omega^{-i} V_i(a,b,c,d,e)$$

(compare the proof of Theorem 10.8, p. 140). Therefore,

$$V_i(b,d,e,c,a)^5 = V_i(a,b,c,d,e)^5.$$

Likewise, the left-hand sides of the equations (11.7) and (11.8) are all equal to $V_i(a, b, c, d, e)^5$. Therefore, summing up all these equations, we obtain

$$5V_i(a, b, c, d, e)^5 = (A + B + C + D + E)(a + b + c + d + e) + 5F.$$

Using (11.6) we conclude

$$V_i(a, b, c, d, e)^5 = \frac{1}{5}(A + B + C + D + E) + F.$$

Since $A$, $B$, ..., $F$ can be rationally calculated from $\omega$, which is already expressed by radicals (see §7.2), it follows that the functions $V_i^5$ can be expressed by radicals. Therefore, $a$, $b$, $c$, $d$, and $e$ can also be expressed by radicals. Using the formula $F_5(a, b, c, d, e)$ of §11.1 (and (11.6)), we obtain

$$a, b, c, d, e = \frac{1}{5}\left(1 + \sum_{i=1}^{4} \sqrt[5]{V_i^5}\right).$$

With hindsight, and with a view towards a possible generalization to cyclotomic equations of higher degree, the crucial steps in the calculations above appear to be

(a) the existence of relations among the roots, which can be used to reduce to degree 1 each polynomial expression in the roots;
(b) the existence of a cyclic permutation of the roots that preserves the relations above.

Given (a) and (b), we number the roots in such a way that the cyclic permutation be

$$x_1 \mapsto x_2 \mapsto x_3 \mapsto \cdots \mapsto x_n \mapsto x_1,$$

and the same arguments as above then show that the $n$-th power of each Lagrange resolvent of the form

$$t(\omega) = x_1 + \omega x_2 + \omega^2 x_3 + \cdots + \omega^{n-1} x_n,$$

for any $n$-th root of unity $\omega$, is a rational expression in $\omega$. Arguing inductively, we may assume that an expression by radicals has been found for $\omega$. Then $t(\omega)^n$ can be expressed by radicals and the roots $x_1, \ldots, x_n$ can also be found by radicals, by Lagrange's formula (Proposition 10.2, p. 130), namely

$$x_i = \frac{1}{n}\left(\sum_{\omega} \omega^{-(i-1)} t(\omega)\right).$$

Now, (a) is clear for the cyclotomic equations $\Phi_p$ for any prime $p$ or, rather, for the equations obtained by de Moivre's change of variable $Y = X + X^{-1}$, whose roots are $2\cos\frac{2k\pi}{p}$ for $k = 1, \ldots, \frac{p-1}{2}$. Indeed, the same trigonometric equation (11.4) yields the required relations. But (b) is very far from clear! Yet Vandermonde simply states

Since, to solve the equation

$$X^m - X^{m-1} - (m-1)X^{m-2} + \text{ etc. } = 0,$$

the question is at most to determine (article VI) the quantity which is indifferently one of its roots, and by no means to arrange it in such a way that it be indifferent to interchange the roots among themselves, this solution will always be very easy [81, p. 416].

And he leaves it at that.

Yet, he had certainly noticed that the relations were not preserved by *every* permutation of the roots. The existence of a cyclic permutation that does preserve the relations is a very remarkable, and quite mysterious, property of cyclotomic equations, which should have awaken Vandermonde's curiosity. If he had investigated this property, he could have developed the theory of cyclotomy about thirty years before Gauss.

Moreover, Vandermonde had pinpointed the very basic idea of Galois theory: in order to determine the "structure" of an equation, deciding eventually whether it is solvable by radicals, and more generally to evaluate its difficulty, one has to look at the permutations of the roots; but one needs only to consider those permutations that preserve the relations between the roots.[2]

This is a conspicuous example of a deep insight that was completely wasted. To conclude this chapter, I could do no better than to quote Lebesgue:

> Surely, any man who discovers something truly important is left behind by his own discovery; he himself hardly understands it, and only by pondering over it for a long time. But Vandermonde never came back to his algebraic investigations because he did not realize their importance in the first place, and if he did not understand them afterwards, it is precisely because he did not reflect deeply on them; he was interested in everything, he was busy with everything; he was not able to go slowly to the bottom of anything. [...] To assess exactly what Vandermonde saw, understood and what he did not catch, one would have to reconstruct not only the mind of a man from the eighteenth century, but Vandermonde's mind, and at the moment when he had a glimpse of genius and went ahead of his age. When trying to do so, one will always give too much or too little credit to Vandermonde [50, pp. 222–223].

---

[2]This restriction does not appear for general equations since their roots are independent indeterminates. In this case, there is no relation to preserve, so every permutation is admissible.

## Exercises

*Exercise 11.1.* List all the terms in the partial types

$$[\alpha\ \beta\ \gamma\ \delta\ \varepsilon],\quad [\alpha\ \varepsilon\ \delta\ \beta\ \gamma],\quad [\alpha\ \gamma\ \beta\ \varepsilon\ \delta],$$
$$\text{v\ iii\ iv\ i\ ii}\qquad\text{v\ iii\ iv\ i\ ii}\qquad\text{v\ iii\ iv\ i\ ii}$$

and $[\alpha\ \delta\ \varepsilon\ \gamma\ \beta]$. Show that the sum of all these partial types is also a
  v iii iv i ii
partial type. What is the subgroup of $S_5$ that leaves this new partial type
invariant? [81, Art. 24, p. 391]

*Exercise 11.2.* Show that the permutation $a \mapsto b \mapsto c \mapsto d \mapsto e \mapsto a$ does
not preserve the relations among $a = 2\cos\frac{2\pi}{11}$, $b = 2\cos\frac{4\pi}{11}$, $c = 2\cos\frac{6\pi}{11}$,
$d = 2\cos\frac{8\pi}{11}$, and $e = 2\cos\frac{10\pi}{11}$.

*Exercise 11.3.* Show that the permutation $a \mapsto b \mapsto c \mapsto a$ preserves the
relations among $a = 2\cos\frac{2\pi}{7}$, $b = 2\cos\frac{4\pi}{7}$, and $c = 2\cos\frac{6\pi}{7}$.

*Exercise 11.4.* Find a permutation of the numbers $2\cos\frac{2k\pi}{13}$ for $k = 1, \ldots,$
6 that preserves the relations among them.

Chapter 12

# Gauss on Cyclotomic Equations

The contributions of Carl Friedrich Gauss (1777–1855) to the theory of equations measure up to the outstanding advances he made in many other research areas. They occupy a special place in his work however, since they were among his earliest achievements. They can be divided up into two main topics: the fundamental theorem of algebra (1799) which we already discussed in Chapter 9, and the solution of cyclotomic equations.

Gauss' results on cyclotomic equations show how to complete Vandermonde's arguments to provide inductively expressions by radicals for the roots of unity (see Corollary 12.29), but they far exceed this goal. In effect, they yield a thorough description of the possible reductions of cyclotomic equations of prime index to equations of smaller degree. Thus, in a brilliant way, they carry out for cyclotomic equations the program envisioned by Lagrange: to solve an equation by determining successively certain functions of the roots. As Gauss shows, the solution of $\Phi_p(X) = 0$ can be reduced to the solution of equations of degree equal to the prime factors of $p - 1$. In particular, the 17-th roots of unity can be determined by solving successively four quadratic equations, since $17 - 1 = 2^4$. As an application of this result, it follows that the regular polygon with 17 sides can be constructed by ruler and compass; this result was obtained by Gauss as early as 1796, and it is said to have been crucial in his decision to pursue a career in mathematics (see Bühler [11, p. 10]). This application will be discussed in an appendix to this chapter.

A definitive account of his results on cyclotomic equations was published by Gauss as the seventh and final section of his epoch-making treatise on number theory, "Disquisitiones Arithmeticae" (1801). The inclusion of such algebraic results in a book on number theory was commented by Gauss himself in the preface [31, p. xii]:

The theory of the division of the circle, or of regular polygons, which is treated in section VII, does not belong *by itself* to Arithmetic, but its *principles* cannot be found but in Higher Arithmetic: this may appear to geometers as unexpected as the new truths that follow from it, and which they will see, I hope, with pleasure.

Accordingly, our review of Gauss' results will be preceded by some number-theoretic preliminaries. Thereafter, we divide the contents of the seventh section of the "Disquisitiones Arithmeticae" into three sections: first, we prove the irreducibility of cyclotomic equations of prime index, which is a key result in Gauss' investigations; next, we discuss the possible reductions of cyclotomic equations and we finish with the solvability by radicals of the cyclotomic equations and auxiliary equations. Some extra results which are needed to justify some of the steps in Gauss' proofs will be found in the final section.

It should be noted that we only review those results of section VII of the "Disquisitiones Arithmeticae" that directly concern the theory of equations. Several details that are meaningful in view of applications to number theory are omitted.

## 12.1 Number-theoretic preliminaries

At the very beginning of the "Disquisitiones Arithmeticae" [31, Art. 2], Gauss introduces the following notation, which has gained wide acceptance and will be used repeatedly in the sequel: if $a$, $b$ and $n$ are integers and $n \neq 0$, one writes

$$a \equiv b \mod n$$

to express the property that $a - b$ is divisible by $n$. The integers $a$ and $b$ are then said to be *congruent modulo n*. The explicit reference to the modulus $n$ is sometimes omitted when no confusion is likely to arise. It is readily verified that this relation is an equivalence relation which is compatible with the sum and the product of integers, i.e., if $a_1 \equiv b_1 \mod n$ and $a_2 \equiv b_2 \mod n$, then $a_1 + a_2 \equiv b_1 + b_2 \mod n$ and $a_1 a_2 \equiv b_1 b_2 \mod n$.

In the sequel, we focus on the case where the modulus $n$ is prime. This case has very distinctive features. We will prove in particular the following result, which plays a key role in Gauss' investigations on cyclotomic equations:

**Theorem 12.1.** *For any prime number p, there exists an integer g whose* *various powers $g^0$, $g^1$, $g^2$, ..., $g^{p-2}$ are congruent to 1, 2, ..., $p-1$ modulo p (not necessarily in that order).*

In the course of proving this theorem, we will see that an integer $g$ satisfies the condition of the theorem if and only if

$$g^{p-1} \equiv 1 \mod p \quad \text{and} \quad g^i \not\equiv 1 \mod p \quad \text{for } i = 1, \ldots, p-2.$$

Therefore, any such integer $g$ is called a *primitive root of p*. (It would be more accurate, although not shorter, to call it a primitive $(p-1)$-st root of 1 modulo $p$.) For instance, 2 is a primitive root of 11, because modulo 11 we have

$$2^0 \equiv 1, \quad 2^1 \equiv 2, \quad 2^2 \equiv 4, \quad 2^3 \equiv 8, \quad 2^4 \equiv 5,$$
$$2^5 \equiv 10, \quad 2^6 \equiv 9, \quad 2^7 \equiv 7, \quad 2^8 \equiv 3, \quad 2^9 \equiv 6.$$

By contrast, 3 is not a primitive root of 11, as $3^5 \equiv 1 \mod 11$, hence $3^6 \equiv 3$, $3^7 \equiv 3^2$, $3^8 \equiv 3^3$, ..., and the powers of 3 take modulo 11 only the values $3^0 \equiv 1$, $3^1 \equiv 3$, $3^2 \equiv 9$, $3^3 \equiv 5$, and $3^4 \equiv 4$.

The proof of Theorem 12.1 occupies the rest of this section. We closely follow one of the two proofs that Gauss included in the "Disquisitiones Arithmeticae" [31, Art. 55].

**Lemma 12.2 (Gauss [31, Art. 14]).** *Let a, b be integers and let p be a prime number. If*

$$ab \equiv 0 \mod p,$$

*then $a \equiv 0 \mod p$ or $b \equiv 0 \mod p$.*

This amounts to the well-known fact that if a prime number divides a product of integers, then it divides one of the factors. To prove it, it suffices to mimic the proof of Lemma 5.9 (p. 49), substituting integers for polynomials.

**Proposition 12.3 (Gauss [31, Art. 43]).** *Let p be a prime number and let $a_0$, $a_1$, ..., $a_d$ be integers. If $a_d \not\equiv 0 \mod p$, then the congruence equation*

$$a_d X^d + a_{d-1} X^{d-1} + \cdots + a_1 X + a_0 \equiv 0 \mod p \tag{12.1}$$

*has at most d incongruent solutions modulo p.*

**Proof.** We argue by induction on $d$. Since the case $d = 0$ is clear, we may assume inductively that every congruence equation of degree $d - 1$ has at most $d - 1$ solutions modulo $p$.

If equation (12.1) has $d + 1$ solutions $x_1, \ldots, x_{d+1}$ pairwise distinct modulo $p$, then the change of variable $Y = X - x_1$ transforms equation (12.1) into another equation of degree $d$,

$$a_d Y^d + a'_{d-1} Y^{d-1} + \cdots + a'_1 Y + a'_0 \equiv 0 \quad \bmod p$$

which has the same leading coefficient $a_d$ and has the $d + 1$ solutions

$$0, \quad x_2 - x_1, \quad \ldots, \quad x_{d+1} - x_1.$$

Since 0 is a root, it follows that $a'_0 \equiv 0 \bmod p$ and the equation in $Y$ can be written

$$Y(a_d Y^{d-1} + a'_{d-1} Y^{d-2} + \cdots + a'_1) \equiv 0 \quad \bmod p.$$

Now, $x_2 - x_1, x_3 - x_1, \ldots, x_{d+1} - x_1$ are non-zero roots of this equation, hence, by Lemma 12.2, they are roots of the second factor $a_d Y^{d-1} + \cdots + a'_1$, which is a polynomial of degree $d - 1$. This contradicts the induction hypothesis. $\qquad\square$

**Remark 12.4.** As with any other equivalence relation, the congruence relation defines a partition of the set on which it is defined into equivalence classes. The congruence class modulo $n$ of an integer $m$ consists of all the integers that are congruent to $m$ modulo $n$ or, in other words, of all the integers of the form $kn + m$, for $k \in \mathbb{Z}$.

Since the congruence relation is compatible with the sum and the product of integers, these operations induce well-defined operations on the set of congruence classes of integers, and it is readily verified that this set inherits the commutative ring structure of $\mathbb{Z}$. The ring of congruence classes modulo $n$ is denoted by $\mathbb{Z}/n\mathbb{Z}$. (Compare §9.1.) This ring has only finitely many elements, namely the congruence classes modulo $n$ of $0, 1, \ldots, n-1$.

Lemma 12.2 asserts that $\mathbb{Z}/p\mathbb{Z}$ is a domain, for $p$ prime, and the arguments in Proposition 12.3 show, more generally, that an equation of degree $d$ with coefficients in a domain has at most $d$ solutions in the domain.

In fact, since $\mathbb{Z}/n\mathbb{Z}$ is finite, it is easily seen that this ring is a field whenever it is a domain. Indeed, if $aX \equiv 0 \bmod n$ implies $X \equiv 0 \bmod n$, then the products $ax$ are pairwise distinct modulo $n$ when $x$ runs over $0$, $1, \ldots, n-1$. Therefore, one of these products is congruent to 1 modulo $n$. This proves that $a$ is invertible modulo $n$. Consequently, for $p$ prime, the ring $\mathbb{Z}/p\mathbb{Z}$ is a field, which is denoted by $\mathbb{F}_p$.

For the proof of Theorem 12.1, we also need the following result, due to Pierre de Fermat (1601–1665), and proved by Gauss in [31, Art. 50]:

**Theorem 12.5 (Fermat).** *Let $p$ be any prime number and let $a$ be an integer. If $a \not\equiv 0 \bmod p$, then $a^{p-1} \equiv 1 \bmod p$.*

**Proof (Euler).** We will prove

$$a^p \equiv a \quad \bmod p \qquad \text{for every integer } a. \tag{12.2}$$

It will then follow that

$$a(a^{p-1} - 1) \equiv 0 \quad \bmod p \qquad \text{for every integer } a,$$

hence, by Lemma 12.2, $a^{p-1} - 1 \equiv 0 \bmod p$ whenever $a \not\equiv 0 \bmod p$.

The basic observation is that the binomial coefficients $\binom{p}{i} = \frac{p!}{i!(p-i)!}$ are all divisible by $p$ for $i = 1, \ldots, p-1$. Therefore,

$$(a + 1)^p \equiv a^p + 1 \quad \bmod p \qquad \text{for every integer } a,$$

and it easily follows by induction on $a$ that (12.2) holds for every positive integer $a$. The property for negative $a$ is then readily proved, since

$$(-a)^p \equiv -a^p \quad \bmod p$$

for every integer $a$ and every prime $p$. (For $p = 2$, observe that $-1 \equiv 1 \bmod 2$.) $\qquad \square$

**Corollary 12.6.** *Let $p$ be a prime number. The following conditions on an integer $g$ are equivalent:*

(a) *$g^{p-1} \equiv 1 \bmod p$ and $g^i \not\equiv 1 \bmod p$ for $i = 1, \ldots, p-2$;*
(b) *the powers $g^0, g^1, \ldots, g^{p-2}$ take the values $1, 2, \ldots, p-1$ modulo $p$.*

**Proof.** (a) $\Rightarrow$ (b) If $g^{p-1} \equiv 1 \bmod p$, then $g \not\equiv 0 \bmod p$ and it follows from Lemma 12.2 that the values modulo $p$ of the powers $g^0, g^1, \ldots, g^{p-2}$ range in $\{1, 2, \ldots, p-1\}$. Therefore, to prove (b), it suffices to show that the powers $g^i$ are pairwise distinct modulo $p$ for $i = 0, 1, \ldots, p-2$.

Assume on the contrary

$$g^i \equiv g^j \quad \bmod p$$

for some integers $i$, $j$ between 0 and $p-2$, and $i < j$. Then,

$$g^i(1 - g^{j-i}) \equiv 0 \quad \bmod p,$$

hence, by Lemma 12.2,

$$g^{j-i} \equiv 1 \quad \bmod p.$$

Since $j - i$ is an integer between 1 and $p - 2$, this congruence contradicts (a).

(b) $\Rightarrow$ (a) From (b), it clearly follows that $g \not\equiv 0 \bmod p$, hence

$$g^{p-1} \equiv 1 \quad \bmod p,$$

by Theorem 12.5. Moreover, condition (b) implies that $g^i \not\equiv g^0$ for $i = 1$, $\ldots, p - 2$, hence

$$g^i \not\equiv 1 \quad \bmod p \qquad \text{for } i = 1, \ldots, p - 2.$$

(Compare Proposition 7.9, p. 85.)                                               $\square$

As previously mentioned, every integer $g$ satisfying the equivalent conditions (a) and (b) in Corollary 12.6 is called a *primitive root of p*.

We now introduce the following technical definition: for any prime number $p$ and any integer $a$ relatively prime to $p$, the *exponent (modulo p) of a* is the smallest positive integer $e$ such that $a^e \equiv 1 \bmod p$. Thus, the integers of exponent $p - 1$, which will be shown to exist for every prime $p$, are the primitive roots of $p$. (Compare Definitions 7.7, p. 85.)

**Lemma 12.7.** *Let $e$ be the exponent (modulo a prime number p) of an integer a relatively prime to p, and let m be an integer. Then $a^m \equiv 1 \bmod p$ if and only if e divides m. In particular, e divides $p - 1$.*

**Proof.** The same arguments as in the proof of Lemma 7.8, p. 85, apply.    $\square$

Of course, it is not clear *a priori* that there exist integers of exponent $e$ for every divisor $e$ of $p - 1$. As a first step in the proof of Theorem 12.1, we now show the existence of such integers in the case where $e$ is the power of a prime number.

**Lemma 12.8 (Gauss [31, Art. 55]).** *Let $p$ and $q$ be prime numbers. If some power $q^m$ of $q$ divides $p - 1$, then there exists an integer of exponent $q^m$ (modulo p).*

**Proof.** By Proposition 12.3, the congruence equation $X^{(p-1)/q} \equiv 1 \bmod p$ has at most $(p - 1)/q$ solutions (modulo $p$). Therefore, one can find an integer $x$ that is not a root of this equation and is relatively prime to $p$. Let then $a = x^{(p-1)/q^m}$. By Fermat's theorem (Theorem 12.5), we have $x^{p-1} \equiv 1 \bmod p$, hence

$$a^{q^m} \equiv 1 \quad \bmod p.$$

Lemma 12.7 then shows that the exponent of $a$ divides $q^m$. On the other hand,

$$a^{q^{m-1}} = x^{(p-1)/q} \not\equiv 1 \mod p,$$

so the exponent of $a$ does not divide $q^{m-1}$. Since $q$ is prime, the only (positive) integer that divides $q^m$ but not $q^{m-1}$ is $q^m$, and the exponent of $a$ is therefore equal to $q^m$. $\qquad\square$

**Proof of Theorem 12.1.** Let

$$p - 1 = q_1^{m_1} \dots q_r^{m_r}$$

be the decomposition of $p - 1$ into a product of prime factors, where the primes $q_1, \dots, q_r$ are pairwise distinct. By Lemma 12.8, one can find integers $a_1, \dots, a_r$ of respective exponents $q_1^{m_1}, \dots, q_r^{m_r}$ modulo $p$. To prove the theorem, we show that the product $a_1 \dots a_r$ has exponent $p - 1$, and is therefore a primitive root of $p$.

Let $e$ be the exponent of $a_1 \dots a_r$. Lemma 12.7 shows that $e$ divides $p - 1$. If $e \neq p - 1$, then $e$ lacks at least one of the prime factors of $p - 1$ and divides therefore $(p-1)/q_i$ for some $i = 1, \dots, r$. Assume for instance $e$ divides $(p-1)/q_1$. Then, by Lemma 12.7,

$$(a_1 \dots a_r)^{(p-1)/q_1} \equiv 1 \mod p. \tag{12.3}$$

Now, since the exponent $q_i^{m_i}$ of $a_i$ divides $(p-1)/q_1$ for $i = 2, \dots, r$, we have

$$a_i^{(p-1)/q_1} \equiv 1 \mod p \qquad \text{for } i = 2, \dots, r.$$

Therefore, (12.3) yields

$$a_1^{(p-1)/q_1} \equiv 1 \mod p.$$

This congruence shows, by Lemma 12.7, that the exponent $q_1^{m_1}$ of $a_1$ divides $(p-1)/q_1$. Therefore, $p - 1$ is divisible by $q_1^{m_1+1}$. This is a contradiction, which shows that is was absurd to assume $e \neq p - 1$. (Alternatively, the claim that $e = p - 1$ can also be established by the same arguments as in the proof of Proposition 7.13, p. 87.) $\qquad\square$

**Remarks 12.9. (a)** If $g$ is a primitive root of an odd prime $p$, then $g^{(p-1)/2} \equiv -1 \mod p$. Indeed, since

$$\left(g^{(p-1)/2}\right)^2 \equiv g^{p-1} \equiv 1 \mod p,$$

it follows that $g^{(p-1)/2}$ is a root of the congruence equation

$$X^2 \equiv 1 \mod p.$$

Now, this equation has two roots modulo $p$ (see Proposition 12.3), which are 1 and $-1$, so

$$g^{(p-1)/2} \equiv \pm 1 \mod p.$$

Since $g$ is primitive, no power of exponent smaller than $p-1$ is congruent to 1; therefore, $g^{(p-1)/2} \equiv 1 \mod p$ is impossible and it follows that $g^{(p-1)/2} \equiv -1 \mod p$.

**(b)** Fermat's theorem (Theorem 12.5) can also be proved by the following elaboration on Lagrange's Theorem (Theorem 10.4, p. 134): for any element $a$ of a (multiplicative) group $G$, define the *order* (or the *exponent*) of $a$ as the smallest positive integer $e$ such that $a^e = 1$ in $G$. If the order of $a$ is finite, and denoted by $e$, then it is easily seen that the set

$$S = \{1, a, a^2, \dots, a^{e-1}\} \subset G$$

is a subgroup of $G$, and the arguments in Corollary 12.6 or Lemma 7.8 (p. 85) show that the elements 1, $a$, ..., $a^{e-1}$ are pairwise distinct, hence $|S| = e$. By Lagrange's Theorem, it follows that $e$ divides $|G|$ (if $G$ is finite). In particular, $a^{|G|} = 1$ for all $a \in G$, if $G$ is finite.

Fermat's theorem follows by applying this result to the multiplicative group $\mathbb{F}_p^\times = \mathbb{F}_p \setminus \{0\}$ of the field with $p$ elements. (See Remark 12.4.)

**(c)** Tracing back through the proof of Theorem 12.1, it appears that the only information on $\mathbb{F}_p$ that was used (besides the fact that $\mathbb{F}_p^\times$ is an abelian group) was that the number of solutions of the equation $X^{(p-1)/q} = 1$ does not exceed $(p-1)/q$. Therefore, the arguments in this proof provide the following result: if a finite abelian group $G$ is such that for every integer $n$ the equation $X^n = 1$ has at most $n$ solutions in $G$, then $G$ is cyclic, i.e., $G$ is generated by a single element,

$$G = \{1, a, a^2, \dots, a^{|G|-1}\} \qquad \text{for some } a \in G.$$

In particular, every finite subgroup of the multiplicative group of a field is cyclic.

## 12.2    Irreducibility of the cyclotomic polynomials of prime index

The aim of this section is to provide a proof and develop some of the consequences of the following theorem:

**Theorem 12.10.** *For every prime $p$, the cyclotomic polynomial*

$$\Phi_p(X) = X^{p-1} + X^{p-2} + \cdots + X + 1$$

*is irreducible over the field of rational numbers.*

This theorem was first proved by Gauss in Article 341 of the "Disquisitiones Arithmeticae." Since then, it has been generalized, and the proofs have been simplified by several mathematicians, including Eisenstein, Dedekind, Kronecker, Mertens, Landau, and Schur. Instead of following Gauss' own proof, which requires a careful analysis of several cases, we will follow Eisenstein's ideas, which are simpler and in some sense more general, in that they provide a useful sufficient condition for the irreducibility of polynomials over $\mathbb{Q}$. In §12.5, we prove some generalizations of this theorem, after Dedekind and Kronecker. The proofs given there yield an alternative proof of Theorem 12.10. The starting point of all the proofs is a result known as "Gauss' lemma."

**Lemma 12.11 (Gauss [31, Art. 42]).** *If a monic polynomial in $\mathbb{Q}[X]$ divides a monic polynomial with integral coefficients, then its coefficients are all integral.*

**Proof.** Let $f = X^n + a_{n-1}X^{n-1} + \cdots + a_1 X + a_0 \in \mathbb{Q}[X]$ be a monic polynomial which divides a monic polynomial $P \in \mathbb{Z}[X]$, and let $g \in \mathbb{Q}[X]$ be the quotient,

$$fg = P \in \mathbb{Z}[X].$$

Since $P$ and $f$ are monic, $g$ is monic too. Let

$$g = X^m + b_{m-1}X^{m-1} + \cdots + b_1 X + b_0 \in \mathbb{Q}[X].$$

We have to prove that the coefficients $a_0$, $a_1$, ..., $a_{n-1}$ are all integers. Suppose the contrary and let $d$ be the least common multiple of the denominators of $a_0$, ..., $a_{n-1}$; then

$$f = \tfrac{1}{d}(dX^n + a'_{n-1}X^{n-1} + \cdots + a'_1 X + a'_0)$$

where $d$, $a'_{n-1}$, ..., $a'_0$ are relatively prime integers. Similarly, let

$$g = \tfrac{1}{e}(eX^m + b'_{m-1}X^{m-1} + \cdots + b'_1 X + b'_0)$$

where $e$, $b'_{m-1}$, ..., $b'_0$ are relatively prime integers. Let $p$ be a prime number which divides $d$. Since $d$, $a'_{n-1}$, ..., $a'_0$ are relatively prime, there is a largest index $k$ such that $p$ does not divide $a'_k$. Let also $\ell$ be the largest index such that $p$ does not divide $b'_\ell$ (if $p$ does not divide $e$, let $\ell = m$ and $b'_\ell = e$). Since $fg = P \in \mathbb{Z}[X]$, it follows that

$$(dX^n + a'_{n-1}X^{n-1} + \cdots + a'_0)(eX^m + b'_{m-1}X^{m-1} + \cdots + b'_0) \in de\mathbb{Z}[X].$$

In particular, the coefficient of $X^{k+\ell}$ in the product is divisible by $p$ since $p$ divides $d$, i.e.,

$$\sum_{i+j=k+\ell} a_i' b_j' = \cdots + a_{k-1}' b_{\ell+1}' + a_k' b_\ell' + a_{k+1}' b_{\ell-1}' + \cdots \equiv 0 \mod p.$$

Since $a_i' \equiv 0 \mod p$ for $i > k$ and $b_j' \equiv 0 \mod p$ for $j > \ell$, we have

$$\sum_{i+j=k+\ell} a_i' b_j' \equiv a_k' b_\ell' \mod p.$$

Therefore, the previous equation yields

$$a_k' b_\ell' \equiv 0 \mod p.$$

This is a contradiction, since $p$ does not divide $a_k'$ nor $b_\ell'$.                     $\square$

**Theorem 12.12 (Eisenstein).** *Let $P$ be a monic polynomial with integral coefficients,*

$$P = X^t + c_{t-1} X^{t-1} + \cdots + c_1 X + c_0 \in \mathbb{Z}[X].$$

*If there is a prime number $p$ which divides $c_i$ for $i = 0, \ldots, t-1$ but such that $p^2$ does not divide $c_0$, then $P$ is irreducible over $\mathbb{Q}$.*

**Proof.** Assume on the contrary that there exist non-constant polynomials $f, g \in \mathbb{Q}[X]$ such that

$$P = fg.$$

Since $P$ is monic, we may assume that $f$ and $g$ are both monic, hence, by Gauss' lemma (Lemma 12.11), that $f$ and $g$ have integral coefficients. Let

$$f = X^n + a_{n-1} X^{n-1} + \cdots + a_1 X + a_0 \in \mathbb{Z}[X]$$

and

$$g = X^m + b_{m-1} X^{m-1} + \cdots + b_1 X + b_0 \in \mathbb{Z}[X].$$

Since $a_0 b_0 = c_0$, it follows that $p$ divides $a_0$ or $b_0$, but not both since $p^2$ does not divide $c_0$. Since $f$ and $g$ are interchangeable, we may assume without loss of generality that $p$ divides $a_0$ but not $b_0$.

Let then $i$ be the largest index such that $p$ divides $a_i$; thus, $p$ divides $a_k$ for $k \leq i$ but not for $k = i + 1$ (possibly, $i = n - 1$; we then let $a_{i+1} = 1$). Now, the idea, as in the proof of Gauss' lemma, is to look at a well-chosen coefficient in the product $fg$; namely, we consider the coefficient of $X^{i+1}$, which is

$$c_{i+1} = a_{i+1} b_0 + a_i b_1 + a_{i-1} b_2 + \cdots + a_0 b_{i+1}. \tag{12.4}$$

(We let $b_m = 1$ and $b_j = 0$ if $j > m$.) Since $i + 1 \leq n < t$, the hypothesis implies that $c_{i+1}$ is divisible by $p$; but since $a_i$, $a_{i-1}$, ..., $a_0$ are all divisible by $p$, it follows from (12.4) above that $p$ divides $a_{i+1} b_0$. This is a contradiction since it was assumed that $p$ does not divide $a_{i+1}$ nor $b_0$.                     $\square$

**Proof of Theorem 12.10.** If $\Phi_p(X)$ is not irreducible, then there is a factorization

$$\Phi_p(X) = f(X)g(X)$$

for some non-constant polynomials $f$, $g$ in $\mathbb{Q}[X]$. The change of variable $X = Y + 1$ converts this equation into

$$\Phi_p(Y + 1) = f(Y + 1)g(Y + 1).$$

Since $f(Y+1)$ and $g(Y+1)$ are non-constant polynomials in $\mathbb{Q}[Y]$, it follows that $\Phi_p(Y + 1)$ is reducible in $\mathbb{Q}[Y]$. But

$$\Phi_p(Y + 1) = \frac{(Y + 1)^p - 1}{(Y + 1) - 1},$$

and by expanding the numerator it follows that

$$\Phi_p(Y + 1) = Y^{p-1} + pY^{p-2} + \binom{p}{2}Y^{p-3} + \cdots + \binom{p}{2}Y + p$$

(where $\binom{p}{i} = \frac{p!}{i!(p-i)!}$ is the binomial coefficient). Eisenstein's criterion shows that $\Phi_p(Y+1)$ is irreducible in $\mathbb{Q}[Y]$. Therefore, $\Phi_p(X)$ is irreducible in $\mathbb{Q}[X]$. $\qquad\square$

The importance of this theorem lies in the fact that it enables us to reduce to a standard form every rational expression in the $p$-th roots of unity. We let $\mu_p$ be the set of $p$-th roots of unity, as in §7.2, and let $\mathbb{Q}(\mu_p)$ denote the set of complex numbers that have a rational expression in these $p$-th roots of unity. Thus, letting

$$\mu_p = \{\rho_1, \ldots, \rho_p\},$$

we have

$$\mathbb{Q}(\mu_p) = \left\{ \frac{f(\rho_1, \ldots, \rho_p)}{g(\rho_1, \ldots, \rho_p)} \mid f, g \in \mathbb{Q}[X_1, \ldots, X_p] \text{ and } g(\rho_1, \ldots, \rho_p) \neq 0 \right\}.$$

This set is clearly a subfield of $\mathbb{C}$, since it is closed under sums, differences, products, and divisions by non-zero elements.

Recall from Proposition 7.9, p. 85 (and Corollary 7.12, p. 86) that if $\zeta$ is any $p$-th root of unity other than 1, then every $p$-th root of unity is a power of $\zeta$ with an exponent between 0 and $p - 1$. Thus

$$\mu_p = \{1, \zeta, \zeta^2, \ldots, \zeta^{p-1}\}$$

and the complex numbers that are denoted by $\rho_1$, ..., $\rho_p$ above are powers of $\zeta$. Therefore, every rational expression in $\rho_1$, ..., $\rho_p$ is a rational expression in $\zeta$, and conversely, so that

$$\mathbb{Q}(\mu_p) = \left\{ \frac{f(\zeta)}{g(\zeta)} \mid f, g \in \mathbb{Q}[X] \text{ and } g(\zeta) \neq 0 \right\}.$$

**Theorem 12.13.** *Every element in* $\mathbb{Q}(\mu_p)$ *can be expressed in one and only one way as a linear combination with rational coefficients of the p-th roots of unity other than* 1,

$$a_1 \zeta + a_2 \zeta^2 + \cdots + a_{p-1} \zeta^{p-1} \qquad (a_i \in \mathbb{Q}).$$

Since some of the arguments used in the proof will be useful in various contexts, we quote them in full generality.

**Lemma 12.14.** *Let* $P$ *and* $Q$ *be polynomials with coefficients in some field* $F$, *and assume* $P$ *is irreducible in* $F[X]$. *If* $P$ *and* $Q$ *have a common root in some field* $K$ *containing* $F$, *then* $P$ *divides* $Q$.

(Compare Gauss [31, Art. 346].)

**Proof.** If $P$ does not divide $Q$, then $P$ and $Q$ are relatively prime, because $P$ is irreducible. Corollary 5.4 (p. 47) then shows that there exist polynomials $U$, $V$ in $F[X]$ such that

$$PU + QV = 1.$$

Substituting in this equation the common root $u$ of $P$ and $Q$ for the indeterminate $X$, we obtain

$$P(u)U(u) + Q(u)V(u) = 1 \qquad \text{in } K.$$

Since $P(u) = Q(u) = 0$, this equation yields $0 = 1$ in $K$, a contradiction. Therefore, $P$ divides $Q$. $\qquad\qquad\qquad\qquad\qquad\qquad\qquad\qquad\qquad\qquad\qquad\square$

Now, consider a field $F$ and an element $u$ in a field $K$ containing $F$. We generalize the above definition of $\mathbb{Q}(\mu_p)$, letting $F(u)$ denote the set of elements in $K$ that have a rational expression in $u$ with coefficients in $F$,

$$F(u) = \left\{ \frac{f(u)}{g(u)} \in K \mid f, g \in F[X] \text{ and } g(u) \neq 0 \right\}. \qquad (12.5)$$

The set $F(u)$ is obviously closed under sums, differences, products, and divisions by non-zero elements, and is therefore a subfield of $K$.

If $u$ is a root of some non-zero polynomial $P \in F[X]$, then it is a root of some monic irreducible factor of $P$. Indeed, if $P = cP_1 \ldots P_r$ is the decomposition of $P$ into prime factors according to Theorem 5.8 (p. 49), then the equation $P(u) = 0$ implies that $P_i(u) = 0$ for at least one index $i$. Therefore, substituting for $P$ a suitable monic irreducible factor if necessary, we may assume $P$ itself is irreducible and monic.

**Proposition 12.15.** *If* $u \in K$ *is a root of an irreducible polynomial* $P \in F[X]$ *of degree* $d$, *then every element in* $F(u)$ *can be uniquely written in the form*

$$a_0 + a_1 u + a_2 u^2 + \cdots + a_{d-1} u^{d-1} \qquad \text{with } a_i \in F.$$

**Proof.** Let $f(u)/g(u)$ be an arbitrary element in $F(u)$. Since $g(u) \neq 0$, the polynomial $g$ is not divisible by $P$, and is therefore relatively prime to $P$, because $P$ is irreducible. By Corollary 5.4 (p. 47), there exist polynomials $h$ and $U$ such that

$$gh + PU = 1 \qquad \text{in } F[X].$$

Substituting $u$ for the indeterminate $X$ in this equation, and taking into account the fact that $P(u) = 0$, we get

$$g(u)h(u) = 1 \qquad \text{in } K.$$

This shows that $f(u)/g(u)$ can be written as a *polynomial* expression in $u$,

$$\frac{f(u)}{g(u)} = f(u)h(u) \qquad \text{in } K.$$

Now, let $R$ be the remainder of the division of $fh$ by $P$,

$$fh = PQ + R \qquad \text{in } F[X], \quad \text{with } \deg R \leq d - 1.$$

Since $P(u) = 0$, it follows that

$$f(u)h(u) = R(u) \qquad \text{in } K,$$

and since $R \in F[X]$ is a polynomial of degree at most $d - 1$, we have converted an arbitrary rational expression $f(u)/g(u) \in F(u)$ into a polynomial expression of the type

$$a_0 + a_1 u + \cdots + a_{d-1}u^{d-1}$$

with $a_i \in F$.

To prove the uniqueness of this expression, assume

$$a_0 + a_1 u + \cdots + a_{d-1}u^{d-1} = b_0 + b_1 u + \cdots + b_{d-1}u^{d-1}$$

for some $a_0, \ldots, a_{d-1}, b_0, \ldots, b_{d-1} \in F$. Collecting all the terms on one side, we see that $u$ is then a root of the polynomial

$$V(X) = (a_0 - b_0) + (a_1 - b_1)X + \cdots + (a_{d-1} - b_{d-1})X^{d-1} \in F[X].$$

From Lemma 12.14, it follows that $P$ divides $V$, but since $\deg V \leq d - 1$, this is impossible unless $V = 0$. Therefore,

$$a_0 - b_0 = a_1 - b_1 = \cdots = a_{d-1} - b_{d-1} = 0. \qquad \square$$

**Remark 12.16.** There is only one monic irreducible polynomial $P \in F[X]$ that has $u$ as a root. To see this, assume $Q \in F[X]$ is another polynomial with the same properties. Lemma 12.14 shows that $P$ divides $Q$ and, reversing the roles of $P$ and $Q$, that $Q$ divides $P$ also. Since $P$ and $Q$ are both monic, it follows that $P = Q$. Moreover, among the non-zero polynomials in $F[X]$ that have $u$ as a root, $P$ is the polynomial of least degree since it divides all the others. Therefore, this (unique) monic irreducible polynomial $P \in F[X]$ that has $u$ as a root is called the *minimum polynomial of $u$ over $F$.*

**Proof of Theorem 12.13.** We already observed above that $\mathbb{Q}(\mu_p) = \mathbb{Q}(\zeta)$. Since $\zeta$ is a root of $\Phi_p$, which is irreducible by Theorem 12.10 and has degree $p - 1$, it follows from Proposition 12.15 that every element $a \in \mathbb{Q}(\mu_p)$ can be uniquely expressed in the form

$$a = a_0 + a_1\zeta + a_2\zeta^2 + \cdots + a_{p-2}\zeta^{p-2} \tag{12.6}$$

for some $a_i \in \mathbb{Q}$. To obtain the required form, it now suffices to use the fact that

$$\Phi_p(\zeta) = 1 + \zeta + \zeta^2 + \cdots + \zeta^{p-1} = 0. \tag{12.7}$$

This equation yields

$$a_0 = -a_0(\zeta + \zeta^2 + \cdots + \zeta^{p-1}),$$

hence, substituting in (12.6) above,

$$a = (a_1 - a_0)\zeta + (a_2 - a_0)\zeta^2 + \cdots + (a_{p-2} - a_0)\zeta^{p-2} + (-a_0)\zeta^{p-1}.$$

The uniqueness of this expression follows from the uniqueness of the expression (12.6): if

$$a_1\zeta + \cdots + a_{p-1}\zeta^{p-1} = b_1\zeta + \cdots + b_{p-1}\zeta^{p-1},$$

then, using (12.7) to eliminate $\zeta^{p-1}$, we obtain

$$-a_{p-1} + (a_1 - a_{p-1})\zeta + \cdots + (a_{p-2} - a_{p-1})\zeta^{p-2}$$
$$= -b_{p-1} + (b_1 - b_{p-1})\zeta + \cdots + (b_{p-2} - b_{p-1})\zeta^{p-2}.$$

By the uniqueness of the expression (12.6), it follows that the coefficients of $1, \zeta, \zeta^2, \ldots, \zeta^{p-2}$ on each side are equal, and therefore

$$a_{p-1} = b_{p-1}, \quad a_1 = b_1, \quad \ldots, \quad a_{p-2} = b_{p-2}. \qquad \square$$

**Remark 12.17.** For later use, we note the expression of rational numbers in the form indicated by Theorem 12.13: it easily follows from (12.7) that any element $a \in \mathbb{Q}$ is expressed as

$$a = (-a)\zeta + (-a)\zeta^2 + \cdots + (-a)\zeta^{p-1}.$$

## 12.3 The periods of cyclotomic equations

Let $p$ be a prime number and let $\zeta \in \mathbb{C}$ be a primitive $p$-th root of unity (i.e., a $p$-th root of unity other than 1, see Corollary 7.12, p. 86). If $m$ and $n$ are integers such that $m \equiv n \bmod p$, then $\zeta^m = \zeta^n$.

Now, let $g$ be a primitive root of $p$. Since the integers $g^0$, $g^1$, $g^2$, ..., $g^{p-2}$ are congruent modulo $p$ to 1, 2, 3, ..., $p-1$ (in some order), it follows that the complex numbers

$$\zeta^{g^0}, \ \zeta^{g^1}, \ \zeta^{g^2}, \ \ldots, \ \zeta^{g^{p-2}}$$

are the same as

$$\zeta, \ \zeta^2, \ \zeta^3, \ \ldots, \ \zeta^{p-1},$$

in some order. Therefore, letting $\zeta_i = \zeta^{g^i}$ for $i = 0, \ldots, p-2$ to simplify the notation, we may write the set of $p$-th roots of unity as

$$\mu_p = \{1, \zeta_0, \zeta_1, \ldots, \zeta_{p-2}\}.$$

It turns out that this new ordering $\zeta_0$, $\zeta_1$, ..., $\zeta_{n-2}$ of the primitive $p$-th roots of unity is exactly the ordering for which Vandermonde's arguments in §11.2 hold. More precisely, the point is that the cyclic permutation

$$\sigma: \ \zeta_0 \mapsto \zeta_1 \mapsto \cdots \mapsto \zeta_{p-2} \mapsto \zeta_0,$$

extended to $\mu_p$ by setting $\sigma(1) = 1$, preserves the relations among the $p$-th roots of unity. This is the essence of the following proposition:

**Proposition 12.18.** *For $\rho$, $\omega \in \mu_p$,*

$$\sigma(\rho\omega) = \sigma(\rho)\sigma(\omega).$$

**Proof.** For $i = 0, \ldots, p-3$, we have $\sigma(\zeta_i) = \zeta_{i+1}$, hence, by definition of $\zeta_i$ and $\zeta_{i+1}$,

$$\sigma(\zeta_i) = \zeta_i^g.$$

This equation also holds for $i = p - 2$, since $\zeta_{p-2}^g = \zeta^{g^{p-1}}$ and $g^{p-1} \equiv g^0 \bmod p$ by Fermat's theorem (Theorem 12.5). It also holds trivially with 1 instead of $\zeta_i$; thus,

$$\sigma(\rho) = \rho^g \qquad \text{for all } \rho \in \mu_p.$$

Since $\rho\omega \in \mu_p$ for $\rho$, $\omega \in \mu_p$, we have

$$\sigma(\rho\omega) = (\rho\omega)^g = \sigma(\rho)\sigma(\omega) \qquad \text{for } \rho, \ \omega \in \mu_p. \qquad \square$$

For example, if $p = 11$, then we can choose $g = 2$, as we observed in the example after the statement of Theorem 12.1. The corresponding ordering of the primitive 11-th roots of unity is the following:

$$\zeta_0 = \zeta, \quad \zeta_1 = \zeta^2, \quad \zeta_2 = \zeta^4, \quad \zeta_3 = \zeta^8, \quad \zeta_4 = \zeta^5,$$
$$\zeta_5 = \zeta^{10}, \quad \zeta_6 = \zeta^9, \quad \zeta_7 = \zeta^7, \quad \zeta_8 = \zeta^3, \quad \zeta_9 = \zeta^6$$

where we can choose for instance $\zeta = \cos \frac{2\pi}{11} + i \sin \frac{2\pi}{11}$.

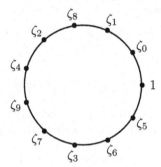

The values of $-2 \cos \frac{2k\pi}{11}$, which were denoted by $a$, $b$, $c$, $d$, $e$ in §11.2, are thus given by

$$a = -2 \cos \frac{2\pi}{11} = -(\zeta_0 + \zeta_5),$$

$$b = -2 \cos \frac{4\pi}{11} = -(\zeta_1 + \zeta_6),$$

$$c = -2 \cos \frac{6\pi}{11} = -(\zeta_3 + \zeta_8),$$

$$d = -2 \cos \frac{8\pi}{11} = -(\zeta_2 + \zeta_7),$$

$$e = -2 \cos \frac{10\pi}{11} = -(\zeta_4 + \zeta_9).$$

The permutation $\sigma \colon \zeta_0 \mapsto \zeta_1 \mapsto \cdots \mapsto \zeta_9 \mapsto \zeta_0$ induces the following permutation of $a, \ldots, e$:

$$a \mapsto b \mapsto d \mapsto c \mapsto e \mapsto a.$$

This is the permutation which played a crucial role in §11.2.

Thus, mimicking the arguments in our comments to Vandermonde's solution of $\Phi_{11}(X) = 0$, it is not hard to see that the cyclotomic equation $\Phi_p(X) = 0$ is solvable by radicals for every prime $p$. However, this result appears as secondary in Gauss' investigations and we postpone its proof to the next section. Gauss' primary concern is to decompose the solution of cyclotomic equations into the simplest steps as possible.

This decomposition is achieved as follows: for any two positive integers $e$, $f$ such that

$$ef = p - 1,$$

Gauss defines $e$ complex numbers, which he calls the *periods of $f$ terms*:

$$\eta_0 = \zeta_0 + \zeta_e + \zeta_{2e} + \cdots + \zeta_{e(f-1)},$$
$$\eta_1 = \zeta_1 + \zeta_{e+1} + \zeta_{2e+1} + \cdots + \zeta_{e(f-1)+1},$$
$$\eta_2 = \zeta_2 + \zeta_{e+2} + \zeta_{2e+2} + \cdots + \zeta_{e(f-1)+2},$$
$$\vdots$$
$$\eta_{e-1} = \zeta_{e-1} + \zeta_{2e-1} + \zeta_{3e-1} + \cdots + \zeta_{p-2}.$$

In particular, the periods of 1 term are the roots $\zeta_0$, $\zeta_1$, ..., $\zeta_{p-2}$, and the (unique) period of $p - 1$ terms is the sum of all the $p$-th roots of unity other than 1 or, in other words, the sum of all the roots of $\Phi_p$. This period is therefore rational: it is the opposite of the first coefficient of $\Phi_p$,

$$\zeta_0 + \zeta_1 + \cdots + \zeta_{p-2} = -1.$$

As a further example, for $p \geq 3$, the periods of two terms can be seen to be the values of $2\cos\frac{2k\pi}{p}$ for $k = 1$, ..., $\frac{p-1}{2}$. This was already shown above for $p = 11$, but can be proved in general by considering the form of these periods,

$$\eta_j = \zeta_j + \zeta_{j+\frac{p-1}{2}}.$$

By definition of the indexing, we have

$$\zeta_{j+\frac{p-1}{2}} = (\zeta_j)^{g^{(p-1)/2}}$$

and since $g^{(p-1)/2} \equiv -1 \bmod p$ by Remark 12.9(a), it follows that

$$\eta_j = \zeta_j + \zeta_j^{-1}.$$

Therefore, the $\frac{p-1}{2}$ periods of two terms are the roots of the equation of degree $\frac{p-1}{2}$ obtained from $\Phi_p(X) = 0$ by setting $Y = X + X^{-1}$. This proves the claim, in view of Remark 7.6 (p. 84).

As Gauss shows, the periods of $f$ terms thus defined have the following remarkable properties.

**Property 12.19.** *Any period of $f$ terms can be determined rationally from any other period of $f$ terms.*

**Property 12.20.** *If $f$ and $g$ are two divisors of $p - 1$ and if $f$ divides $g$, then any period of $f$ terms is a root of an equation of degree $g/f$ whose coefficients are rational expressions of a period of $g$ terms.*

These properties will be proved below, see Corollary 12.24 and Corollary 12.26.

Thus, the periods can be used to provide remarkable examples of the step-by-step solution of equations as envisioned by Lagrange. Fix a sequence of integers:

$$f_0 = p - 1, \quad f_1, \quad \ldots, \quad f_{r-1}, \quad f_r = 1$$

such that $f_i$ divides $f_{i-1}$ for $i = 1, \ldots, r$, and define $V_i$ to be a period of $f_i$ terms (arbitrarily chosen) for $i = 0, \ldots, r$. Then $V_0$ is rational, and for $i = 1, \ldots, r$ the complex number $V_i$ can be determined by solving an equation of degree $f_{i-1}/f_i$ whose coefficients are rational expressions in $V_{i-1}$. Since $V_r$ is a period of 1 term, this process eventually yields a primitive $p$-th root of unity. The other $p$-th roots of unity are then readily obtained as powers of this one.

The choice of $V_i$ among the periods of $f_i$ terms does not affect essentially the solution, because Property 12.19 shows that the periods of $f_i$ terms are rational expressions of each other.

Of course, it is not clear *a priori* that the equation used to determine $V_i$ from $V_{i-1}$ is solvable by radicals, since $f_{i-1}/f_i$ might exceed 5, but Gauss further proves that these equations are indeed solvable by radicals for any value of $f_{i-1}/f_i$, including $p - 1$. (This case occurs if $r = 1$.) If one wants to deal with equations of the least possible degrees, one can choose the sequence $f_0, f_1, \ldots, f_r$ in such a way that the successive quotients $f_{i-1}/f_i$ be the prime factors of $p - 1$, but this is in no way compulsory.

Take for instance $p = 37$ and look at the lattice of divisors of $p - 1 = 2^2 \cdot 3^2$:

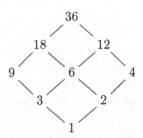

(In this diagram, a straight line indicates a relation of divisibility.) To every path going down from 36 to 1 (without going up at any step) corresponds a pattern of solution of $\Phi_{37}(X) = 0$ by successive equations, whose degrees are the successive quotients. For instance, if we choose the path 36, 12, 6, 1, then we first determine a period of 12 terms by an equation of degree $36/12 = 3$, next a period of 6 terms by an equation of degree $12/6 = 2$ and finally a period of 1 term, i.e., a primitive 37-th root of unity, by an equation of degree 6.

Instead of solving directly this last equation, one could determine a period of 3 terms by an equation of degree $6/3 = 2$ and a period of 1 term by an equation of degree 3. This amounts to refine the proposed path into 36, 12, 6, 3, 1.

For $p = 17$, the lattice of divisors of $p - 1 = 2^4$ is much simpler, it is

Thus, a primitive 17-th root of unity can be determined by solving successively four quadratic equations. This is the key fact that leads to the construction of the regular polygon with 17 sides by ruler and compass (see the appendix).

We now turn to the proof of Properties 12.19 and 12.20, which we adapt from Gauss' own arguments with the added thrust of some elementary linear algebra.[1] First, we define a map from the field $\mathbb{Q}(\mu_p)$ onto itself, extending by linearity the map $\sigma$ defined on $\mu_p$. (See the definition of $\sigma$ before Proposition 12.18.) We thus set

$$\sigma(a_0\zeta_0 + \cdots + a_{p-2}\zeta_{p-2}) = a_0\sigma(\zeta_0) + \cdots + a_{p-2}\sigma(\zeta_{p-2}),$$

i.e.,

$$\sigma(a_0\zeta_0 + a_1\zeta_1 + \cdots + a_{p-3}\zeta_{p-3} + a_{p-2}\zeta_{p-2})$$
$$= a_0\zeta_1 + a_1\zeta_2 + \cdots + a_{p-3}\zeta_{p-2} + a_{p-2}\zeta_0.$$

Theorem 12.13 shows that this definition is not ambiguous, and that it is sufficient to define $\sigma$ on the whole of $\mathbb{Q}(\mu_p)$.

**Proposition 12.21.** *The map $\sigma$ is a field automorphism of $\mathbb{Q}(\mu_p)$, which leaves every element of $\mathbb{Q}$ invariant.*

---

[1] Gauss' original arguments also use linear algebra, but expressed in an elementary way via systems of linear equations, see [31, Art. 346].

**Proof.** That $\sigma$ is bijective and that

$$\sigma(ua + vb) = u\sigma(a) + v\sigma(b)$$

for $a$, $b \in \mathbb{Q}(\mu_p)$ and $u$, $v \in \mathbb{Q}$ (i.e., that $\sigma$ is $\mathbb{Q}$-linear) readily follow from the definition of $\sigma$. Moreover, since by Remark 12.17 the rational numbers $a \in \mathbb{Q}$ are written as

$$a = (-a)\zeta_0 + (-a)\zeta_1 + \cdots + (-a)\zeta_{p-2},$$

the definition also shows that every rational number is invariant under $\sigma$.

Thus, it only remains to prove

$$\sigma(ab) = \sigma(a)\sigma(b) \qquad \text{for all } a, \, b \in \mathbb{Q}(\mu_p).$$

This was already proved in Proposition 12.18 in the particular case where $a$, $b \in \mu_p$. From this case, the general case can be derived by linearity: let

$$a = \sum_{i=0}^{p-2} a_i\zeta_i \quad \text{and} \quad b = \sum_{j=0}^{p-2} b_j\zeta_j$$

with $a_i$, $b_j \in \mathbb{Q}$ for all $i$, $j$. Then $ab = \sum_{i,j=0}^{p-2} a_i b_j \zeta_i \zeta_j$ hence, since $\sigma$ is $\mathbb{Q}$-linear,

$$\sigma(ab) = \sum_{i,j=0}^{p-2} a_i b_j \sigma(\zeta_i \zeta_j).$$

On the other hand, we have

$$\sigma(a)\sigma(b) = \sum_{i,j=0}^{p-2} a_i b_j \sigma(\zeta_i)\sigma(\zeta_j).$$

Therefore, Proposition 12.18 shows that $\sigma(ab) = \sigma(a)\sigma(b)$. $\qquad \square$

**Remark.** The irreducibility of $\Phi_p$ was used above in an essential, but rather implicit, way. Indeed, that the map $\sigma$ is well-defined on $\mathbb{Q}(\mu_p)$ results from the fact that the expression $a_0\zeta_0 + \cdots + a_{p-2}\zeta_{p-2}$ for the elements in $\mathbb{Q}(\mu_p)$ is unique; the proof of this fact, in Theorem 12.13, ultimately relies on the irreducibility of $\Phi_p$.

Now, let $e$ and $f$ be (positive) integers such that

$$ef = p - 1.$$

Let $K_f$ denote the set of elements in $\mathbb{Q}(\mu_p)$ that are invariant under $\sigma^e$. Since $\sigma$, hence also $\sigma^e$, is a field automorphism of $\mathbb{Q}(\mu_p)$ which is the identity on $\mathbb{Q}$, the set $K_f$ is clearly closed under sums, differences, products, and divisions by non-zero elements, and contains $\mathbb{Q}$. In other words, $K_f$ is a subfield of $\mathbb{Q}(\mu_p)$ containing $\mathbb{Q}$. Using the standard form of the elements in $\mathbb{Q}(\mu_p)$, a standard form for the elements of $K_f$ is easily found, as the next proposition shows.

**Proposition 12.22.** *Every element in $K_f$ can be written in a unique way as a linear combination with rational coefficients of the $e$ periods of $f$ terms.*

**Proof.** Let $a$ be an arbitrary element in $\mathbb{Q}(\mu_p)$, which we write as follows:

$$a = a_0\zeta_0 + a_1\zeta_1 + \cdots + a_{e-1}\zeta_{e-1}$$
$$+ a_e\zeta_e + a_{e+1}\zeta_{e+1} + \cdots + a_{2e-1}\zeta_{2e-1}$$
$$+ \cdots$$
$$+ a_{e(f-1)}\zeta_{e(f-1)} + a_{e(f-1)+1}\zeta_{e(f-1)+1} + \cdots + a_{p-2}\zeta_{p-2}.$$

Then, by definition of $\sigma$, we have

$$\sigma^e(a) = a_0\zeta_e + a_1\zeta_{e+1} + \cdots + a_{e-1}\zeta_{2e-1}$$
$$+ a_e\zeta_{2e} + a_{e+1}\zeta_{2e+1} + \cdots + a_{2e-1}\zeta_{3e-1}$$
$$+ \cdots$$
$$+ a_{e(f-1)}\zeta_0 + a_{e(f-1)+1}\zeta_1 + \cdots + a_{p-2}\zeta_{e-1}.$$

If $\sigma^e(a) = a$, then, by Theorem 12.13, the coefficients of $\zeta_i$ in the two expressions above are the same, for $i = 0, \ldots, p-2$, hence

$$\begin{aligned}
a_0 &= a_e = a_{2e} = \cdots = a_{e(f-1)}, \\
a_1 &= a_{e+1} = a_{2e+1} = \cdots = a_{e(f-1)+1}, \\
&\quad\vdots \\
a_{e-1} &= a_{2e-1} = a_{3e-1} = \cdots = a_{p-2}.
\end{aligned}$$

Therefore, every element $a \in K_f$ can be written as

$$a = a_0(\zeta_0 + \zeta_e + \cdots + \zeta_{e(f-1)})$$
$$+ a_1(\zeta_1 + \zeta_{e+1} + \cdots + \zeta_{e(f-1)+1})$$
$$+ \cdots$$
$$+ a_{e-1}(\zeta_{e-1} + \zeta_{2e-1} + \cdots + \zeta_{p-2}).$$

This proves that $a$ is a linear combination of the periods, since the expressions between brackets are the periods of $f$ terms.

The uniqueness of this expression of $a$ readily follows from Theorem 12.13, which shows that every element in $\mathbb{Q}(\mu_p)$ can be written in only one way as a linear combination of $\zeta_0, \ldots, \zeta_{p-2}$. $\quad\square$

**Proposition 12.23.** *Let $\eta$ be a period of $f$ terms. Every element in $K_f$ can be written as*

$$a_0 + a_1\eta + a_2\eta^2 + \cdots + a_{e-1}\eta^{e-1}$$

*for some $a_0, \ldots, a_{e-1} \in \mathbb{Q}$.*

**Proof.** Since $K_f$ is a field containing $\mathbb{Q}$, it can be considered as a vector space over $\mathbb{Q}$ in a natural way: the vector space operations are induced by the operations in the field. To prove the proposition, it obviously suffices to show that $1, \eta, \ldots, \eta^{e-1}$ is a basis of $K_f$ over $\mathbb{Q}$. In fact, it even suffices to prove that $1, \eta, \ldots, \eta^{e-1}$ are linearly independent over $\mathbb{Q}$, since Proposition 12.22 shows that the $e$ periods of $f$ terms form a basis of $K_f$ over $\mathbb{Q}$, hence that $\dim_{\mathbb{Q}} K_f = e$.

In order to prove this linear independence, suppose

$$a_0 + a_1\eta + \cdots + a_{e-1}\eta^{e-1} = 0 \tag{12.8}$$

for some rational numbers $a_0, \ldots, a_{e-1}$. Then $\eta$ is a root of the polynomial

$$P(X) = a_0 + a_1 X + \cdots + a_{e-1} X^{e-1}.$$

Applying $\sigma$, next $\sigma^2$, $\sigma^3$ and so on until $\sigma^{e-1}$ to each side of (12.8), and taking into account the fact that the coefficients $a_i$ are invariant under $\sigma$, we observe that $\sigma(\eta)$, $\sigma^2(\eta)$, $\ldots$, $\sigma^{e-1}(\eta)$ also are roots of $P$. Now, $\eta$, $\sigma(\eta)$, $\ldots$, $\sigma^{e-1}(\eta)$ are the $e$ periods of $f$ terms, which are pairwise distinct by Proposition 12.22. Since the polynomial $P$ has degree at most $e - 1$, it cannot have as roots the $e$ periods of $f$ terms, unless it is the zero polynomial. Therefore,

$$a_0 = \cdots = a_{e-1} = 0,$$

and this proves the linear independence of $1, \eta, \ldots, \eta^{e-1}$. $\qquad\square$

**Corollary 12.24.** *If $\eta$ and $\eta'$ are periods of $f$ terms, then*

$$\eta' = a_0 + a_1\eta + \cdots + a_{e-1}\eta^{e-1}$$

*for some rational numbers $a_0, \ldots, a_{e-1}$.*

**Proof.** This readily follows from Proposition 12.23, since $\eta' \in K_f$. $\qquad\square$

This corollary proves Property 12.19 of the periods. In order to prove Property 12.20, we now introduce another pair of integers $g$, $h$ such that

$$gh = p - 1,$$

and assume that $f$ divides $g$. Then, denoting $k = g/f = e/h$, we have

$$\sigma^e = (\sigma^h)^k.$$

Therefore, every element invariant under $\sigma^h$ is also invariant under $\sigma^e$, which means that

$$K_g \subset K_f.$$

**Proposition 12.25.** *Let $f$ and $g$ be divisors of $p - 1$. If $f$ divides $g$, then every element in $K_f$ is a root of a polynomial of degree $g/f$ with coefficients in $K_g$.*

**Proof.** For $a \in K_f$, we consider the polynomial

$$P(X) = (X - a)\big(X - \sigma^h(a)\big)\big(X - \sigma^{2h}(a)\big) \ldots \big(X - \sigma^{h(k-1)}(a)\big).$$

This polynomial has degree $k = g/f$, and its coefficients are the elementary symmetric polynomials in $a$, $\sigma^h(a)$, $\sigma^{2h}(a)$, $\ldots$, $\sigma^{h(k-1)}(a)$. Since

$$\sigma^h(\sigma^{h(k-1)}(a)) = \sigma^e(a) = a,$$

the map $\sigma^h$ permutes $a$, $\sigma^h(a)$, $\ldots$, $\sigma^{h(k-1)}(a)$ among themselves and leaves therefore the coefficients of $P$ invariant. This shows that the coefficients of $P$ are in $K_g$. The polynomial $P$ thus satisfies the required properties. $\square$

The proof of Property 12.20 can now be completed.

**Corollary 12.26.** *Let $f$ and $g$ be divisors of $p-1$ and let $\eta$ and $\xi$ be periods of $f$ and $g$ terms respectively. If $f$ divides $g$, then $\eta$ is a root of a polynomial of degree $g/f$ whose coefficients are rational expressions of $\xi$.*

**Proof.** Since $\xi \in K_g$ and $\eta \in K_f$, this corollary readily follows from Propositions 12.25 and 12.23. $\square$

It is instructive to note, with a view towards the modern framework of Galois theory, that the subfields $K_f$ form a lattice of subfields of $\mathbb{Q}(\mu_p)$, which is anti-isomorphic to the lattice of divisors of $p - 1$, since $K_g \subset K_f$ if and only if $f$ divides $g$. Thus, if for instance $p = 37$, the periods define the following lattice of subfields of $\mathbb{Q}(\mu_{37})$:

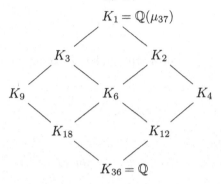

(A straight line indicates a relation of inclusion.)

## 12.4  Solvability by radicals

After his careful analysis of the periods of cyclotomic equations and their properties, Gauss shows in Art. 359–360 of "Disquisitiones Arithmeticae" that the equations by which the periods are determined can be solved by radicals. His exposition in this part is more sketchy and slurs at some point over a difficulty which will be pinpointed below.

We use the notation of the preceding section. In particular, we let $e$, $f$ and $g$, $h$ be two pairs of integers such that

$$ef = gh = p - 1.$$

We assume that $f$ divides $g$ and set

$$k = \frac{g}{f} = \frac{e}{h}.$$

We write $\eta_0, \ldots, \eta_{e-1}$ (resp. $\xi_0, \ldots, \xi_{h-1}$) for the periods of $f$ (resp. $g$) terms,

$$\eta_i = \zeta_i + \zeta_{e+i} + \zeta_{2e+i} + \cdots + \zeta_{e(f-1)+i},$$
$$\xi_j = \zeta_j + \zeta_{h+j} + \zeta_{2h+j} + \cdots + \zeta_{h(g-1)+j}.$$

In Corollary 12.26 we have seen that, when the periods $\xi_0, \ldots, \xi_{h-1}$ are considered as known, then any period $\eta_i$ can be determined by an equation of degree $g/f$. Our goal in this section is to show that this equation is solvable by radicals.

Consider for instance the equation that yields $\eta_0$. (The arguments for the other periods is exactly the same, but the notation is more complicated.)

We let $P(X) = 0$ denote this equation of degree $k$. Since the coefficients of $P$ are in $K_g$, they are invariant under $\sigma^h$; hence, by repeatedly applying $\sigma^h$ to each side of the equation $P(\eta_0) = 0$ we find

$$P\big(\sigma^h(\eta_0)\big) = 0, \quad P\big(\sigma^{2h}(\eta_0)\big) = 0, \quad \ldots, \quad P\big(\sigma^{h(k-1)}(\eta_0)\big) = 0.$$

Therefore, the roots of $P$ are $\eta_0$ and its images under $\sigma^h$, $\sigma^{2h}$, $\ldots$, $\sigma^{h(k-1)}$, which are $\eta_h$, $\eta_{2h}$, $\ldots$, $\eta_{h(k-1)}$.

In order to prove that $P(X) = 0$ is solvable by radicals, it suffices, after Lagrange's formula (Proposition 10.2, p. 130), to show that the $k$-th power of the Lagrange resolvent

$$t(\omega) = \eta_0 + \omega\eta_h + \omega^2\eta_{2h} + \cdots + \omega^{k-1}\eta_{h(k-1)}$$

(where $\omega$ is a $k$-th root of unity) can be calculated from the periods of $g$ terms.

**Proposition 12.27.** *For every $k$-th root of unity $\omega$, the complex number $t(\omega)^k$ has a rational expression in terms of $\omega$ and of the periods of $g$ terms.*

**Proof.** First, we observe that, by Proposition 12.22, the product of any two periods of $f$ terms can be expressed as a linear combination of the periods of $f$ terms. We thus obtain relations among the periods, which can be used to reduce to 1 the degree of any polynomial expression in the periods. In particular,

$$\begin{aligned}
t(\omega)^k &= \big(\eta_0 + \omega\eta_h + \cdots + \omega^{k-1}\eta_{h(k-1)}\big)^k \\
&= a_0\eta_0 + \cdots + a_{h-1}\eta_{h-1} \\
&\quad + a_h\eta_h + \cdots + a_{2h-1}\eta_{2h-1} \\
&\quad + \cdots \\
&\quad + a_{h(k-1)}\eta_{h(k-1)} + \cdots + a_{e-1}\eta_{e-1}
\end{aligned} \tag{12.9}$$

where the coefficients $a_0$, $\ldots$, $a_{e-1}$ are rational (in fact polynomial) expressions in $\omega$ over $\mathbb{Q}$.

Since the relations among the periods $\eta_0$, $\ldots$, $\eta_{e-1}$ are preserved under $\sigma^h$, by Proposition 12.21, we can replace $\eta_0$ by $\sigma^h(\eta_0) = \eta_h$, $\eta_1$ by $\sigma^h(\eta_1) = \eta_{h+1}$, etc. in the above calculation of $t(\omega)^k$. We thus find

$$\begin{aligned}
\big(\eta_h + \omega\eta_{2h} + \cdots + \omega^{k-1}\eta_0\big)^k &= a_0\eta_h + \cdots + a_{h-1}\eta_{2h-1} \\
&\quad + a_h\eta_{2h} + \cdots + a_{2h-1}\eta_{3h-1} \\
&\quad + \cdots \\
&\quad + a_{h(k-1)}\eta_0 + \cdots + a_{e-1}\eta_{h-1}.
\end{aligned} \tag{12.10}$$

This yields an expression of $(\sigma^h(t(\omega)))^k$. However, since

$$\sigma^h(t(\omega)) = \omega^{-1}t(\omega),$$

we have

$$(\sigma^h(t(\omega)))^k = t(\omega)^k,$$

so that (12.9) and (12.10) are two expressions of $t(\omega)^k$. Replacing in the initial calculation of $t(\omega)^k$ the period $\eta_i$ by $\sigma^{2h}(\eta_i)$, next by $\sigma^{3h}(\eta_i)$, ..., $\sigma^{h(k-1)}(\eta_i)$ (for $i = 0, \ldots, e - 1$), we still find $k - 2$ other expressions of $t(\omega)^k$. Inspection shows that the coefficients of a given period $\eta_i$ in these various expressions are $a_i, a_{i+h}, a_{i+2h}, \ldots, a_{i+h(k-1)}$. Therefore, if we sum up all these expressions, we obtain

$$\begin{aligned}
kt(\omega)^k &= (a_0 + \cdots + a_{h(k-1)})(\eta_0 + \cdots + \eta_{h(k-1)}) \\
&\quad + (a_1 + \cdots + a_{h(k-1)+1})(\eta_1 + \cdots + \eta_{h(k-1)+1}) \\
&\quad + \cdots \\
&\quad + (a_{h-1} + \cdots + a_{e-1})(\eta_{h-1} + \cdots + \eta_{e-1}).
\end{aligned}$$

Since $\eta_i + \eta_{h+i} + \cdots + \eta_{h(k-1)+i} = \xi_i$ for $i = 0, \ldots, h - 1$, it follows that $t(\omega)^k$ is rationally expressed in terms of $\omega$ and $\xi_0, \ldots, \xi_{h-1}$, in the form

$$t(\omega)^k = \frac{1}{k}\big((a_0 + \cdots + a_{h(k-1)})\xi_0 + \cdots + (a_{h-1} + \cdots + a_{e-1})\xi_{h-1}\big). \quad \Box$$

**Remark 12.28.** The above proof is quite similar to that of Gauss, but the final arguments are different. Gauss argues as follows: after observing that the right-hand sides of (12.9) and (12.10) are equal, since both are expressions of $t(\omega)^k$, he draws the conclusion that the coefficients of any given period are the same in both expressions, hence

$$\begin{aligned}
a_0 &= a_h = a_{2h} = \cdots = a_{h(k-1)}, \\
a_1 &= a_{h+1} = a_{2h+1} = \cdots = a_{h(k-1)+1}, \\
&\qquad\qquad \vdots \\
a_{h-1} &= a_{2h-1} = a_{3h-1} = \cdots = a_{e-1}.
\end{aligned}$$

These equations can be used to simplify (12.9) to

$$\begin{aligned}
t(\omega)^k &= a_0(\eta_0 + \eta_h + \cdots + \eta_{h(k-1)}) \\
&\quad + a_1(\eta_1 + \eta_{h+1} + \cdots + \eta_{h(k-1)+1}) \\
&\quad + \cdots \\
&\quad + a_{h-1}(\eta_{h-1} + \eta_{2h-1} + \cdots + \eta_{e-1}).
\end{aligned}$$

This completes the proof of the proposition, since the expressions between brackets in the right-hand side are the periods of $g$ terms.

However, the comparison of coefficients, which was also used in the proof of Proposition 12.22 above, is justified only insofar as the expression of an element as a linear combination of $\eta_0, \ldots, \eta_{e-1}$ (or, more generally, of $\zeta_0, \ldots, \zeta_{p-2}$) is known to be unique. This was shown in Theorem 12.13 for linear combinations with rational coefficients, which was sufficient to prove Proposition 12.22, but here the scalars are rational expressions of a $k$-th root of unity $\omega$, so new arguments are needed.

From the proof of Theorem 12.13, it is clear that the crucial fact on which this uniqueness property ultimately relies is the irreducibility of $\Phi_p$. Therefore, in order to justify Gauss' argument, we need to prove the irreducibility of $\Phi_p$ not only over the field $\mathbb{Q}$ of rational numbers, but over $\mathbb{Q}(\omega)$, where $\omega$ is a $k$-th root of unity for some integer $k$ dividing $p - 1$. This will be done in the next section, see Corollary 12.33.

To complete this section, we observe with Gauss [31, Art. 360] that the full generality of periods is not needed if we only aim to show that the roots of unity can be expressed by radicals.

**Corollary 12.29.** *For every integer $n$, the $n$-th roots of unity have expressions by radicals.*

**Proof.** We argue by induction on $n$. The corollary is trivial if $n = 1$ or 2, so we may assume that for every integer $k < n$ the $k$-th roots of unity are expressible by radicals. If $n$ is not prime, then Theorem 7.3 (p. 82) and the induction hypothesis readily show that the $n$-th roots of unity can be expressed by radicals. We may thus assume that $n$ is prime. We then order the $n$-th roots of unity other than 1 as at the beginning of §12.3 with the aid of a primitive root of $n$ and we consider the Lagrange resolvent

$$t(\omega) = \zeta_0 + \omega\zeta_1 + \cdots + \omega^{n-2}\zeta_{n-2}$$

(where $\omega$ is an $(n-1)$-th root of unity). By the induction hypothesis, $\omega$ can be expressed by radicals. Proposition 12.27 (with $k = g = n - 1$) shows that $t(\omega)^{n-1}$ has a rational expression in terms of $\omega$, hence an expression by radicals. Lagrange's formula (Proposition 10.2, p. 130) now yields expressions by radicals for the $n$-th roots of unity,

$$\zeta_i = \frac{1}{n-1}\left(\sum_\omega \omega^{-i} \sqrt[n-1]{t(\omega)^{n-1}}\right). \qquad \square$$

## 12.5   Irreducibility of the cyclotomic polynomials

The purpose of this section is to justify Gauss' argument (see Remark 12.28), by proving the irreducibility of the cyclotomic polynomial $\Phi_p$ over $\mathbb{Q}(\mu_k)$, when $p$ is a prime number and $k$ is an integer that is relatively prime to $p$. (This result will not be used in subsequent chapters.)

A proof of this result was first published by Kronecker in 1854. The proof we give is inspired by some ideas of Dedekind (see van der Waerden [77, §60], Weber [85, §174]). It holds in fact for any integer $n$ instead of $p$. Its essential step is to prove the irreducibility of $\Phi_n$ over $\mathbb{Q}$, which was first established for non-prime $n$ by Gauss in 1808 (see Bühler [11, p. 74]).

**Lemma 12.30.** *Let $f$ be a monic irreducible factor of $\Phi_n$ in $\mathbb{Q}[X]$ and let $p$ be a prime number that does not divide $n$. If $\omega \in \mathbb{C}$ is a root of $f$, then $\omega^p$ also is a root of $f$, so*

$$f(\omega) = 0 \Rightarrow f(\omega^p) = 0.$$

**Proof.** Assume on the contrary that $f(\omega) = 0$ but $f(\omega^p) \neq 0$. Since $\Phi_n$ divides $X^n - 1$, we have

$$X^n - 1 = fg \tag{12.11}$$

for some monic polynomial $g \in \mathbb{Q}[X]$. Since $f(\omega) = 0$, it follows that $\omega^n = 1$, hence also, raising each side to the $p$-th power,

$$(\omega^p)^n = 1.$$

In other words, $\omega^p$ is a root of $X^n - 1$. Since on the other hand it was assumed that $f(\omega^p) \neq 0$, equation (12.11) implies

$$g(\omega^p) = 0.$$

This last equation shows that $\omega$ is a root of $g(X^p)$. Therefore, by Lemma 12.14, $f(X)$ divides $g(X^p)$. Let $h(X) \in \mathbb{Q}[X]$ be a monic polynomial such that

$$g(X^p) = f(X)h(X). \tag{12.12}$$

Gauss' lemma (Lemma 12.11) and equations (12.11) and (12.12) show that $f$, $g$, and $h$ have integral coefficients. Therefore, we may consider the polynomials $\overline{f}$, $\overline{g}$, and $\overline{h}$ whose coefficients are the congruence classes modulo $p$ of the coefficients of $f$, $g$, and $h$ respectively, i.e., the images of these coefficients in $\mathbb{F}_p$ ($= \mathbb{Z}/p\mathbb{Z}$, see Remark 12.4). By reduction modulo $p$, equations (12.11) and (12.12) yield

$$X^n - 1 = \overline{f}(X)\overline{g}(X) \qquad \text{in } \mathbb{F}_p[X] \tag{12.13}$$

and

$$\overline{g}(X^p) = \overline{f}(X)\overline{h}(X) \quad \text{in } \mathbb{F}_p[X]. \tag{12.14}$$

Now, Fermat's theorem (Theorem 12.5) shows that $a^p = a$ for all $a \in \mathbb{F}_p$. Therefore, if

$$\overline{g}(X) = a_0 + a_1 X + \cdots + a_{r-1} X^{r-1} + X^r,$$

we also have

$$\overline{g}(X) = a_0^p + a_1^p X + \cdots + a_{r-1}^p X^{r-1} + X^r,$$

hence

$$\overline{g}(X^p) = a_0^p + a_1^p X^p + \cdots + a_{r-1}^p X^{p(r-1)} + X^{pr}.$$

Since $(u + v)^p = u^p + v^p$ in $\mathbb{F}_p$ (because the binomial coefficients $\binom{p}{i}$ are divisible by $p$ for $i = 1, \ldots, p - 1$), it follows that

$$\overline{g}(X^p) = (a_0 + a_1 X + \cdots + a_{r-1} X^{r-1} + X^r)^p = \overline{g}(X)^p \quad \text{in } \mathbb{F}_p[X].$$

Thus, (12.14) can be rewritten as

$$\overline{g}^p = \overline{f}\,\overline{h}$$

and this shows that $\overline{f}$ and $\overline{g}$ are not relatively prime. Let $\varphi(X) \in \mathbb{F}_p[X]$ be a non-constant common factor of $\overline{f}$ and $\overline{g}$. Equation (12.13) shows that $\varphi^2$ divides $X^n - 1$. Let

$$X^n - 1 = \varphi^2 \psi \quad \text{in } \mathbb{F}_p[X].$$

Comparing the derivatives of both sides, we obtain

$$nX^{n-1} = \varphi \cdot (2\partial\varphi \cdot \psi + \varphi \cdot \partial\psi),$$

hence $\varphi$ divides $X^n - 1$ and $nX^{n-1}$. This is impossible since $X^n - 1$ and $nX^{n-1}$ are relatively prime in $\mathbb{F}_p[X]$. (It is here that the hypothesis that $p$ does not divide $n$ is needed.) This contradiction shows that the hypothesis $f(\omega^p) \neq 0$ was absurd. $\qquad\square$

**Theorem 12.31.** *For every integer $n \geq 1$, the cyclotomic polynomial $\Phi_n$ is irreducible over $\mathbb{Q}$.*

**Proof.** Let $f$ be a monic irreducible factor of $\Phi_n$ in $\mathbb{Q}[X]$. We will prove that every root of $\Phi_n$ in $\mathbb{C}$ is a root of $f$. Since the roots of $\Phi_n$ are simple, it will then follow from Proposition 5.10 (p. 50) that $\Phi_n$ divides $f$, hence that $\Phi_n = f$, since $f$ and $\Phi_n$ divide each other and are both monic.

Let $\zeta$ be a root of $f$. Then $\zeta$ is a root of $\Phi_n$, which means that $\zeta$ is a primitive $n$-th root of unity. From Proposition 7.10 (p. 86) we recall that any other primitive $n$-th root of unity has the form $\zeta^k$, where $k$ is an integer relatively prime to $n$ between 0 and $n$. Factoring $k$ into (not necessarily distinct) prime factors

$$k = p_1 \dots p_s,$$

we find, by successive applications of Lemma 12.30,

$$f(\zeta) = 0 \Rightarrow f(\zeta^{p_1}) = 0 \Rightarrow f(\zeta^{p_1 p_2}) = 0 \Rightarrow \cdots$$
$$\Rightarrow f(\zeta^{p_1 \cdots p_{s-1}}) = 0 \Rightarrow f(\zeta^k) = 0.$$

Thus, $f$ has as root every primitive $n$-th root of unity, i.e., every root of $\Phi_n$. $\qquad\square$

**Theorem 12.32.** *If $m$ and $n$ are relatively prime integers, then $\Phi_n$ is irreducible over $\mathbb{Q}(\mu_m)$.*

**Proof.** Let $f$ be a monic irreducible factor of $\Phi_n$ in $\mathbb{Q}(\mu_m)[X]$ and let $\zeta \in \mathbb{C}$ be a root of $f$. Arguing as above, we see that it suffices to prove

$$f(\zeta^k) = 0$$

for every integer $k$ relatively prime to $n$ between 0 and $n$.

Let $\eta$ be a primitive $m$-th root of unity. As observed before Theorem 12.13, we have $\mathbb{Q}(\mu_m) = \mathbb{Q}(\eta)$, hence, by Proposition 12.15, every coefficient of $f$ is a polynomial expression in $\eta$ with rational coefficients. Therefore,

$$f(X) = \varphi(\eta, X)$$

for some polynomial $\varphi(Y, X) \in \mathbb{Q}[Y, X]$.

Let now $\rho = \zeta \eta$. Since $m$ and $n$ are relatively prime, it follows from Proposition 7.13 (p. 87) that $\rho$ is a primitive $mn$-th root of unity. Moreover, since $m$ and $n$ are relatively prime, by Remark 5.5(c) there exist integers $r$ and $s$ such that

$$mr + ns = 1.$$

As $\zeta^n = 1$ and $\eta^m = 1$, this equation implies that

$$\zeta = \zeta^{mr} = \rho^{mr} \quad \text{and} \quad \eta = \eta^{ns} = \rho^{ns}.$$

Since $f(\zeta) = 0$, we have $\varphi(\eta, \zeta) = 0$, or

$$\varphi(\rho^{ns}, \rho^{mr}) = 0.$$

Lemma 12.14 and Theorem 12.31 then show that $\Phi_{mn}(X)$ divides $\varphi(X^{ns}, X^{mr})$. It follows that

$$\varphi(\omega^{ns}, \omega^{mr}) = 0 \qquad (12.15)$$

for every primitive $mn$-th root of unity $\omega$.

For any integer $k$ relatively prime to $n$ between 0 and $n$, let

$$\ell = kmr + ns.$$

Since $mr + ns = 1$, we have $mr \equiv 1 \bmod n$ and $ns \equiv 1 \bmod m$, hence

$$\ell \equiv k \quad \bmod n \qquad \text{and} \qquad \ell \equiv 1 \quad \bmod m. \qquad (12.16)$$

It follows that $\zeta^{\ell} = \zeta^{k}$ and $\eta^{\ell} = \eta$, and since we already observed that $\zeta = \rho^{mr}$ and $\eta = \rho^{ns}$, we have

$$\rho^{\ell mr} = \zeta^{k} \qquad \text{and} \qquad \rho^{\ell ns} = \eta.$$

On the other hand, the congruences in (12.16) also show that $\ell$ is relatively prime to $mn$. Therefore, $\rho^{\ell}$ is a primitive $mn$-th root of unity, and equation (12.15) yields

$$\varphi(\rho^{\ell ns}, \rho^{\ell mr}) = 0,$$

i.e., $\varphi(\eta, \zeta^{k}) = 0$, or $f(\zeta^{k}) = 0$. $\qquad \square$

**Corollary 12.33.** *Let $p$ be a prime number and let $k$ be an integer that divides $p - 1$. Let also $\zeta \in \mathbb{C}$ be a primitive $p$-th root of unity. Then every element in $\mathbb{Q}(\mu_k)(\mu_p)$ can be uniquely written in the form*

$$a_1\zeta + a_2\zeta^2 + \cdots + a_{p-1}\zeta^{p-1},$$

*for some $a_1, \ldots, a_{p-1}$ in $\mathbb{Q}(\mu_k)$.*

**Proof.** The hypothesis on $k$ implies that $k$ is relatively prime to $p$, hence $\Phi_p$ is irreducible over $\mathbb{Q}(\mu_k)$, by Theorem 12.32. The corollary then follows by the same arguments as in the proof of Theorem 12.13. $\qquad \square$

## Appendix: Ruler and compass construction of regular polygons

Our goal is to construct regular polygons in the plane, using ruler and compass only. This construction should be possible whenever the center of the polygon (i.e., the center of the circumscribed circle) and one of its vertices are arbitrarily chosen. Therefore, we may regard the center $O$ and one of the vertices $A$ as given, and we have to determine the other vertices. From the two given points $O$ and $A$, new points can be constructed by a (finite) sequence of operations of the following types:

(1) draw a line through two points already determined,
(2) draw a circle with center a point already determined and radius the distance between two points already determined.

New points are determined as intersection points of the lines or circles drawn according to (1) and (2). The points that can be thus determined are called *constructible points*.

The problem is to decide for which values of $n$ the vertices of the regular polygon with $n$ sides, with center $O$, and $A$ as one of the vertices, are constructible. To solve this problem, we first give an algebraic characterization of the constructible points, via their coordinates in a suitable basis, which we construct as follows: we consider the perpendicular to $OA$ through $O$ and write $B$ for one of the intersection points of this perpendicular and the circle with center $O$ and radius $OA$:

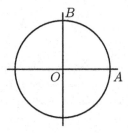

(Observe that the point $B$ is constructible.)

**Proposition.** *A point in the plane can be constructed by ruler and compass from $O$ and $A$ if and only if its coordinates in the basis $(OA, OB)$ can be obtained from 0 and 1 by a (finite) sequence of operations of the following types:*

(i) *rational operations,*
(ii) *extraction of square roots.*

**Proof.** First, we show that the points whose coordinates satisfy the condition above are constructible.

Since the perpendicular through a given point to a given line can be constructed by ruler and compass, a point with coordinates $(a, b)$ is constructible if (and only if) the points $(a, 0)$ and $(0, b)$ are constructible. Moreover, since $(0, b)$ is the intersection of the axis $OB$ with the circle with center $O$ passing through $(b, 0)$, it suffices to consider points with co-ordinates $(u, 0)$. So, we have to prove that a point with coordinates $(u, 0)$ is constructible with ruler and compass if $u$ is obtained from 0 and 1 by a sequence of operations (i) and (ii) above.

We argue inductively on the number of operations. Thus, we will prove that if $(u, 0)$ and $(v, 0)$ are constructible, then $(u + v, 0)$, $(u - v, 0)$, $(uv, 0)$, $(uv^{-1}, 0)$ (assuming $v \neq 0$) and $(\sqrt{u}, 0)$ (assuming $u \geq 0$) are constructible.

This is clear for $(u + v, 0)$ and $(u - v, 0)$. In order to construct $(uv, 0)$ and $(uv^{-1}, 0)$, consider the following figure:

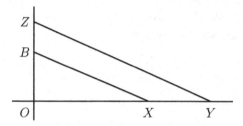

Since $BX$ and $YZ$ are parallel, we have

$$\frac{OX}{OB} = \frac{OY}{OZ}.$$

Let $X = (x, 0)$, $Y = (y, 0)$, and $Z = (0, z)$. Since $B = (0, 1)$, it follows that

$$x = yz^{-1} \quad \text{and} \quad y = xz.$$

Therefore, if we regard $X$ and $Z$ as given, we can construct $Y$ by drawing the parallel to $BX$ through $Z$. This construction yields $(xz, 0)$ from $(x, 0)$ and $(0, z)$ (or, equivalently, $(z, 0)$). On the other hand, if we regard $Y$ and $Z$ as given, then we can obtain $X$ by drawing the parallel to $YZ$ through $B$; this yields $(yz^{-1}, 0)$ from $(y, 0)$ and $(0, z)$ (or $(z, 0)$).

To complete the proof of the "if" part, it only remains to show that $(\sqrt{u}, 0)$ can be constructed from $(u, 0)$ (assuming $u \geq 0$). This can be done as follows:

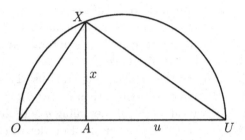

Let $U$ be the point with coordinates $(1 + u, 0)$ and let $X$ be one of the intersection points of the perpendicular to $OU$ through $A$ with the circle with diameter $OU$. We thus have $X = (1, x)$ for some $x$. Since the triangles $OAX$ and $XAU$ are similar, we have

$$\frac{AX}{OA} = \frac{AU}{AX},$$

hence

$$\frac{x}{1} = \frac{u}{x}, \quad \text{so} \quad x = \sqrt{u}.$$

Since the point $(x, 0)$ can be easily determined from $(1, x)$, this construction yields $(\sqrt{u}, 0)$ from $(u, 0)$.

We have thus proved that the points whose coordinates are obtained from 0 and 1 by rational operations and extraction of square roots are constructible.

To prove the converse, we first observe that if a line passes through two points $(a_1, b_1)$ and $(a_2, b_2)$, then its equation has the form

$$\alpha X + \beta Y = \gamma$$

where $\alpha$, $\beta$ and $\gamma$ are rational expressions in $a_1$, $a_2$, $b_1$ and $b_2$. (Specifically, $\alpha = b_2 - b_1$, $\beta = a_1 - a_2$ and $\gamma = a_1 b_2 - b_1 a_2$.) Likewise, the equation of a circle with center $(a_1, b_1)$ and radius the distance between $(a_2, b_2)$ and $(a_3, b_3)$ is

$$(X - a_1)^2 + (Y - b_1)^2 = (a_2 - a_3)^2 + (b_2 - b_3)^2,$$

hence it has the form

$$X^2 + Y^2 = \alpha X + \beta Y + \gamma$$

where $\alpha$, $\beta$, $\gamma$ are rational expressions in $a_1$, $a_2$, $a_3$, $b_1$, $b_2$, $b_3$. Now, direct calculations show that the coordinates of the intersection point of two lines

$$\alpha_1 X + \beta_1 Y = \gamma_1 \qquad \text{and} \qquad \alpha_2 X + \beta_2 Y = \gamma_2$$

are rational expressions in $\alpha_1$, $\beta_1$, $\gamma_1$, $\alpha_2$, $\beta_2$, $\gamma_2$. Thus, if a point is constructed as the intersection of two lines passing through given points, its coordinates are rational expressions in the coordinates of the given points.

Similarly, it can be seen that the coordinates of the intersection points of a line and a circle

$$\alpha_1 X + \beta_1 Y = \gamma_1 \qquad \text{and} \qquad X^2 + Y^2 = \alpha_2 X + \beta_2 Y + \gamma_2$$

are obtained by rational operations and extraction of a square root from $\alpha_1$, $\beta_1$, $\gamma_1$, $\alpha_2$, $\beta_2$, $\gamma_2$. Therefore, if a point is constructed as the intersection of a line through given points and a circle with given center and with radius the distance between given points, then its coordinates are obtained from the coordinates of the given points by rational operations and extraction of a square root.

Finally, the intersection of two circles

$$X^2 + Y^2 = \alpha_1 X + \beta_1 Y + \gamma_1 \qquad \text{and} \qquad X^2 + Y^2 = \alpha_2 X + \beta_2 Y + \gamma_2$$

can be obtained as the intersection of the circle

$$X^2 + Y^2 = \alpha_1 X + \beta_1 Y + \gamma_1$$

and the line

$$\alpha_1 X + \beta_1 Y + \gamma_1 = \alpha_2 X + \beta_2 Y + \gamma_2,$$

hence the same conclusion as for the preceding case holds. These arguments show that the coordinates of the constructible points are obtained by operations (i) and (ii) from the coordinates of $O$ and $A$, i.e., from 0 and 1. $\qquad\square$

This constructibility criterion seems to have been first published by Pierre Laurent Wantzel (1814–1848) in 1837, but it was undoubtedly known to Gauss (and presumably also to others) around 1796.

**Theorem.** *If $p$ is a prime number of the form $p = 2^m + 1$ (with $m \in \mathbb{N}$) then the regular polygon with $p$ sides can be constructed with ruler and compass.*

**Proof.** Since $p-1$ is a power of 2, the lattice of divisors of $p-1$ is a chain

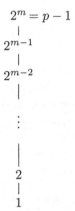

$$2^m = p - 1$$
$$|$$
$$2^{m-1}$$
$$|$$
$$2^{m-2}$$
$$|$$
$$\vdots$$
$$|$$
$$2$$
$$|$$
$$1$$

Therefore, the results of §12.3 show that the periods of two terms can be determined by solving a sequence of quadratic equations. Since, as observed p. 171, the periods of two terms are the values $2\cos\frac{2k\pi}{p}$, for $k = 1, \ldots, \frac{p-1}{2}$, and since the solution of a quadratic equation only requires rational operations and extraction of square roots, it follows that $\cos\frac{2\pi}{p}$ can be obtained from the integers (or even from 0 and 1) by rational operations and extraction of square roots. The preceding proposition then shows that the point with coordinates $(\cos\frac{2\pi}{p}, 0)$ is constructible.

The point $P = (\cos\frac{2\pi}{p}, \sin\frac{2\pi}{p})$ can then be obtained as the intersection of the circle with center $O$ and radius $OA$ with the perpendicular to $OA$ through $(\cos\frac{2\pi}{p}, 0)$. The point $P$ is a vertex of the regular polygon with $p$ sides. In fact, it is one of the two vertices that are closest to $A$, and the other vertices can be found by reproducing the distance $AP$ on the circle. □

If a prime number $p$ has the form $2^m + 1$, then it is easily seen that $m$ is a power of 2. Indeed, if $m$ is divisible by some odd integer $k$, then $2^m + 1$ is divisible by $2^{m/k} + 1$, as can be seen by letting $X = 2^{m/k}$ in the equation

$$X^k + 1 = (X+1)(X^{k-1} - X^{k-2} + \cdots - X + 1).$$

Thus, the prime numbers that satisfy the hypothesis of the proposition are in fact of the form $p = 2^{2^n} + 1$ for some integer $n$. These prime numbers are called *Fermat primes*, after Pierre de Fermat, who conjectured that the number $F_n = 2^{2^n} + 1$ is prime for every integer $n$. For $n = 0, 1, 2, 3, 4$, this formula yields 3, 5, 17, 257, and 65537, which are indeed prime, but in 1732 Euler showed that $F_5 = 641 \cdot 6700417$. Since then, the numbers

$F_n$ have been shown to be composite for various values of $n$, and no new Fermat prime has been found. Although it has not been proved that no other Fermat prime exists, it is at least known that there is no such prime between 65538 and $10^{39456}$ (i.e., $F_n$ is not prime for $5 \leq n \leq 16$).

**Corollary.** *The regular polygon with $n$ sides can be constructed with ruler and compass if $n$ is a product of distinct Fermat primes and of a power of 2.*

**Proof.** Since the regular polygon with $n$ sides is constructible when $n$ is a power of 2 (by repeated bisections of angles) or when $n$ is a Fermat prime (by the preceding theorem), it suffices to show that when $n_1$ and $n_2$ are relatively prime integers such that the regular polygons with $n_1$ and $n_2$ sides are constructible, then the regular polygon with $n_1 n_2$ sides is constructible.

If $n_1$ and $n_2$ are relatively prime, then there exist integers $m_1$ and $m_2$ such that $m_1 n_1 + m_2 n_2 = 1$ (see Remark 5.5(c)). Multiplying each side by $\frac{2\pi}{n_1 n_2}$, we obtain

$$m_1 \frac{2\pi}{n_2} + m_2 \frac{2\pi}{n_1} = \frac{2\pi}{n_1 n_2}.$$

Therefore, the arc $\frac{2\pi}{n_1 n_2}$ can be constructed by reproducing a certain number of times the arcs $\frac{2\pi}{n_1}$ and $\frac{2\pi}{n_2}$, and it readily follows that the regular polygon with $n_1 n_2$ sides can be constructed from the regular polygons with $n_1$ and $n_2$ sides. □

**Remark.** It can be proved that the converses of the theorem and of the corollary above also hold. Thus, the regular polygon with $n$ sides can be constructed with ruler and compass if and only if $n$ is a product of distinct Fermat primes and of a power of 2. This result is explicitly stated (without proof) by Gauss [31, Art. 366], but a smooth proof of these converse statements requires a detailed analysis of field degrees, which would carry us too far afield. Therefore, we refer the interested reader to Carrega [14, Chap. 4] or Stewart [70, Chap. 17] for a proof. We also refer to Hardy and Wright [36, §5.8] for an explicit geometric construction of the 17-gon with ruler and compass.

## Exercises

*Exercise 12.1.* Recall from Exercise 7.7 (p. 92) that for every integer $n \geq 2$, the number of integers that are relatively prime to $n$ between 0 and $n$ is denoted by $\varphi(n)$. Prove the following generalization (due to Euler) of Fermat's theorem (Theorem 12.5): $a^{\varphi(n)} \equiv 1 \bmod n$ for every integer $n \geq 2$ and every integer $a$ relatively prime to $n$.

*Exercise 12.2.* Show that Girard's theorem on rational roots (Theorem 6.1, p. 67) readily follows from Gauss' lemma (Lemma 12.11).

*Exercise 12.3.* Prove that the periods with an even number of terms are real numbers.

*Exercise 12.4.* Prove that the set of periods of $f$ terms does not depend on the choice of a primitive root of $p$ nor of a primitive $p$-th root of unity. More precisely, let $\zeta$ and $\zeta' \in \mathbb{C}$ be primitive $p$-th roots of unity, let $g$, $g' \in \mathbb{Z}$ be primitive roots of $p$ and let $p - 1 = ef$ for some positive integers $e$, $f$. Write $\zeta_i = \zeta^{g^i}$ and $\zeta'_i = \zeta'^{g'^i}$ for $i = 0, \ldots, p - 2$. Show that for any $i = 0, \ldots, f - 1$, there is an integer $j$ between 0 and $f - 1$ such that

$$\zeta_i + \zeta_{e+i} + \cdots + \zeta_{e(f-1)+i} = \zeta'_j + \zeta'_{e+j} + \cdots + \zeta'_{e(f-1)+j}.$$

*Exercise 12.5.* With the notation of Proposition 12.22, prove that if $K_g \subset K_f$, then $f$ divides $g$. Moreover, show that in this case, $\dim_{K_g} K_f = g/f$.

*Exercise 12.6. The following exercise provides complementary observations to the proof of Corollary 12.29.* Let the notation be as in Corollary 12.29. Show that $t(\omega) \neq 0$ and that $t(\omega^k)t(\omega)^{-k}$ has a rational expression in terms of $\omega$, for $k \in \mathbb{Z}$. (Compare Exercise 10.4, p. 142.) Conclude that it suffices to extract a single $(n - 1)$-th root to determine $\zeta_i$.

*Exercise 12.7.* By looking at the algebraic expression of 5-th roots of unity, find a construction of regular pentagons by ruler and compass. Find also a construction of regular 20-gons.

# Chapter 13

# Ruffini and Abel on General Equations

Lagrange's investigations were primarily aimed at the solution of "general" equations, i.e., equations with letters as coefficients, such as

$$X^n - s_1 X^{n-1} + s_2 X^{n-2} - \cdots + (-1)^n s_n = 0$$

(see Definition 8.1, p. 94). At about the same time when Gauss completed the solution of the class of particular equations that arise from the division of the circle (known as cyclotomic equations), Lagrange's line of investigation bore new fruits in the hands of Paolo Ruffini (1765–1822). In 1799, Ruffini published a massive two-volume treatise: "Teoria Generale delle Equazioni" [64, t. 1, pp. 1–324], in which he proves that the general equations of degree at least 5 are not solvable by radicals.

Ruffini's proof was received with skepticism by the mathematical community. Indeed, the proof was rather hard to follow through the 516 pages of his books. A few years after the publication, negative comments were made but, to Ruffini's dismay, no clear, focused objection was raised. Vague criticism was denying Ruffini the credit of having validly proved his claim. Negative reactions prompted Ruffini to simplify his proof, and he eventually came up with very clean arguments, but distrust of Ruffini's work did not subside. Typical in this respect is the following anecdote: in order to get a clear, motivated pronouncement from the French Academy of Sciences, Ruffini submitted a paper to the Academy in 1810. A year later, the referees (Lagrange, Lacroix and Legendre) had not yet given their conclusions. Ruffini then wrote to Delambre, who was secretary of the Academy, to withdraw his paper. In his reply, Delambre explains the referees' attitude:

> Whatever decision Your Referees would have reached, they had to work considerably either to motivate their approval or to refute Your proof. You know how precious is time to realize also how reluctant most geometers are to occupy themselves for a long time with the works of each

other, and if they would have happened not to be of Your opinion, they
would have had to be moved by a quite powerful motive to enter the
lists against a geometer so learned and so skillful [64, t. 3, p. 59].

At least, unconvincing as it was, Ruffini's proof seems to have completed
the reversal of the current opinion towards general equations: while the
works of Bézout and Euler around the middle of the eighteenth century
were grounded on the opinion that general equations were solvable, and that
finding the solution of the fifth degree equations was only a matter of clever
transformations, the opposite view became common in the beginning of the
nineteenth century (see Ayoub [4, p. 274]). Some comments of Gauss may
also have been influential in this respect. In his proof of the fundamental
theorem of algebra, [30, §9], Gauss writes:

> After the works of many geometers left very little hope of ever arriving
> at the resolution of the general equation algebraically, it appears more
> and more likely that this resolution is impossible and contradictory.

He voiced again the same skepticism in Article 359 of "Disquisitiones Arithmeticae."

Ruffini's credit also includes advances in the theory of permutations,
which was crucial for his proof. Ruffini's results in this direction were soon
generalized by Cauchy. Incidentally, it is noteworthy that Cauchy was very
appreciative of Ruffini's work and that he supported Ruffini's claim that
his proof was valid (see [64, t. 3, pp. 88–89]). In fact, it now appears that
Ruffini's proofs do have a significant gap, which we will point out below.

In 1824, a new proof was found by Niels-Henrik Abel (1802–1829) [1,
n° 3], independently of Ruffini's work. An expanded version of Abel's proof
was published in 1826 in the first issue of Crelle's journal (the "Journal für
die reine und angewandte Mathematik") [1, n° 7]. This proof also contains
some minor flaws (see [1, vol. 2, pp. 292–293]), but it essentially settled the
issue of solvability of general equations.

Abel's approach is remarkably methodical. He explains it in some detail
in the introduction to a subsequent paper: "Sur la résolution algébrique des
équations" (1828) [1, n° 18].

> To solve these equations [of degree at most 4], a uniform method has
> been found, and it was believed that it could be applied to equations
> of arbitrary degree; but in spite of the efforts of a Lagrange and other
> distinguished geometers, one was not able to reach this goal. This led
> to the presumption that the algebraic solution of general equations was
> impossible; but that could not be decided, since the method that was

used could not lead to definite conclusions except in the case where the equations were solvable. Indeed, the purpose was to solve equations, without knowing whether this was possible. In this case, one might get the solution, although that was not sure at all; but if unfortunately the solution happened to be impossible, one could have sought it for ever without finding it. In order to obtain unfailingly something in this matter, it is therefore necessary to take another way. One has to cast the problem in such a form that it be always possible to solve, which can be done with any problem. Instead of seeking a relation of which it is not known whether it exists or not, one has to seek whether such a relation is indeed possible. For instance, in the integral calculus, instead of trying by a kind of divination or by trial and error to integrate differential formulas, one has to look rather whether it is possible to integrate them in this or that way. When a problem is thus presented, the statement itself contains the seed of the solution and shows the way that is to be taken; and I think that there will be few cases where one could not reach more or less important propositions, even when one could not completely solve the question because the calculations would be too complicated.

The method thus advocated by Abel can be interpreted in the realm of algebraic equations as a kind of generic method. One has to find the most general form of the expected solution and work on it to investigate what kind of information can be obtained on this expression if it is a root of the general equation. Abel thus proves, by an intricate inductive argument, that if an expression by radicals is a root of the general equation of some degree, then every function of which it is composed is a rational expression of the roots (see Theorem 13.12 for a precise statement). This fills the gap in Ruffini's proofs. Some delicate arguments involving the number of values of functions under permutations of the variables and, in particular, a theorem of Cauchy generalizing earlier results of Ruffini, complete the proof. This last part of the proof can be significantly streamlined by using arguments from the last of Ruffini's proofs, as Wantzel later noticed. In the following sections, we present this easy version, but we point out that this approach unfortunately downplays the advances in the theory of permutations (i.e., in the study of the symmetric group $S_n$) that were prompted by Ruffini's earlier work.

## 13.1  Radical extensions

Abel's calculations with expressions by radicals, which we discuss in this section and the following as a first step in the proof that general equations of degree higher than 4 are not solvable, can be adequately cast into the vocabulary of field extensions. This point of view will be used throughout since it is probably more enlightening for the modern reader.

An expression by radicals is constructed from some quantities that are regarded as known (usually the coefficients of an equation, in this context) by the four usual operations of arithmetic and the extraction of roots. This means that any such expression lies in a field obtained from the field of rational expressions in the known quantities by successive adjunctions of roots of some orders. In fact, it is clearly sufficient to consider roots of prime order, since if $n = p_1 \ldots p_r$ is the factorization of a positive integer $n$ into prime factors, then

$$a^{1/n} = \left(\ldots \left((a^{1/p_1})^{1/p_2}\right)\ldots\right)^{1/p_r}.$$

This shows that an $n$-th root of any element $a$ can be obtained by extracting a $p_1$-th root $a^{1/p_1}$ of $a$, next a $p_2$-th root of $a^{1/p_1}$ and so on. Moreover, it obviously suffices to extract $p$-th roots of elements that are not $p$-th powers, otherwise the base field is not enlarged. We thus come to the notion of a radical field extension. Before spelling out this notion in mathematical terms, we note that, in order to avoid some technical difficulties, we restrict attention throughout the chapter to fields of characteristic zero; in other words, we assume that $1 + 1 + \cdots + 1 \neq 0$ (for any number of terms) or that every field under consideration contains (an isomorphic copy of) the field $\mathbb{Q}$ of rational numbers. This is of course the classical case, which was the only case considered by Ruffini and Abel.

**Definitions 13.1.** A field $R$ containing a field $F$ of characteristic zero is called a *radical extension of height* 1 of $F$ if there exist a prime number $p$, an element $a \in F$ that is not a $p$-th power in $F$, and an element $u \in R$ such that

$$R = F(u) \qquad \text{and} \qquad u^p = a.$$

(Recall from (12.5) (p. 166) that $F(u)$ is the set of all rational expressions in $u$ with coefficients in $F$.) The prime $p$ is called the *degree* of the radical extension of height 1. (The degree does not depend on the representation of $R$ as $F(u)$: see Exercise 13.2.) An element $u$ as above is sometimes denoted by $a^{1/p}$ or $\sqrt[p]{a}$, and, accordingly, one sometimes writes

$$R = F(a^{1/p}) \qquad \text{or} \qquad R = F(\sqrt[p]{a}).$$

This is in fact an abuse of notation, since the element $u$ is not uniquely determined by $a$ and $p$: there are indeed $p$ different $p$-th roots of $a$. Worse still, the field $R$ itself is in general not uniquely determined by $F$, $a$, and $p$. For instance, there are three subfields of $\mathbb{C}$ which qualify as $\mathbb{Q}(2^{1/3})$. (See however Exercises 13.5 and 13.6.) Therefore, the notation above will be used with caution.

Radical extensions of height $h$, for any positive integer $h$, are defined inductively as radical extensions of height 1 of radical extensions of height $h - 1$. More precisely, a field $R$ containing a field $F$ is called a *radical extension of height $h$* of $F$ if there is a field $R_1$ between $R$ and $F$ such that $R$ is a radical extension of height 1 of $R_1$, and $R_1$ is a radical extension of height $h - 1$ of $F$. Thus, in this case we can find a tower of extensions between $R$ and $F$,

$$R \supset R_1 \supset R_2 \supset \cdots \supset R_{h-1} \supset F,$$

such that, letting $R = R_0$ and $F = R_h$, we have for $i = 1, \ldots, h$

$$R_{i-1} = R_i\left(a_i^{1/p_i}\right)$$

for some prime number $p_i$ and some element $a_i \in R_i$ that is not a $p_i$-th power in $R_i$.

We simply term *radical extension* any radical extension of some (finite) height and, for completeness, we say that any field is a *radical extension of height 0* of itself.

The definitions above are quite convenient to translate into mathematically amenable terms questions concerning expressions by radicals. For instance, to say that a complex number $z$ has an expression by radicals means that there is a radical extension of the field $\mathbb{Q}$ of rational numbers containing $z$. More generally, we say that an element $v$ of a field $L$ *has an expression by radicals over some field $F \subset L$* if there is a radical extension of $F$ containing $v$. Likewise, we say that a polynomial equation $P(X) = 0$ over some field $F$ is *solvable by radicals over $F$* if there is a radical extension of $F$ containing a root of $P$. In the case of general equations

$$P(X) = (X - X_1) \ldots (X - X_n) = X^n - s_1 X^{n-1} + \cdots + (-1)^n s_n = 0,$$

we are concerned with radical expressions involving only the coefficients $s_1$, $\ldots$, $s_n$, so the base field $F$ will be the field of rational functions in $s_1, \ldots, s_n$ (which can be considered as independent indeterminates, according to Remark 8.9(a), p. 101). To be more precise, we have to specify a field of reference in which the rational functions are allowed to take their coefficients. A logical choice is of course the field $\mathbb{Q}$ of rational numbers, but in fact, since we are aiming at a negative result, the reference field can be chosen arbitrarily large. Indeed, we will prove that if an equation is solvable by radicals over some field $F$, then it is solvable by radicals over every field $L$ containing $F$; therefore, if the general equation of degree $n$ is not solvable over $\mathbb{C}(s_1, \ldots, s_n)$, it is not solvable over $\mathbb{Q}(s_1, \ldots, s_n)$ either.

Of course, Ruffini and Abel did not address in these terms the problem of assigning a reference field, but their free use of roots of unity suggests that all the roots of unity are at their disposal in the base field. The choice $F = \mathbb{C}(s_1, \ldots, s_n)$ seems therefore close in spirit to Ruffini's and Abel's work.

The hypothesis that the base field contains all the roots of unity also has a technical advantage, in that it allows more flexibility in the treatment of radical extensions, as the next result shows:

**Proposition 13.2.** *Let $R$ be a field containing a field $F$. If $R$ has the form $R = F(u)$ for some element $u$ such that $u^n \in F$ for some integer $n$, and if $F$ contains a primitive $n$-th root of unity (hence all the $n$-th roots of unity, since the other roots are powers of this one), then $R$ is a radical extension of $F$. Moreover, there is a tower of radical extensions of height 1*

$$R = R_0 \supset R_1 \supset R_2 \supset \cdots \supset R_{h-1} \supset R_h = F$$

*such that the degree of each radical extension $R_{i-1} \supset R_i$ of height 1 divides $n$ (unless $R = F$).*

In other words, in the definition of radical extensions, we need not require that the exponent $n$ be a prime number, nor that $u^n$ be not the $n$-th power of an element in $F$, provided that $F$ contains a primitive $n$-th root of unity.

**Proof.** We argue by induction on $n$. If $n = 1$, then $u \in F$, hence $R = F$ and $R$ is then a radical extension of height 0 of $F$. We may thus assume that $n \geq 2$ and that the proposition holds when the exponent of $u$ is at most $n - 1$.

If $n$ is not prime, let $n = rs$ for some (positive) integers $r$, $s < n$. By the induction hypothesis, $F(u)$ is a radical extension of $F(u^r)$ and $F(u^r)$ is a radical extension of $F$, since $u^r$ satisfies $(u^r)^s \in F$. Therefore, $F(u)$ is a radical extension of $F$, since it is clear from the definition that the property of being radical is transitive, namely, in a tower of extensions $L \supset K \supset F$, if $L$ is a radical extension of $K$ and $K$ is a radical extension of $F$, then $L$ is a radical extension of $F$. More precisely, by induction we may assume there is a tower of radical extensions between $F(u)$ and $F(u^r)$ (resp. between

$F(u^r)$ and $F$) such that the degree of each intermediate radical extension of height 1 divides $r$ (resp. divides $s$). Thus, there is between $F(u)$ and $F$ a tower of radical extensions in which the degree of each intermediate extension of height 1 divides $n$.

If $n$ is prime, we consider two cases, according to whether $u^n$ is or is not the $n$-th power of an element in $F$. If it is not, then $R$ is a radical extension of $F$, by definition, and its degree is $n$. If it is, let

$$u^n = b^n$$

for some $b \in F$. If $b = 0$, then $u = 0$ and $R = F$, a radical extension of height 0 of $F$. If $b \neq 0$, then the preceding equation yields

$$\left(\frac{u}{b}\right)^n = 1,$$

hence $u/b$ is an $n$-th root of unity. Since the $n$-th roots of unity are all in $F$, it follows that $u/b \in F$, hence $u \in F$ and again $R = F$, a radical extension of height 0 of $F$. □

So far, we have dealt only with the case where roots of unity are in the base field. In order to reduce more general situations to this case, we have to use Gauss' result that every root of unity has an expression by radicals. Since we now have a formal definition for "expression by radicals," it seems worthwhile to spell out how Gauss' arguments actually fit in this framework.

**Proposition 13.3.** *For any integer $n$ and any field $F$, the $n$-th roots of unity lie in a radical extension $R$ of $F$. This radical extension can be chosen to satisfy the following condition: there is a tower of radical extensions*

$$R = R_0 \supset R_1 \supset R_2 \supset \cdots \supset R_h = F$$

*in which the degree of each intermediate radical extension $R_{i-1} \supset R_i$ of height 1 is at most $n - 1$ (unless $R = F$).*

**Proof.** It suffices to show that a primitive $n$-th root of unity $\zeta$ lies in a radical extension of $F$ as above, since the other $n$-th roots of unity are powers of $\zeta$ and therefore lie in the same radical extension as $\zeta$.

We argue by induction on $n$. For $n = 1$, we have $\zeta = 1$, hence $\zeta$ lies in $F$, which is a radical extension of height 0 of itself. We may thus assume that $n \geq 2$ and that the proposition holds for roots of unity of exponent less than $n$.

If $n$ is not prime, let $n = rs$ for some (positive) integers $r$, $s < n$. Then $\zeta^r$ is an $s$-th root of unity. By the induction hypothesis, we can find a radical extension $K$ of $F$ containing $\zeta^r$, and assume that there is a tower of radical extensions between $K$ and $F$ in which the degree of each intermediate extension of height 1 is at most $s - 1$, hence less than $n - 1$. By the induction hypothesis again, we can find a radical extension $L$ of $K$ (hence also of $F$) that contains a primitive $r$-th root of unity. Moreover, we may assume that $L$ is obtained from $K$ by a tower of radical extensions in which every intermediate radical extension of height 1 has degree at most $r - 1$. Then, since $\zeta^r \in L$, it follows from Proposition 13.2 that $L(\zeta)$ is a radical extension of $L$, hence of $F$, and that there is a tower of radical extensions between $L(\zeta)$ and $L$ in which the degree of every intermediate extension of height 1 divides $r$, and is therefore less than $n - 1$. The proposition is thus proved in this case.

If $n$ is prime, then we have to use Gauss' results. First, we can find a radical extension $K$ of $F$ that contains the $(n - 1)$-th roots of unity, by the induction hypothesis, and assume there is a tower of radical extensions between $K$ and $F$ in which every intermediate extension of height 1 has degree at most $n-2$. We then consider the Lagrange resolvents $t(\omega) \in K(\zeta)$ as in the proof of Corollary 12.29 (p. 181). By Proposition 12.27 (p. 179), we have

$$t(\omega)^{n-1} \in K$$

for every $(n - 1)$-th root of unity $\omega$. Therefore Proposition 13.2 shows that $K(t(\omega))$ is a radical extension of $K$, with a tower of radical extensions in which each intermediate extension of height 1 has degree at most $n - 1$. Adjoining successively all the Lagrange resolvents $t(\omega)$, we find in $K(\zeta)$ a radical extension $R$ of $K$, hence of $F$, which contains $t(\omega)$ for all $\omega \in \mu_{n-1}$. From Lagrange's formula (Proposition 10.2, p. 130) it follows that $\zeta$ can be rationally calculated from the Lagrange resolvents, hence $\zeta \in R$ (so in fact $R = K(\zeta)$) and the proof is complete.                                          $\square$

We now aim to prove the afore-mentioned fact that solvability of an equation by radicals over some field $F$ implies solvability by radicals over any larger field $L$. This fact may seem obvious, since every expression by radicals involving elements of $F$ is an expression by radicals involving elements of $L$. However, it requires a careful justification. The point is that, in building radical extensions or expressions by radicals, we allow only extractions of $p$-th roots of elements that are not $p$-th powers in $F$, but these elements could become $p$-th powers in the larger field $L$.

**Lemma 13.4.** *Let $L$ be a field containing a field $F$, and let $R$ be a radical extension of $F$.*

(i) *There is a radical extension $S$ of $L$ such that $R$ can be identified with a subfield of $S$.*

(ii) *Suppose $R$ is obtained from $F$ by a tower of radical extensions in which the intermediate radical extensions of height 1 have degree $p_1, \ldots, p_h$, and $F$ contains a primitive $p_i$-th root of unity for $i = 1, \ldots, h$. If $L$ and $R$ are both contained in a field $K$, there is a radical extension $S$ of $L$ in $K$ containing $R$.*

**Proof.** We argue by induction on the height $h$ of $R$. If $h = 0$, then $R = F$ and we can choose $S = L$ in each case.

If $h = 1$, let $R = F(u)$ where $u$ is such that $u^p = a$ for some prime $p$ and some element $a \in F$ that is not a $p$-th power in $F$. In case (ii), we have $u \in K$. If $a$ is not a $p$-th power in $L$, then $L(u)$ is a radical extension of height 1 of $L$, and this extension contains $R$ since it contains $F$ and $u$. We may then take $S = L(u)$. If $a$ is a $p$-th power in $L$, let $b \in L$ be such that $b^p = a$. Since the $p$-th powers of $u$ and $b$ are equal, it follows that $(u/b)^p = 1$. Then $u/b \in F$ because $F$ contains all the $p$-th roots of unity by hypothesis, hence $u \in L$ and we may take $S = L$.

In case (i), we extend $L$ into a field $E$ over which the polynomial $X^p - a$ splits into a product of linear factors. (The existence of such a field $E$ follows from Girard's theorem (Theorem 9.3, p. 110).) Since $u$ is one of the roots of $X^p - a$, it can be identified with an element in $E$, and every rational function in $u$ with coefficients in $F$, i.e., every element in $R$, is then identified with an element in $E$. We may thus henceforth assume that $R$ is contained in $E$. We then consider the same alternative as in case (ii): if $a$ is not a $p$-th power in $L$, then $L(u)$ is a radical extension of height 1 of $L$, and we may take $S = L(u)$. If there exists $b \in L$ such that $b^p = a$, then $u/b$ is a $p$-th root of unity, and Proposition 13.3 shows that there is a radical extension $S$ of $L$ which contains $u/b$. Since $b \in L$, it follows that $u \in S$, hence $R \subset S$ and the proof is complete in the case where the height $h$ of $R$ is 1.

If $h \geq 2$, the lemma follows from the case $h = 1$ and the induction hypothesis. Indeed, we can find in $R$ a subfield $R_1$ that is a radical extension of height $h - 1$ of $F$ and such that $R$ is a radical extension of height 1 of $R_1$. By the induction hypothesis, we may assume that $R_1$ is contained in a radical extension $S_1$ of $L$ and, by the case $h = 1$ already considered, $R$ can be identified with a subfield of a radical extension $S$ of $S_1$. The field $S$ is then a radical extension of $L$ satisfying the condition in (i). For (ii), we may assume $S_1 \subset K$, and find $S$ in $K$ as in the case $h = 1$. $\square$

**Corollary 13.5.** *Let* $v_1, \ldots, v_n$ *be elements in a field* $K$ *containing a field* $F$. *If each* $v_i$ *lies in a radical extension of* $F$ *contained in* $K$, *then there is a radical extension* $R$ *of* $F$ *containing all the elements* $v_1, \ldots, v_n$. *Moreover, if* $F$ *contains the field* $\mathbb{C}$ *of complex numbers, so that all the roots of unity lie in* $F$, *then the radical extension* $R$ *can be chosen inside* $K$.

**Proof.** For $i = 1, \ldots, n$, let $K_i \subset K$ be a radical extension of $F$ containing $v_i$. Consider a tower or radical extensions

$$K_i = R_{i,0} \supset R_{i,1} \supset \cdots \supset R_{i,h_i} = F$$

with each intermediate extension $R_{i,j-1} \supset R_{i,j}$ radical of height 1: for $i = 1, \ldots, n$ and $j = 1, \ldots, h_i$, we have

$$R_{i,j-1} = R_{i,j}(u_{i,j}) \qquad \text{with} \quad u_{i,j}^{p_{i,j}} = a_{i,j}$$

for some prime $p_{i,j}$ and some $a_{i,j} \in R_{i,j}$ that is not a $p_{i,j}$-th power in $R_{i,j}$. Let $m$ be a common multiple of all the $p_{i,j}$'s for $i = 1, \ldots, n$ and $j = 1, \ldots, h_i$. By Proposition 13.3, we may find a radical extension $F'$ of $F$ containing all the $m$-th roots of unity. Lemma 13.4(i) yields a radical extension $S$ of $K$ containing a copy of $F'$. We identify $F'$ with a subfield of $S$, let $R'_{i,h_i} = F'$ for $i = 1, \ldots, n$, and define by decreasing induction on $j$

$$R'_{i,j-1} = R'_{i,j}(u_{i,j}) \qquad \text{for } i = 1, \ldots, n \text{ and } j = 1, \ldots, h_i.$$

Let also $K'_i = R'_{i,0}$ for $i = 1, \ldots, n$. Thus, $R_{i,j} \subset R'_{i,j}$ for all $i$, $j$, and in particular $v_i \in K'_i$. Since $u_{i,j}^{p_{i,j}} = a_{i,j} \in R'_{i,j}$ and $R'_{i,j}$ contains $F'$, hence also a primitive $p_{i,j}$-th root of unity, Proposition 13.2 shows that $R'_{i,j-1}$ is a radical extension of $R'_{i,j}$. Therefore, $K'_i$ is a radical extension of $F'$.

We now show that there is inside $S$ a radical extension $R$ of $F'$ containing $v_1, \ldots, v_n$. Since $F'$ is a radical extension of $F$, the field $R$ is also a radical extension of $F$, which satisfies the conditions of the corollary.

We argue by induction on $n$. If $n = 1$, we may take $R = K'_1$. Suppose we have $L \subset S$ a radical extension of $F'$ containing $v_1, \ldots, v_{n-1}$. Since $F'$ contains a primitive $p_{n,j}$-th root of unity for $j = 1, \ldots, h_n$, we may apply Lemma 13.4(ii) to find inside $S$ a radical extension $R$ of $L$ containing $K'_n$, hence also $v_n$. The field $R$ thus satisfies the requirements: it is a radical extension of $F'$ in $S$ containing $v_1, \ldots, v_n$.

In the particular case where $F$ contains $\mathbb{C}$, we have $F' = F$, hence we may take $S = K$. The field $R$ then lies in $K$. $\qquad \square$

**Theorem 13.6.** *Let* $P$ *be a polynomial with coefficients in a field* $F$ *(of characteristic zero). If* $P(X) = 0$ *is solvable by radicals over* $F$, *then it is solvable by radicals over every field* $L$ *containing* $F$.

**Proof.** Let $R$ be a radical extension of $F$ containing a root $r$ of $P$. Lemma 13.4(i) shows that we may assume $R$ is contained in some radical extension $S$ of $L$. The radical extension $S$ then contains the root $r$, hence $P(X) = 0$ is solvable by radicals over $L$. $\qquad\square$

The following special case of the theorem is particularly relevant for this chapter:

**Corollary 13.7.** *If the general equation of degree $n$*

$$P(X) = (X - x_1) \ldots (X - x_n) = X^n - s_1 X^{n-1} + \cdots + (-1)^n s_n = 0$$

*is not solvable by radicals over* $\mathbb{C}(s_1, \ldots, s_n)$, *then it is not solvable by radicals over* $\mathbb{Q}(s_1, \ldots, s_n)$ *either.*

We may thus henceforth assume that the base field contains all the roots of unity.

## 13.2 Abel's theorem on natural irrationalities

Any proof that the general equation of some degree is not solvable by radicals obviously proceeds *ad absurdum*. Thus, we assume by way of contradiction that there is a radical extension $R$ of $\mathbb{C}(s_1, \ldots, s_n)$ which contains a root $x_i$ of the general equation

$$(X - x_1) \ldots (X - x_n) = X^n - s_1 X^{n-1} + s_2 X^{n-2} - \cdots + (-1)^n s_n = 0.$$

The first step in Abel's proof (which was missing in Ruffini's proofs) is to show that $R$ can be supposed to lie inside $\mathbb{C}(x_1, \ldots, x_n)$. This means that the irrationalities that occur in an expression by radicals for a root of the general equation of degree $n$ can be chosen to be *natural*, as opposed to *accessory* irrationalities, which designate the elements of extensions of $\mathbb{C}(s_1, \ldots, s_n)$ outside $\mathbb{C}(x_1, \ldots, x_n)$ (see Ayoub [4, p. 268]). (The terms "natural" and "accessory" irrationalities were coined by Kronecker.)

The purpose of this section is to give a proof of this result, following Abel's approach in [1, n° 7, §2]. Recall that all the fields in this chapter are assumed to be of characteristic zero.

**Lemma 13.8.** *Let $p$ be a prime number and let $a$ be an element of some field $F$. Assume $a$ is not a $p$-th power in $F$.*

(a) *For $k = 1, \ldots, p-1$, the $k$-th power $a^k$ is not a $p$-th power in $F$ either.*

(b) *The polynomial $X^p - a$ is irreducible over $F$.*

**Proof.** (a) If $k$ is an integer between 1 and $p - 1$, then it is relatively prime to $p$, hence we can find integers $\ell$ and $q$ such that $pq + k\ell = 1$ (see Remark 5.5(c)). Then

$$a = (a^q)^p (a^k)^\ell.$$

Therefore, if $a^k = b^p$ for some $b \in F$, then we have

$$a = (a^q b^\ell)^p,$$

in contradiction with the hypothesis that $a$ is not a $p$-th power in $F$. This contradiction proves (a).

(b) Let $P$ and $Q$ be polynomials in $F[X]$ such that

$$X^p - a = PQ.$$

We may assume that $P$ and $Q$ are monic, and we have to prove that $P$ or $Q$ is the constant polynomial 1. Let $K$ be an extension of $F$ over which $X^p - a$ splits into a product of linear factors. (The existence of such a field follows from Girard's theorem (Theorem 9.3, p. 110).) Since the roots of $X^p - a$ are the $p$-th roots of $a$, which are obtained from any of them by multiplication by the various $p$-th roots of unity (see §7.2), we have in $K[X]$

$$\prod_{\omega \in \mu_p} (X - \omega u) = PQ$$

where $u \in K$ is one of the $p$-th roots of $a$ in $K$. This equation shows that $P$ and $Q$ split in $K[X]$ into products of factors $X - \omega u$. More precisely, $\mu_p$ decomposes into a union of disjoint subsets $I$ and $J$ such that

$$P = \prod_{\omega \in I} (X - \omega u) \qquad \text{and} \qquad Q = \prod_{\omega \in J} (X - \omega u).$$

Let $b$ be the constant term of $P$, and write $k$ for the degree of $P$, which is the number of elements in $I$. The above factorization of $P$ shows that

$$b = \Big( \prod_{\omega \in I} \omega \Big) (-u)^k.$$

Since $\omega^p = 1$ for any $\omega \in I$, we obtain by raising each side of the preceding equation to the $p$-th power

$$b^p = (-1)^{kp} a^k,$$

hence $a^k = \big( (-1)^k b \big)^p$ is a $p$-th power in $F$. Part (a) of the lemma then shows that $k = 0$ or $k = p$. In the first case $P = 1$ and in the second case $P = X^p - a$, hence $Q = 1$. $\qquad \square$

Let now $R$ be a radical extension of height 1 of some field $F$. By definition, this means that there exists an element $u \in R$ such that $R = F(u)$ and $u^p = a$ for some prime $p$ and some element $a \in F$ that is not a $p$-th power in $F$. Using Lemma 13.8, we can give a standard form to the elements of $R$.

**Corollary 13.9.** *Every element $v \in R$ can be written in a unique way as*

$$v = v_0 + v_1 u + v_2 u^2 + \cdots + v_{p-1} u^{p-1},$$

*for some elements $v_0, v_1, \ldots, v_{p-1} \in F$.*

**Proof.** This readily follows from Proposition 12.15 (p. 166), by Lemma 13.8(b). □

In fact, when $v \in R$ is given beforehand outside $F$, the element $u$ can be chosen in such a way that $v_1 = 1$ in the expression above, as we now show:

**Lemma 13.10.** *Let $R$ be a radical extension of height 1 of some field $F$ and let $v \in R$. If $v \notin F$, then the element $u \in R$ such that $R = F(u)$ and $u^p \in F$ can be chosen in such a way that*

$$v = v_0 + u + v_2 u^2 + \cdots + v_{p-1} u^{p-1}$$

*for some $v_0, v_2, \ldots, v_{p-1} \in F$.*

**Proof.** Let $u'$ be an element of $R$ such that $R = F(u')$ and $u'^p = a'$ for some element $a' \in F$ that is not a $p$-th power in $F$. By Corollary 13.9, we may write

$$v = v_0' + v_1' u' + v_2' u'^2 + \cdots + v_{p-1}' u'^{p-1}$$

for some $v_0', \ldots, v_{p-1}' \in F$. The coefficients $v_1', \ldots, v_{p-1}'$ are not all zero since $v \notin F$. Let $k$ be an index between 1 and $p-1$ such that $v_k' \neq 0$, and let

$$u = v_k' u'^k. \tag{13.1}$$

Raising each side of this equation to the $p$-th power, we obtain

$$u^p = v_k'^p a'^k.$$

This shows that $u$ satisfies the equation $u^p = a$ with $a = v_k'^p a'^k \in F$.

If $a$ is the $p$-th power of an element in $F$, then from $a = v_k'^p a'^k$ we see that $a'^k$ also is a $p$-th power in $F$. But then it follows from Lemma 13.8(a) that $a'$ itself is a $p$-th power in $F$, which contradicts the hypothesis on $u'$. Therefore, $a$ is not a $p$-th power in $F$.

Since $u \in R$, we obviously have $F(u) \subset R$. In order to prove that $R = F(u)$, it thus suffices to show that every element in $R$ has a rational expression in $u$ with coefficients in $F$. We first show that the powers of $u'$ have such expressions. For any $i = 0, \ldots, p - 1$, we obtain by raising each side of (13.1) to the $i$-th power

$$u^i = v_k'^{\,i} u'^{ki}. \tag{13.2}$$

Now, recall the permutation $\sigma_k$ of $\{0, 1, \ldots, p - 1\}$ that maps every integer $i$ between 0 and $p - 1$ to the unique integer $\sigma_k(i)$ between 0 and $p - 1$ such that

$$\sigma_k(i) \equiv ik \mod p$$

(see Proposition 10.7, p. 139). By definition of $\sigma_k(i)$, there is an integer $m$ such that $ik - \sigma_k(i) = pm$, hence

$$u'^{ik} = (u'^p)^m u'^{\sigma_k(i)}.$$

Therefore, recalling that $u'^p = a'$ and letting $b_i = (v_k'^{\,i} a'^m)^{-1} \in F$ for $i = 0$, $\ldots, p - 1$, we obtain from (13.2)

$$b_i u^i = u'^{\sigma_k(i)}$$

for $i = 0, \ldots, p - 1$. Now, every $x \in R$ has an expression $x = \sum_{i=0}^{p-1} x_i u'^i$, with $x_i \in F$ for $i = 0, \ldots, p - 1$, which can be rewritten as

$$x = \sum_{i=0}^{p-1} x_{\sigma_k(i)} u'^{\sigma_k(i)},$$

as $\sigma_k$ is a permutation of $\{0, \ldots, p - 1\}$. Substituting $b_i u^i$ for $u'^{\sigma_k(i)}$, we obtain

$$x = \sum_{i=0}^{p-1} x_{\sigma_k(i)} b_i u^i.$$

This shows that every element in $R$ has a rational expression in $u$ with coefficients in $F$, hence $R = F(u)$. For the given $v \in R$, we have

$$v = \sum_{i=0}^{p-1} v_{\sigma_k(i)}' b_i u^i. \tag{13.3}$$

Since $\sigma_k(1) = k$, the coefficient of $u$ in this expression is $v_k' b_1$. But taking $i = 1$ in the calculations above, we see that $b_1 = (v_k')^{-1}$. Therefore, the coefficient of $u$ in the expression (13.3) of $v$ is 1. This completes the proof. $\qquad\square$

**Lemma 13.11.** *As in Lemma 13.10, let $R$ be a radical extension of height 1 of some field $F$, let $v \in R$, and let $u \in R$ be such that $R = F(u)$, $u^p \in F$ for some prime $p$, and*

$$v = v_0 + u + v_2 u^2 + \cdots + v_{p-1} u^{p-1}$$

*for some $v_0$, $v_2$, ..., $v_{p-1} \in F$. Assume that $F$ contains a primitive $p$-th root of unity $\zeta$ (hence all the $p$-th roots of unity, since the others are powers of $\zeta$). If $v$ is a root of an equation with coefficients in $F$, then $R$ contains $p$ roots of this equation, and $u$, $v_0$, $v_2$, ..., $v_{p-1}$ have rational expressions in these roots with coefficients in $\mathbb{Q}(\zeta)$.*

**Proof.** Let $P \in F[X]$ be such that $P(v) = 0$. Define a polynomial $Q$ with coefficients in $F$ by

$$Q(Y) = P(v_0 + Y + v_2 Y^2 + \cdots + v_{p-1} Y^{p-1}) \in F[Y].$$

Because $P(v) = 0$, it follows that $Q(u) = 0$. On the other hand, $u$ is also a root of the polynomial $Y^p - a$, which is irreducible by Lemma 13.8(b). Therefore, Lemma 12.14 (p. 166) shows that $Y^p - a$ divides $Q(Y)$, and it follows that every root of $Y^p - a$ is a root of $Q(Y)$. Since the roots of $Y^p - a$ are the $p$-th roots of $a$, which are of the form $\zeta^i u$, for $i = 0, \ldots, p - 1$, we have

$$Q(\zeta^i u) = 0 \qquad \text{for } i = 0, \ldots, p - 1. \tag{13.4}$$

For $i = 0, \ldots, p - 1$, let

$$z_i = v_0 + \zeta^i u + v_2 \zeta^{2i} u^2 + \cdots + v_{p-1} \zeta^{(p-1)i} u^{p-1}.$$

(Thus, $z_0 = v$.) Equation (13.4) yields

$$P(z_i) = 0 \qquad \text{for } i = 0, \ldots, p - 1,$$

which proves that $R$ contains $p$ roots of $P$. To complete the proof, we now show that $u$, $v_0$, $v_2$, ..., $v_{p-1}$ have rational expressions in $z_0$, ..., $z_{p-1}$, by calculations that are reminiscent of Lagrange's formula (Proposition 10.2, p. 130). Grouping the terms that contain a given factor $v_j u^j$ in the sum of $\zeta^{-ik} z_i$, we have

$$\sum_{i=0}^{p-1} \zeta^{-ik} z_i = \sum_{j=0}^{p-1} \left( \sum_{i=0}^{p-1} \zeta^{(j-k)i} \right) v_j u^j \qquad \text{for } k = 0, \ldots, p - 1, \tag{13.5}$$

where we have let $v_1 = 1$. If $j \neq k$, then $\zeta^{j-k}$ is a $p$-th root of unity other than 1, hence a root of $\Phi_p(X) = \sum_{i=0}^{p-1} X^i$. Therefore,

$$\sum_{i=0}^{p-1} \zeta^{(j-k)i} = 0.$$

Hence, all the terms with index $j \neq k$ vanish in the right-hand side of (13.5), and it remains

$$\sum_{i=0}^{p-1} \zeta^{-ik} z_i = pv_k u^k \qquad \text{for } k = 0, \ldots, p - 1.$$

This proves that $v_k u^k$ has a rational expression (indeed a linear expression) in $z_0, \ldots, z_{p-1}$ with coefficients in $\mathbb{Q}(\zeta_p)$. In particular, for $k = 1$, we see that $u$ has such an expression, and since $v_k = (v_k u^k)u^{-k}$, it follows that $v_0$, $v_2, \ldots, v_{p-1}$ also have rational expressions in $z_0, \ldots, z_{p-1}$ with coefficients in $\mathbb{Q}(\zeta_p)$.                                                                                     $\square$

Now, let

$$K = \mathbb{C}(x_1, \ldots, x_n),$$

where $x_1, \ldots, x_n$ are independent indeterminates over $\mathbb{C}$, and let $F$ be the subfield of symmetric functions. By Theorem 8.3 (p. 95), we have

$$F = \mathbb{C}(s_1, \ldots, s_n),$$

where $s_1, \ldots, s_n$ are the elementary symmetric polynomials in $x_1, \ldots, x_n$.

**Theorem 13.12 (of natural irrationalities).** *If an element $v \in K$ lies in a radical extension of $F$, then there is inside $K$ a radical extension of $F$ containing $v$.*

**Proof.** We argue by induction on the height of the radical extension $R$ of $F$ containing $v$, which is assumed to exist. There is nothing to prove if the height of $R$ is 0 (i.e., if $R = F$) since in this case $R$ lies inside $K$. We may thus assume the height of $R$ is $h \geq 1$ and consider $R$ as a radical extension of height 1 of some subfield $R_1$, which is a radical extension of $F$ of height $h - 1$.

If $v \in R_1$, then we are done by the induction hypothesis. For the rest of the proof, we may thus assume that $v$ lies outside $R_1$. Lemma 13.10 then shows that

$$R = R_1(u)$$

for some element $u$ such that $u^p \in R_1$ (for some prime $p$) and

$$v = v_0 + u + v_2 u^2 + \cdots + v_{p-1} u^{p-1} \tag{13.6}$$

for some elements $v_0, v_2, \ldots, v_{p-1} \in R_1$. Now, since $v \in K$ we may consider the polynomial

$$P(X) = \prod_{\sigma \in S_n} (X - \sigma(v)).$$

This polynomial lies in $F[X]$ because its coefficients are symmetric functions. Since $v$ is a root of $P$, Lemma 13.11 shows that $R$ contains $p$ roots of $P$, and that $u$, $v_0$, $v_2$, ..., $v_{p-1}$ have rational expressions in these roots. Since all the roots of $P$ lie in $K$, it follows that $u$, $v_0$, $v_2$, ..., $v_{p-1} \in K$.

But $u^p$, $v_0$, $v_2$, ..., $v_{p-1}$ also lie in $R_1$, which is a radical extension of height $h - 1$ of $F$. By the induction hypothesis, $u^p$, $v_0$, $v_2$, ..., $v_{p-1}$ all lie in radical extensions of $F$ inside $K$ and, by Corollary 13.5, we can find a single radical extension $R'$ of $F$ inside $K$ containing $u^p$, $v_0$, $v_2$, ..., $v_{p-1}$. Since $u^p \in R'$, the field $R'(u)$ is a radical extension of $R'$, hence a radical extension of $F$. But we have already observed above that $u \in K$, hence $R'(u) \subset K$. Since (13.6) shows that $v \in R'(u)$, the proof is complete.    $\square$

## 13.3  Proof of the unsolvability of general equations of degree higher than 4

In order to prove that general equations of degree higher than 4 are not solvable by radicals, we have to show, according to Definitions 13.1 above, that for $n \geq 5$ there is no radical extension of $\mathbb{C}(s_1, \ldots, s_n)$ containing a root $x_i$ of the general equation of degree $n$

$$(X - x_1) \ldots (X - x_n) = X^n - s_1 X^{n-1} + \cdots + (-1)^n s_n = 0.$$

The proof we give below is based upon Ruffini's last proof (1813) [64, vol. 2, pp. 162–170]. It is sometimes called the Wantzel modification of Abel's proof (see [64, vol. 2, p. 505] and Serret [66, n° 516]), although Wantzel was relying on Ruffini's papers (see Ayoub [4, p. 270]).

**Lemma 13.13.** *Let $u$ and $a$ be elements of $\mathbb{C}(x_1, \ldots, x_n)$ such that*

$$u^p = a$$

*for some prime number $p$, and assume $n \geq 5$. If $a$ is invariant under the permutations*

$$\sigma \colon x_1 \mapsto x_2 \mapsto x_3 \mapsto x_1; \quad x_i \mapsto x_i \quad \text{for } i > 3$$

*and*

$$\tau \colon x_3 \mapsto x_4 \mapsto x_5 \mapsto x_3; \quad x_i \mapsto x_i \quad \text{for } i = 1, 2 \text{ and } i > 5,$$

*then so is $u$.*

**Proof.** Applying $\sigma$ to each side of the equation $u^p = a$, we get $\sigma(u)^p = a$, hence

$$\sigma(u)^p = u^p.$$

Since the lemma is trivial if $u = 0$, we may assume $u \neq 0$ and divide each side of the preceding equation by $u^p$. We thus obtain

$$\left(\frac{\sigma(u)}{u}\right)^p = 1,$$

hence

$$\sigma(u) = \omega_\sigma u$$

for some $p$-th root of unity $\omega_\sigma$. Applying $\sigma$ to each side of this last equation, we get $\sigma^2(u) = \omega_\sigma^2 u$, next $\sigma^3(u) = \omega_\sigma^3 u$. But $\sigma^3$ is the identity map, hence $\sigma^3(u) = u$, and therefore

$$\omega_\sigma^3 = 1. \tag{13.7}$$

Arguing similarly with $\tau$ instead of $\sigma$, we find

$$\tau(u) = \omega_\tau u$$

with

$$\omega_\tau^3 = 1. \tag{13.8}$$

From these equations, we also deduce

$$\sigma\big(\tau(u)\big) = \omega_\sigma \omega_\tau u \qquad \text{and} \qquad \sigma^2\big(\tau(u)\big) = \omega_\sigma^2 \omega_\tau u. \tag{13.9}$$

However, since the composition of $\sigma$ and $\tau$ yields

$$\sigma \circ \tau \colon\ x_1 \mapsto x_2 \mapsto x_3 \mapsto x_4 \mapsto x_5 \mapsto x_1; \quad x_i \mapsto x_i \quad \text{for } i > 5$$

and since

$$\sigma^2 \circ \tau \colon\ x_1 \mapsto x_3 \mapsto x_4 \mapsto x_5 \mapsto x_2 \mapsto x_1; \quad x_i \mapsto x_i \quad \text{for } i > 5,$$

we have $(\sigma \circ \tau)^5 = (\sigma^2 \circ \tau)^5 = \text{Id}$ (the identity map). Therefore, we derive from (13.9)

$$\left(\omega_\sigma \omega_\tau\right)^5 = \left(\omega_\sigma^2 \omega_\tau\right)^5 = 1. \tag{13.10}$$

Now, observe that

$$\omega_\sigma = \omega_\sigma^6 (\omega_\sigma \omega_\tau)^5 (\omega_\sigma^2 \omega_\tau)^{-5},$$

hence (13.7) and (13.10) yield

$$\omega_\sigma = 1.$$

From (13.10), we then deduce $\omega_\tau^5 = 1$, and since

$$\omega_\tau = \omega_\tau^6 \omega_\tau^{-5},$$

it follows from (13.8) that $\omega_\tau = 1$. This shows that $u$ is invariant under $\sigma$ and $\tau$. $\qquad\square$

**Corollary 13.14.** *Let $R$ be a radical extension of $\mathbb{C}(s_1, \ldots, s_n)$ contained in $\mathbb{C}(x_1, \ldots, x_n)$. If $n \geq 5$, then every element of $R$ is invariant under the permutations $\sigma$ and $\tau$ of Lemma 13.13.*

**Proof.** We argue by induction on the height of $R$, which we denote by $h$. If $h = 0$, then $R = \mathbb{C}(s_1, \ldots, s_n)$ and the corollary is obvious. If $h \geq 1$, then there is an element $u \in R$ and a radical extension $R_1$ of height $h - 1$ of $\mathbb{C}(s_1, \ldots, s_n)$ such that

$$R = R_1(u) \qquad \text{and} \qquad u^p \in R_1$$

for some prime number $p$. By induction, we may assume that every element of $R_1$ is invariant under $\sigma$ and $\tau$. Lemma 13.13 then shows that $u$ is also invariant under $\sigma$ and $\tau$, and, since the elements in $R$ have rational expressions in $u$, it follows that every element in $R$ is invariant under $\sigma$ and $\tau$. $\qquad\square$

We thus reach the conclusion:

**Theorem 13.15.** *If $n \geq 5$, the general equation of degree $n$*

$$P(X) = (X - x_1) \ldots (X - x_n) = X^n - s_1 X^{n-1} + \cdots + (-1)^n s_n = 0$$

*is not solvable by radicals over $\mathbb{Q}(s_1, \ldots, s_n)$, nor over $\mathbb{C}(s_1, \ldots, s_n)$.*

**Proof.** According to Corollary 13.7, it suffices to show that $P(X) = 0$ is not solvable by radicals over $\mathbb{C}(s_1, \ldots, s_n)$. Assume on the contrary that there is a radical extension $R$ of $\mathbb{C}(s_1, \ldots, s_n)$ containing a root $x_i$ of $P$. Changing the numbering of $x_1$, $\ldots$, $x_n$ if necessary, we may assume that $i = 1$. Moreover, by the theorem of natural irrationalities (Theorem 13.12), this radical extension $R$ may be assumed to lie within $\mathbb{C}(x_1, \ldots, x_n)$. Then, Corollary 13.14 shows that every element of $R$ is invariant under $\sigma$ and $\tau$. But $x_1 \in R$ and $x_1$ is not invariant under $\sigma$. This is a contradiction. $\qquad\square$

## Exercises

*Exercise 13.1.* Show that over $\mathbb{R}$ and over $\mathbb{C}$, every equation of any degree is solvable by radicals.

*Exercise 13.2.* Let $R$ be a field containing a field $F$ of characteristic zero. Assume $R = F(u) = F(v)$ where $u$, $v$ are subject to $u^p = a$, $v^q = b$ for some primes $p$, $q$, and some $a$, $b \in F$ that are not $p$-th or $q$-th powers in

$F$ respectively. Show that $p = q$ and $a^k b^{-1}$ is a $p$-th power in $F$ for some $k \in \{1, \ldots, p-1\}$. [Hint: For the last part, reduce to the case where $F$ contains a primitive $p$-th root of unity.]

*Exercise 13.3.* Show that the general cubic equation

$$(X - x_1)(X - x_2)(X - x_3) = X^3 - s_1 X^2 + s_2 X - s_3 = 0$$

is solvable by radicals over $\mathbb{Q}(s_1, s_2, s_3)$. Construct explicitly a radical extension of $\mathbb{Q}(s_1, s_2, s_3)$ containing one of the roots of this cubic and show that this radical extension is not contained in $\mathbb{Q}(x_1, x_2, x_3)$. Thus, the solution of the general cubic equation by radicals over $\mathbb{Q}(s_1, s_2, s_3)$ involves accessory irrationalities.

Same questions for the general equation of degree four.

*Exercise 13.4.* Let $\zeta_7$ (resp. $\zeta_3$) be a primitive 7-th (resp. cube) root of unity. Show that $\mathbb{Q}(\zeta_7)$ is not a radical extension of $\mathbb{Q}$, but that $\mathbb{Q}(\zeta_7, \zeta_3)$ is a radical extension of $\mathbb{Q}$.

*Exercise 13.5.* Let $R$ be a radical extension of a field $F$, of the form $R = F(a^{1/p})$, for some $a \in F$ that is not a $p$-th power in $F$. Find an isomorphism which is the identity on $F$

$$F[X]/(X^p - a) \xrightarrow{\sim} R.$$

Conclude that all the fields of the form $F(a^{1/p})$ are isomorphic, under isomorphisms leaving $F$ invariant.

*Exercise 13.6.* Show that there are three different subfields of $\mathbb{C}$ of the form $\mathbb{Q}(2^{1/3})$. Show that if $F$ is a subfield of $\mathbb{C}$ containing a primitive $p$-th root of unity, then for any $a \in F$ that is not a $p$-th power in $F$ there is only one subfield of $\mathbb{C}$ of the form $F(a^{1/p})$.

*Exercise 13.7. To make up partially for the lack of details on the early stages of the theory of groups in Ruffini's and Cauchy's works, the following exercise presents a result of Cauchy on the number of values of rational functions under permutations of the indeterminates, which was used in Abel's proof that general equations are not solvable by radicals.*

Let $n$ be an integer, $n \geq 3$, let $\Delta = \Delta(x_1, \ldots, x_n)$ be the polynomial defined in §8.2 and let $I(\Delta) \subset S_n$ be the isotropy group of $\Delta$, i.e.,

$$I(\Delta) = \{\sigma \in S_n \mid \sigma(\Delta) = \Delta\}.$$

(This subgroup of $S_n$ is called the *alternating group* on $\{1, \ldots, n\}$, and denoted by $A_n$.)

(a) Show that any permutation of $n$ elements is a composition of permutations that interchange two elements and leave the other elements invariant. (Permutations of this type are called *transpositions*.)

(b) Show that a permutation leaves $\Delta$ invariant if and only if it is a composition of an even number of transpositions.

(c) Let $p$ be an odd prime, $p \leq n$. Show that the cyclic permutations of length $p$

$$i_1 \mapsto i_2 \mapsto \cdots \mapsto i_p \mapsto i_1$$

(where $i_1, \ldots, i_p \in \{1, \ldots, n\}$) generate $I(\Delta)$.

[Hint: By (b), it suffices to show that the composition of any two transpositions is a composition of cycles of length $p$.]

(d) Let again $p$ be an odd prime, $p \leq n$, and let $V$ be a rational function in $x_1, \ldots, x_n$ which takes strictly less than $p$ values under the permutations of $x_1, \ldots, x_n$. Show that $V$ has the form $V = R + \Delta S$ where $R$ and $S$ are symmetric rational functions, hence that the number of values of $V$ is 1 or 2.

[Hint: Show that $V$ is invariant under the cyclic permutations of length $p$.]

(e) Translate the result above in the following purely group-theoretic terms: if $G \subset S_p$ is a subgroup of index $< p$ (with $p$ prime), then $G$ contains the alternating group $A_p$.

[Hint: Use Proposition 10.6, p. 139.]

# Chapter 14

# Galois

After Gauss, Ruffini, and Abel, two major classes of equations have been treated thoroughly, with divergent results: the cyclotomic equations are solvable by radicals in any degree, while general equations of degree at least five are not. Thus, the obvious question arises: *which* equations are solvable by radicals?

Abel himself addressed this question, returning several times to the theory of equations, which he called his *thème favori* [1, t. II, p. 260]. Following a clue from Gauss, he discovered a large class of solvable equations, containing the cyclotomic equations as particular cases. In the introduction to the seventh chapter of "Disquisitiones Arithmeticae," in which he discusses cyclotomic equations, Gauss had written [31, Art. 335]:

> Moreover, the principles of the theory that we are about to explain extend much farther than we let it see here. Indeed, they apply not only to circular functions, but also with the same success to numerous other transcendental functions, e.g. to those which depend on the integral $\int \frac{dx}{\sqrt{1-x^4}}$ and also to various kinds of congruences.

The integral $\int \frac{dx}{\sqrt{1-x^4}}$ occurs in the calculation of the arc length of a lemniscate, just like the integral $\int \frac{dx}{\sqrt{1-x^2}} = \sin^{-1} x$ occurs for the arc length of a circle.

Following this clue, Abel realized that Gauss' method for cyclotomic equations could also be applied to the equations that arise from the division of a lemniscate. In complete analogy with Gauss' results on the constructibility of regular polygons with ruler and compass, Abel even proved that a lemniscate can be divided into $2^n + 1$ equal parts by ruler and compass, whenever $2^n + 1$ is a prime number (see [1, t. II, p. 261], Rosen [62]). Pushing his investigations further, Abel eventually developed the following grand generalization (published in 1829) [1, t. I, p. 479]:

**Theorem 14.1 (Abel).** *Let $P$ be a polynomial with roots $r_1, \ldots, r_n$. If the roots $r_2, \ldots, r_n$ can be rationally expressed in terms of $r_1$, i.e., if there exist rational functions $\theta_2, \ldots, \theta_n$ such that*

$$r_i = \theta_i(r_1) \quad \text{for } i = 2, \ldots, n,$$

*and if moreover*

$$\theta_i \theta_j(r_1) = \theta_j \theta_i(r_1) \quad \text{for all } i, j,$$

*then the equation $P(X) = 0$ is solvable by radicals.*

This theorem applies in particular to cyclotomic equations

$$\Phi_p(X) = X^{p-1} + X^{p-2} + \cdots + X + 1 = 0,$$

for $p$ prime. This follows because the roots of $\Phi_p$ are the primitive $p$-th roots of unity and, denoting one of the roots by $\zeta$, the other roots are powers of $\zeta$. Rational functions $\theta_2, \ldots, \theta_{p-1}$ as above can thus be chosen to be

$$\theta_i(X) = X^i \quad \text{for } i = 2, \ldots, p - 1.$$

Abel's condition obviously holds, since

$$\theta_i \theta_j(\zeta) = \zeta^{ij} = \theta_j \theta_i(\zeta).$$

Elaborating again on these results, Abel was closing in on general necessary and sufficient conditions for an equation to be solvable by radicals. He was working on a comprehensive memoir on this subject [1, t. II, n° 18], when he was prematurely carried off by tuberculosis in 1829.

The honor of finding a complete solution to the problem eventually fell to another young genius, Évariste Galois (1811–1832), who was only 18 in 1830 when he submitted to the Paris Academy of Sciences a memoir on the theory of equations. This memoir was lost in the academy. A new version was submitted by Galois in 1831, but was rejected because it was not sufficiently developed. Galois died in a duel the following year, without having had the occasion to submit a more thorough (or rather, a less sketchy) exposition of his ideas. His "Mémoire sur les conditions de résolubilité des équations par radicaux" [55, Ch. IV] is indeed very dense, as it proposes a groundbreaking new approach to the solution of equations, but it is also quite terse and makes for rather difficult reading. Fortunately, Joseph Liouville (1809–1882) generously took the trouble to decipher Galois' memoir, publishing it in 1846 in his *Journal de Mathématiques pures et appliquées*, thus rescuing Galois theory from complete oblivion.

Galois' basic observation is that the roots of an irreducible polynomial are interchangeable: there is no way to distinguish one from the other.[1] Solving an equation consists in assigning a value to each root, hence in breaking a symmetry. What is needed in order to evaluate the difficulty of an equation is a measure of the symmetry of the system of roots. Galois obtains such a tool by associating to any equation without multiple roots a permutation group, which consists of all the permutations of the roots that preserve the relations among them. Galois' brilliant insight is that this group provides an effective measure of symmetry; in particular, the solvability of the equation by radicals can be translated in terms of the associated group. This is achieved by describing the behavior of the group under extension of the base field. Of these fertile new ideas, Galois offers a single application, proving that irreducible equations of prime degree are solvable by radicals if and only if all the roots can be rationally expressed in terms of any two of them.

This brief summary of Galois' memoir does not do justice to the novelty of the ideas contained therein. Indeed, it is not at all clear how to characterize the permutations that preserve the relations among the roots in this general context. (For the particular case of cyclotomic equations, see §11.2 and §12.3.) This difficulty appears to be insurmountable when one realizes that the notions of field and group, which are central in the modern version of Galois theory, were not available at the time of Galois. The only technical notion Galois relies on is the irreducibility of polynomials. Whether or not a polynomial is irreducible depends upon the field from which the coefficients of factors may be drawn, so the idea of a base field faintly emerges in Galois' discussion of irreducibility:

Definitions. An equation is said to be reducible if it admits rational divisors; irreducible in the contrary case.

It is necessary to explain here what should be understood by the word rational, because it will often reappear.

When the equation has all its coefficients rational numbers, it should mean simply that the equation can be decomposed into factors which have rational numbers as their coefficients.

But when the coefficients of an equation are not *all* rational numbers,

---

[1] By contrast, when a polynomial without multiple roots factors as $P_1 P_2$, its roots are partitioned into two sets, as they are either roots of $P_1$ or $P_2$, and roots of $P_1$ can be distinguished from roots of $P_2$ by the condition $P_1(r) = 0$. See also Proposition 14.35.

then one should understand by rational divisor a divisor whose coefficients may be expressed as rational functions of the coefficients of the proposed equation; [and] in general by a rational quantity, a quantity which may be expressed as a rational function of the coefficients of the proposed equation.

Further, one could agree to regard as rational every rational function of a certain number of determined quantities, supposed known a priori. For example, one can choose a certain root of a whole number, and regard as rational every rational function of this radical.

When we thus agree to regard certain quantities as known, we shall say that we *adjoin* them to the equation which it is required to solve. We shall say that these quantities are *adjoined* to the equation.

That having been said, we shall call *rational* every quantity which can be expressed as a rational function of the coefficients of the equation together with a certain number of quantities *adjoined* to the equation and agreed arbitrarily. [...]

One sees moreover that the properties and the difficulties of an equation can be quite different according to the quantities which are adjoined to it. For example, the adjunction of a quantity can render an irreducible equation reducible [55, p. 109].

The collection of quantities that Galois decides to regard as rational is of course a field containing $\mathbb{Q}$ (i.e., a field of characteristic zero). Besides the rational quantities, Galois also computes with the roots of the equation to be solved, which means he has a notion of splitting field of a polynomial, as provided by Girard's theorem (Theorem 9.3, p. 110). But there is more: he also allows himself to compute with roots of arbitrary auxiliary equations. Therefore, he implicitly assumes that there is a field containing the base field of rational quantities, the roots of the proposed equation, and also the roots of all the auxiliary equations one might need to solve. This hypothesis raises questions of a set-theoretic nature; it can be justified from a modern perspective by proving the existence of an algebraic closure of any base field, but this kind of construction clearly falls beyond the scope of Galois' interests. In the (most common) case where the coefficients of the proposed equation are rational numbers (or, more generally, complex numbers), the question does not arise since all the computations can be assumed to take place within $\mathbb{C}$, as a consequence of the fundamental theorem of algebra (Theorem 9.1, p. 109).

The invention of groups is Galois' major breakthrough. Lagrange, Ruffini, and Cauchy had already made elaborate computations with permutations, but they had not considered distinguished sets of permutations as an object of study. By contrast, groups of permutations occupy a prominent place in Galois' memoir, but they appear in a form that could be disorienting to the modern reader. Firstly, Galois' terminology deviates from the modern usage: he calls *substitutions* the maps that we call permutations, and *permutations* the arrangements of letters in a sequence. But there is a more substantial difference: his groups consist of arrangements of letters, not of permutations in the modern sense. Yet, he is fully aware that substitutions are the more essential notion:

> Substitutions are the passage from one permutation to another. The permutation from which one starts in order to indicate substitutions is completely arbitrary, as far as functions are concerned; for there is no reason at all why a letter should occupy one place rather than another in a function of several letters.
>
> Nevertheless, since it is impossible to grasp the idea of a substitution without that of a permutation, we will make frequent use of permutations in the language, and we shall not consider substitutions other than as the passage from one permutation to another [55, p. 115].

To avoid confusion, in our discussion of Galois' contributions we will stick to the modern usage of calling permutations the bijections from a (finite) set to itself, and we will call *arrangements* of a set $R$ of $n$ elements the various $n$-tuples that can be obtained by arranging the elements in a sequence. Thus, the arrangements of $R = \{a, b, c\}$ are

| | | |
|---|---|---|
| $(a, b, c)$ | | $a\ b\ c$ |
| $(a, c, b)$ | | $a\ c\ b$ |
| $(b, a, c)$ | or simply, in Galois' notation: | $b\ a\ c$ |
| $(b, c, a)$ | | $b\ c\ a$ |
| $(c, a, b)$ | | $c\ a\ b$ |
| $(c, b, a)$ | | $c\ b\ a$ |

The permutations of $R$ act in an obvious way on the arrangements of $R$: given two arrangements $\alpha$ and $\beta$ of the same letters, there is a unique permutation of the letters that transforms $\alpha$ into $\beta$. This permutation is often[2] denoted by $\left(\begin{smallmatrix}\alpha\\\beta\end{smallmatrix}\right)$. For example, the permutation $\left(\begin{smallmatrix}a\ b\ c\\b\ c\ a\end{smallmatrix}\right)$ is the one that moves the letters as follows: $a \mapsto b \mapsto c \mapsto a$.

With appropriate notation, it is easy to translate all of Galois' statements about arrangements into the language of permutations; this was done early on: see, for instance, the 1866 edition of Serret's textbook [66, Ch. 5].

---

[2]Not in Galois' memoir: Galois does not have a notation for the permutation that transforms a given arrangement into another.

Yet, besides its character of authenticity, the viewpoint of arrangements offers some definite advantages in terms of exposition. It turns out to be quite flexible and very well suited to studying the solvability of equations by radicals. To illustrate this point, our discussion of Galois' results will rely on his original approach through arrangements. In the first section, we introduce the notion of a group of arrangements and set up correspondences with groups of permutations. The subsequent sections survey the results in Galois' memoir, following Galois' own ordering of propositions. We thus begin with the definition of the Galois group (of arrangements) of an equation, next investigate the behavior of the Galois group under extension of the base field, and deduce a necessary and sufficient condition for an equation to be solvable by radicals, in terms of its Galois group. In the final section, the application of this condition to irreducible equations of prime degree is described. Moreover, we use Galois' solvability criterion to give a proof of Abel's theorem (Theorem 14.1); this application is not mentioned by Galois.

We emphasize that our discussion is by no means a literally faithful presentation of Galois' memoir: not only do we use modern notation and concepts like fields and permutation groups, but we also supply detailed proofs for all of the statements, whereas Galois wrote very little by way of proofs. (Some arguments below are taken from Liouville [55, p. 159] and Serret [66, Section V].) Moreover, when revising his memoir (probably as late as the very eve of his duel) Galois replaced some of his propositions by more general (and somewhat confusing) statements: see [55, pp. 119–121]. Our exposition is closer in spirit to the initial version, which is already of sufficient richness to establish the solvability criterion.

## 14.1    Arrangements and permutations

Arguments involving permutations are a key ingredient of Galois' memoir. In preparation for the discussion of their highly innovative use in relation to the solution of equations by radicals, we review the basic notions from a modern (abstract) point of view.

For an arbitrary set $X$, we let $\mathrm{Sym}(X)$ denote the group of permutations of $X$, i.e., the set of bijections $X \to X$, which is a group under the composition of maps. An arbitrary group $\Gamma$ is said to *act on* $X$ if to each $\sigma \in \Gamma$ is attached a permutation $\pi_\sigma \in \mathrm{Sym}(X)$ in such a way that $\pi_{\sigma\tau}(x) = \pi_\sigma\big(\pi_\tau(x)\big)$ for all $x \in X$ and all $\sigma, \tau \in \Gamma$. Thus, an *action of* $\Gamma$

*on* $X$ is just a group homomorphism $\pi\colon \Gamma \to \mathrm{Sym}(X)$. When the action is fixed we abuse notations and write simply $\sigma(x)$ for $\pi_\sigma(x)$, so the condition on $\pi$ reads: $\sigma\tau(x) = \sigma\big(\tau(x)\big)$ for $\sigma$, $\tau \in \Gamma$ and $x \in X$. In the sequel, the action will always[3] be fixed and injective, and may therefore be used to identify $\Gamma$ with a subgroup of $\mathrm{Sym}(X)$; our abuse of notations is thus completely harmless.

Besides the functional notation above, which we will use most frequently, we will also have cause to use the exponential notation, where the image of $x \in X$ under the action of $\sigma \in \Gamma$ is denoted by $x^\sigma$. In this case of course we demand that[4] $x^{\sigma\tau} = (x^\sigma)^\tau$ for $x \in X$ and $\sigma$, $\tau \in \Gamma$.

**Definition 14.2.** For $\Gamma$ a group acting on an arbitrary set $X$, the *orbit* of an element $x \in X$ is the set of images of $x$ under the action of $\Gamma$:
$$\Gamma(x) = \{\sigma(x) \mid \sigma \in \Gamma\}.$$

Note that for $x$, $y \in X$ the orbits $\Gamma(x)$ and $\Gamma(y)$ are disjoint unless they coincide: if $\Gamma(x)$ and $\Gamma(y)$ are not disjoint, let $z \in \Gamma(x) \cap \Gamma(y)$, say $z = \rho(x) = \theta(y)$. Since multiplication on the right by $\rho$ (resp. by $\theta$) defines a permutation of the elements of $\Gamma$, we have
$$\Gamma = \{\sigma\rho \mid \sigma \in \Gamma\} = \{\sigma\theta \mid \sigma \in \Gamma\},$$
hence
$$\Gamma(z) = \{\sigma\rho(x) \mid \sigma \in \Gamma\} = \Gamma(x) \quad \text{and} \quad \Gamma(z) = \{\sigma\theta(y) \mid \sigma \in \Gamma\} = \Gamma(y).$$
Alternatively, the property follows because orbits are equivalence classes under the relation $x \equiv y \bmod \Gamma$ defined by the existence of $\sigma \in \Gamma$ such that $y = \sigma(x)$. Therefore, orbits form a partition of $X$: if $X$ is finite, there are $x_1, \ldots, x_t \in X$ such that
$$X = \Gamma(x_1) \cup \cdots \cup \Gamma(x_t).$$

For $\Gamma$ a group acting on an arbitrary set $X$, and for $x \in X$, we also define
$$I(x) = \{\sigma \in \Gamma \mid \sigma(x) = x\}.$$
This set is a subgroup of $\Gamma$, sometimes called the *isotropy group of* $x$ (see §10.3). Lagrange's arguments in [49, Art. 97] (see Theorem 10.3, p. 131) readily yield the following result (which also deserves the distinction of being called "Lagrange's theorem").

**Proposition 14.3.** *For $\Gamma$ a finite group acting on a set $X$, and for $x \in X$,*
$$|\Gamma| = |\Gamma(x)| \cdot |I(x)|.$$

---

[3]Except in Appendix 2 of Chapter 15, where non-injective actions are considered.

[4]Punctilious readers will distinguish actions *on the left*, which are homomorphisms $\Gamma \to \mathrm{Sym}(X)$, from actions *on the right*, which are antihomomorphisms $\Gamma \to \mathrm{Sym}(X)$.

**Proof.** Let $\Gamma(x) = \{x_1, \ldots, x_m\}$ (with $x = x_1$, say). For $i = 1, \ldots, m$, define

$$I(x \mapsto x_i) = \{\sigma \in \Gamma \mid \sigma(x) = x_i\} \subset \Gamma.$$

Because each $\sigma \in \Gamma$ maps $x$ to a unique element in $\Gamma(x)$, we have a decomposition of $\Gamma$ into a disjoint union $\Gamma = \bigcup_{i=1}^{m} I(x \mapsto x_i)$, hence

$$|\Gamma| = \sum_{i=1}^{m} |I(x \mapsto x_i)|. \tag{14.1}$$

For $\sigma \in I(x \mapsto x_i)$, left multiplication by $\sigma$ defines a bijection $I(x) \overset{\sim}{\to} I(x \mapsto x_i)$ (compare the proof of Theorem 10.3, p. 131), hence all the terms of the sum on the right-hand side of (14.1) are equal to $|I(x)|$. The proposition follows. $\qquad\square$

Now, let $R$ be a finite set of $n$ elements, with $n \geq 2$. We attach to $R$ two sets of $n!$ elements:

$\mathrm{Sym}(R)$ is (as per the general definition above) the group of permutations of $R$, i.e., the set of bijections $R \to R$;

$\mathsf{A}(R)$ is the set of arrangements of $R$, i.e., the set of $n$-tuples of elements of $R$ without repetition. Every arrangement can be viewed as a bijection $\{1, \ldots, n\} \to R$: the arrangement $(r_1, \ldots, r_n)$ corresponds to the bijection $i \mapsto r_i$.

As noted in the introduction, the group $\mathrm{Sym}(R)$ acts in an obvious way on $\mathsf{A}(R)$: for $\sigma \in \mathrm{Sym}(R)$ and $\alpha \in \mathsf{A}(R)$, we let $\sigma(\alpha)$ denote the arrangement obtained from $\alpha$ by letting $\sigma$ act on its entries, so for $\alpha = (r_1, \ldots, r_n)$ we have $\sigma(\alpha) = \big(\sigma(r_1), \ldots, \sigma(r_n)\big)$. This action has the remarkable property that for any two arrangements $\alpha$, $\beta \in \mathsf{A}(R)$ there is a unique permutation $\binom{\alpha}{\beta} \in \mathrm{Sym}(R)$ that transforms $\alpha$ into $\beta$. In particular, fixing an arbitrary arrangement $\varpi \in \mathsf{A}(R)$ we obtain a bijection between $\mathsf{A}(R)$ and $\mathrm{Sym}(R)$ by mapping $\alpha \in \mathsf{A}(R)$ to $\binom{\varpi}{\alpha} \in \mathrm{Sym}(R)$.

**Definition 14.4.** A *group of arrangements* of $R$ is defined by Galois to be a non-empty subset $\mathsf{G} \subset \mathsf{A}(R)$ with the following property:

for any $\alpha$, $\beta$, $\gamma \in \mathsf{G}$, the permutation $\binom{\alpha}{\beta}$ carries $\gamma$ to another arrangement of $\mathsf{G}$.

In other words: for any $\alpha$, $\beta$, $\gamma \in \mathsf{G}$, there exists $\delta \in \mathsf{G}$ such that

$$\binom{\alpha}{\beta} = \binom{\gamma}{\delta}.$$

(As a visual aid to distinguish groups of arrangements from groups of permutations, we use sans serif letters (A, G, H ...) to designate groups of arrangements.)

To any group of arrangements $G \subset A(R)$ we associate the set of permutations

$$\Pi(G) = \left\{ \begin{pmatrix} \alpha \\ \beta \end{pmatrix} \mid \alpha, \beta \in G \right\} \subset \mathrm{Sym}(R). \tag{14.2}$$

Clearly, $\Pi(G)$ contains the identity, which is $\begin{pmatrix} \alpha \\ \alpha \end{pmatrix}$ for any $\alpha \in G$, and the inverse of each of its permutations $\begin{pmatrix} \alpha \\ \beta \end{pmatrix}$, which is $\begin{pmatrix} \beta \\ \alpha \end{pmatrix}$. The condition that $G$ is a group implies that $\Pi(G)$ is stable under the composition of maps, and is therefore a subgroup (in the modern sense) of $\mathrm{Sym}(R)$: for $\alpha, \beta, \gamma, \delta \in G$ we may find $\varepsilon \in G$ such that $\begin{pmatrix} \gamma \\ \delta \end{pmatrix} = \begin{pmatrix} \beta \\ \varepsilon \end{pmatrix}$, and then

$$\begin{pmatrix} \gamma \\ \delta \end{pmatrix} \circ \begin{pmatrix} \alpha \\ \beta \end{pmatrix} = \begin{pmatrix} \beta \\ \varepsilon \end{pmatrix} \circ \begin{pmatrix} \alpha \\ \beta \end{pmatrix} = \begin{pmatrix} \alpha \\ \varepsilon \end{pmatrix} \in \Pi(G).$$

From Definition 14.4, it also follows that if $\varpi \in G$ is a fixed arrangement, then every permutation $\sigma \in \Pi(G)$ can be represented in a unique way in the form $\sigma = \begin{pmatrix} \varpi \\ \alpha \end{pmatrix}$ for some $\alpha \in G$. Therefore, we also have

$$\Pi(G) = \left\{ \begin{pmatrix} \varpi \\ \alpha \end{pmatrix} \mid \alpha \in G \right\},$$

and it follows that $\Pi(G)$ and $G$ have the same number of elements. Moreover, this equation shows that $G$ is the orbit of $\varpi$ under the action of $\Pi(G)$:

$$G = \{ \sigma(\varpi) \mid \sigma \in \Pi(G) \} = \Pi(G)(\varpi).$$

For any group of arrangements $G \subset A(R)$, the set $A(R)$ decomposes into orbits under the action of $\Pi(G)$: we may find arrangements $\alpha_1, \ldots, \alpha_t$ such that

$$A(R) = \Pi(G)(\alpha_1) \cup \cdots \cup \Pi(G)(\alpha_t).$$

Each part $\Pi(G)(\alpha_i)$ is a group of arrangements with permutation group $\Pi(G)$, and $\Pi(G)(\alpha_i) = G$ if $\alpha_i \in G$ (see Exercise 14.1). Therefore, the correspondence between groups of arrangements and groups of permutations is *not* one-to-one, since different groups of arrangements may have the same group of permutations. On the other hand, every subgroup $\Gamma \subset \mathrm{Sym}(R)$ is the group of permutations of some group of arrangements: for an arbitrary $\varpi \in A(R)$, the orbit $\Gamma(\varpi)$ is a group of arrangements with $\Pi(\Gamma(\varpi)) = \Gamma$ (see Exercise 14.1).

**Proposition 14.5.** *Let* $G \subset A(R)$ *be a group of arrangements and let* $\sigma \in \mathrm{Sym}(R)$. *The set*

$$\sigma(G) = \{ \sigma(\alpha) \mid \alpha \in G \} \subset A(R)$$

*is a group of arrangements such that* $\Pi(\sigma(G)) = \sigma \circ \Pi(G) \circ \sigma^{-1}$.

**Proof.** For $\alpha$, $\beta \in \mathsf{A}(R)$ the permutations $\sigma \circ \binom{\alpha}{\beta}$ and $\binom{\sigma(\alpha)}{\sigma(\beta)} \circ \sigma$ are identical, since they both transform $\alpha$ into $\sigma(\beta)$, hence

$$\sigma \circ \binom{\alpha}{\beta} \circ \sigma^{-1} = \binom{\sigma(\alpha)}{\sigma(\beta)}. \tag{14.3}$$

Therefore, if $\alpha$, $\beta$, $\gamma \in \mathsf{G}$ and $\binom{\alpha}{\beta}$ transforms $\gamma$ into $\delta \in \mathsf{G}$, then $\binom{\sigma(\alpha)}{\sigma(\beta)}$ transforms $\sigma(\gamma)$ into $\sigma(\delta) \in \sigma(\mathsf{G})$. It follows that $\sigma(\mathsf{G})$ is a group of arrangements, and $\Pi(\sigma(\mathsf{G})) = \sigma \circ \Pi(\mathsf{G}) \circ \sigma^{-1}$ readily follows from (14.3). $\square$

Besides the action of $\mathrm{Sym}(R)$, the set of arrangements $\mathsf{A}(R)$ also carries an action of the symmetric group $S_n = \mathrm{Sym}(\{1, \ldots, n\})$ (as is clear from the description of arrangements as bijections $\{1, \ldots, n\} \to R$): for any arrangement $\alpha = (r_1, \ldots, r_n) \in \mathsf{A}(R)$ and any permutation $\sigma \in S_n$, define

$$\alpha^\sigma = (r_{\sigma(1)}, \ldots, r_{\sigma(n)}) \in \mathsf{A}(R).$$

It is readily verified that for $\sigma$, $\tau \in S_n$ we have $(\alpha^\sigma)^\tau = \alpha^{\sigma\tau}$, and that for any two arrangements $\alpha$, $\beta$ of $R$ there is a unique permutation $\sigma \in S_n$ such that $\alpha^\sigma = \beta$. The actions of $\mathrm{Sym}(R)$ and $S_n$ are related as follows:

**Lemma 14.6.** *For* $\alpha$, $\beta \in \mathsf{A}(R)$ *and* $\sigma \in S_n$,

$$\binom{\alpha}{\beta} = \binom{\alpha^\sigma}{\beta^\sigma} \quad \text{in } \mathrm{Sym}(R).$$

*Equivalently, for* $\alpha \in \mathsf{A}(R)$, $\sigma \in S_n$, *and* $\tau \in \mathrm{Sym}(R)$,

$$\tau(\alpha^\sigma) = \tau(\alpha)^\sigma \quad \text{in } \mathsf{A}(R).$$

**Proof.** Let $\alpha = (r_1, \ldots, r_n)$ and $\beta = (r'_1, \ldots, r'_n)$. Then $\binom{\alpha^\sigma}{\beta^\sigma}$ maps $r_{\sigma(i)}$ to $r'_{\sigma(i)}$ for $i = 1, \ldots, n$, hence it maps $r_j$ to $r'_j$ for $j = 1, \ldots, n$. $\square$

It is clear from the lemma that if $\mathsf{G} \subset \mathsf{A}(R)$ is a group of arrangements and $\sigma \in S_n$, then the set $\mathsf{G}^\sigma = \{\alpha^\sigma \mid \alpha \in \mathsf{G}\}$ is also a group of arrangements, with $\Pi(\mathsf{G}^\sigma) = \Pi(\mathsf{G})$. For instance, the following groups

$$\mathsf{G} = \left\{ \begin{matrix} a\ b\ c \\ b\ c\ a \\ c\ a\ b \end{matrix} \right\} \quad \text{and} \quad \mathsf{G}' = \left\{ \begin{matrix} a\ c\ b \\ b\ a\ c \\ c\ b\ a \end{matrix} \right\}$$

are related by $\mathsf{G}' = \mathsf{G}^\sigma$ with $\sigma \in S_3$ the permutation that fixes 1 and interchanges 2 and 3. Clearly, $\Pi(\mathsf{G}) = \Pi(\mathsf{G}')$.

Mimicking the construction of $\Pi(\mathsf{G})$, we may also attach to any group of arrangements a subgroup of $S_n$:

**Proposition 14.7.** *For any group of arrangements* $\mathsf{G} \subset \mathsf{A}(R)$, *the set*

$$\Pi'(\mathsf{G}) = \{\sigma \in S_n \mid \alpha^\sigma \in \mathsf{G} \text{ for some } \alpha \in \mathsf{G}\}$$

*is a subgroup of $S_n$. For any fixed $\varpi \in G$, we have*

$$\Pi'(G) = \{\sigma \in S_n \mid \varpi^\sigma \in G\} \quad and \quad G = \varpi^{\Pi'(G)} \ (= \{\varpi^\sigma \mid \sigma \in \Pi'(G)\}).$$

**Proof.** It is clear that $\Pi'(G)$ contains the identity of $S_n$. If $\sigma \in \Pi'(G)$, let $\alpha \in G$ be such that $\alpha^\sigma \in G$; then $\sigma^{-1} \in \Pi'(G)$ because $(\alpha^\sigma)^{\sigma^{-1}} = \alpha \in G$. To see that $\Pi'(G)$ is stable under the composition of maps, let $\sigma, \tau \in \Pi'(G)$ and let $\alpha, \beta \in G$ be such that $\alpha^\sigma, \beta^\tau \in G$. By Lemma 14.6 we have $\binom{\beta}{\alpha} = \binom{\beta^\tau}{\alpha^\tau}$, hence $\alpha^\tau \in G$ because $G$ is a group of arrangements. Likewise, $\binom{\alpha}{\alpha^\sigma} = \binom{\alpha^\tau}{\alpha^{\sigma\tau}}$, hence $\alpha^{\sigma\tau} \in G$. Therefore, $\sigma\tau \in \Pi'(G)$. This completes the proof that $\Pi'(G)$ is a subgroup of $S_n$.

Now, fix $\varpi \in G$, and let $\sigma \in \Pi'(G)$. We may then find $\alpha \in G$ such that $\alpha^\sigma \in G$. By Lemma 14.6 we have $\binom{\alpha}{\varpi} = \binom{\alpha^\sigma}{\varpi^\sigma}$, hence $\varpi^\sigma \in G$ because $G$ is a group of arrangements. This proves

$$\Pi'(G) \subset \{\sigma \in S_n \mid \varpi^\sigma \in G\} \quad and \quad \varpi^{\Pi'(G)} \subset G.$$

The reverse of the first inclusion is clear from the definition of $\Pi'(G)$. To see that $G \subset \varpi^{\Pi'(G)}$, observe that for every $\alpha \in G$ there is a unique $\sigma \in S_n$ such that $\alpha = \varpi^\sigma$. The definition of $\Pi'(G)$ shows that $\sigma$ lies in $\Pi'(G)$. □

Every arrangement $\varpi$ in a group $G$ defines an isomorphism $\Pi(G) \to \Pi'(G)$ which maps $\sigma \in \Pi(G)$ to the permutation $\tau \in \Pi'(G)$ defined by the equation $\sigma(\varpi) = \varpi^\tau$ (see Exercise 14.2), hence the action of $\mathrm{Sym}(R)$ may seem to be sufficient for all purposes. Yet, the action of $S_n$ on $\mathsf{A}(R)$ will be a key ingredient in our exposition (starting with the proofs of Lemma 14.16 and Proposition 14.17 below), when functions of several variables are evaluated on arrangements, because the action of $S_n$ on the variables is related to its action on the arrangements by

$$\sigma(f)(\alpha) = f(\alpha^\sigma);$$

see the proof of Proposition 14.17.

## 14.2 The Galois group of an equation

We define in this section the main new notion introduced by Galois, which is the cornerstone of his theory: the Galois group of an equation. As pointed out in the introduction, this group actually consists of *arrangements* of the roots, and not of permutations. Its construction is bewildering, as it relies on multiple choices and looks therefore unlikely to reveal any intrinsic property of the equation. The key observation is that it can be characterized by the condition that the rational functions that take a rational value when evaluated on an arrangement of the roots are exactly those that are constant on the arrangements of the Galois group: see Property (GP) in Theorem 14.9 below. (A modern equivalent would assert that the Galois group of a field extension consists of the automorphisms that leave the base field fixed.) Strictly speaking, this property does not determine the group uniquely, for it also depends on the choice of a reference arrangement. A major result of this section is that the associated group of *permutations* of the roots is uniquely determined: see Proposition 14.17.

For the convenience of present-day readers, we use the modern set-theoretic notation and write $F$ for the base field consisting of Galois' "rational quantities." As should be clear from the passage of Galois' memoir quoted in the introduction, $F$ is an arbitrary field of characteristic zero. Let $P \in F[X]$ be a polynomial of arbitrary degree $n$ without multiple roots in any extension of $F$.[5] The equation to be solved is $P(X) = 0$. Let $R$ be the set of its roots (in some field containing $F$, see Girard's theorem (Theorem 9.3, p. 110)). We write $F(R)$ for the field of rational functions of the roots of $P$ with coefficients in $F$, and let $\mathsf{A}(R)$ be the set of arrangements of $R$, i.e., the set of $n$-tuples $(r_1, \ldots, r_n)$ such that $R = \{r_1, \ldots, r_n\}$. Each polynomial in $n$ variables $\varphi(x_1, \ldots, x_n) \in F[x_1, \ldots, x_n]$ can be evaluated on an arrangement $\alpha \in \mathsf{A}(R)$: if $\alpha = (r_1, \ldots, r_n)$, we let $\varphi(\alpha) = \varphi(r_1, \ldots, r_n) \in F(R)$.

**Lemma 14.8.** *There exists a polynomial $\varphi(x_1, \ldots, x_n) \in F[x_1, \ldots, x_n]$ such that all the evaluations $\varphi(\alpha)$ for $\alpha \in \mathsf{A}(R)$ yield different results: for $\alpha \neq \alpha'$ in $\mathsf{A}(R)$, we have $\varphi(\alpha) \neq \varphi(\alpha')$ in $F(R)$.*

This property may be regarded as obvious. It is Lemma II in Galois' memoir. For a proof, Galois just states that one can take $\varphi(x_1, \ldots, x_n) = b_1 x_1 + \cdots + b_n x_n$ for suitably chosen integers $b_1, \ldots, b_n$, see [55, p. 111].

To the reader in want of a more detailed explanation, we can offer the following: let $L(x_1, \ldots, x_n, y_1, \ldots, y_n) = x_1 y_1 + \cdots + x_n y_n$ where $y_1, \ldots, y_n$ are indeterminates. For $\alpha, \alpha' \in \mathsf{A}(R)$ with $\alpha \neq \alpha'$, we have $L(\alpha, y_1, \ldots, y_n) \neq L(\alpha', y_1, \ldots, y_n)$ in $F(R)[y_1, \ldots, y_n]$, hence the product $\prod_{\alpha \neq \alpha' \in \mathsf{A}(R)} \big( L(\alpha, y_1, \ldots, y_n) - L(\alpha', y_1, \ldots, y_n) \big)$ is a non-zero polynomial

---

[5]The basic Lemma 14.8 below obviously does not hold if multiple roots are allowed. Note that the restriction to equations without multiple roots is not serious, because Hudde's method transforms any equation into an equation with simple roots, see Theorem 5.21 (p. 54) and Remark 5.23 (p. 55).

in $F(R)[y_1, \ldots, y_n]$. Because $F$ is infinite (as its characteristic is assumed to be zero), one can find $b_1, \ldots, b_n \in F$ such that

$$\prod_{\alpha \neq \alpha' \in A(R)} \big(L(\alpha, b_1, \ldots, b_n) - L(\alpha', b_1, \ldots, b_n)\big) \neq 0.$$

Then $\varphi(x_1, \ldots, x_n) = L(x_1, \ldots, x_n, b_1, \ldots, b_n) \in F[x_1, \ldots, x_n]$ satisfies the condition of Lemma 14.8. $\quad\square$

For $\varphi$ as in Lemma 14.8, it is clear that all the polynomials $\varphi(x_{\sigma(1)}, \ldots, x_{\sigma(n)})$ obtained from $\varphi$ by permuting the indeterminates in all possible ways are different, so that $\varphi$ takes $n!$ different "values" under the permutations of $x_1, \ldots, x_n$, in Lagrange's parlance. But the condition in Lemma 14.8 is more stringent. For example, take $P = X^3 + 3X^2 + 6X + 4$, which has $r_1 = -1$, $r_2 = -1 + \sqrt{-3}$, $r_3 = -1 - \sqrt{-3}$ as roots. The polynomial $\varphi(x_1, x_2, x_3) = x_1 + 2x_2 + 3x_3$ yields 6 different polynomials when $x_1$, $x_2$, $x_3$ are permuted among themselves, and yet $\varphi(r_1, r_2, r_3) = \varphi(r_2, r_3, r_1)$.

Until the end of this section, we fix a polynomial $\varphi$ as in Lemma 14.8. To define the Galois group of the polynomial $P$, consider the polynomial

$$\Omega(Y) = \prod_{\alpha \in A(R)} \big(Y - \varphi(\alpha)\big). \tag{14.4}$$

Its coefficients are the elementary symmetric polynomials in the $\varphi(\alpha)$'s, hence they have symmetric polynomial expressions in the roots of $P$, and can therefore be expressed in terms of the coefficients of $P$, in view of the fundamental theorem of symmetric polynomials (Theorem 8.4, p. 95). Thus, $\Omega(Y) \in F[Y]$. Let $\Phi \in F[Y]$ be a monic irreducible factor of this polynomial. If we fix an arrangement $\varpi \in A(R)$, we may specifically choose for $\Phi$ the monic irreducible factor that has $\varphi(\varpi)$ as a root. In any case, we have

$$\Phi(Y) = \prod_{\alpha \in G} \big(Y - \varphi(\alpha)\big) \quad \text{for some subset } G \subset A(R). \tag{14.5}$$

The proof of the main theorem below requires an extended preparation; we postpone it until after Lemma 14.16.

**Theorem 14.9.** *The set* $G$ *is a group of arrangements of $R$. It satisfies the following property:*

**(GP)** *for $f \in F[x_1, \ldots, x_n]$ and $\varpi \in G$, we have $f(\varpi) \in F$ if and only if $f$ is constant on $G$, i.e., $f(\alpha) = f(\varpi)$ for all $\alpha \in G$.*

The group $\mathsf{G}$ is the *Galois group of $P$* (or *of the equation $P(X) = 0$*). Note that the group $\mathsf{G}$ is not uniquely determined since it depends on the choice of $\Phi$ or, equivalently, on the choice of an arrangement $\varpi \in \mathsf{A}(R)$. The number of its elements is the degree of $\Phi$, but the relation between $P$ and $\Phi$ is fairly subtle, as will be clear from the examples below. In particular, the only general relation between the degree of $\Phi$ and the degree $n$ of $P$ is that $\deg \Phi$ divides $n!$ (see Exercise 14.6).

We next give a few examples of Galois group constructions. The first one yields an explicit illustration of the various steps in a specific (easy) case:

**Example 14.10.** Let $P(X) = (X - 1)(X^2 - 2)(X^2 - 3)$. The roots of $P$ are

$$r_1 = 1, \quad r_2 = \sqrt{2}, \quad r_3 = -\sqrt{2}, \quad r_4 = \sqrt{3}, \quad r_5 = -\sqrt{3}.$$

In order to determine the Galois group of $P(X) = 0$ over the field $\mathbb{Q}$ of rational numbers, we first choose a polynomial $\varphi$ as in Lemma 14.8. We may take for instance

$$\varphi(x_1, x_2, x_3, x_4, x_5) = x_1 + 2x_2 + 4x_3 + 8x_4 + 13x_5.$$

(Substituting $r_1, \ldots, r_5$ for $x_1, \ldots, x_5$ in all possible ways in $\varphi(x_1, \ldots, x_5)$ yields 5! real numbers. To check that they are all different, it helps to observe that 1, $\sqrt{2}$, and $\sqrt{3}$ are linearly independent over $\mathbb{Q}$, and that all the differences between any two of the coefficients 1, 2, 4, 8, 13 are different. By contrast, the polynomial $x_1 + 2x_2 + 3x_3 + 4x_4 + 5x_5$ does not satisfy the condition of Lemma 14.8 because $r_1 + 2r_2 + 3r_3 + 4r_4 + 5r_5 = r_1 + 2r_4 + 3r_5 + 4r_2 + 5r_3$.)

We then have to choose an irreducible factor of the polynomial $\Omega(Y)$, which has degree $5! = 60$. Heuristically, one can intuit that if an expression in $r_1, \ldots, r_5$ has a rational value, then $r_2, r_3$ on one hand and $r_4, r_5$ on the other hand should appear symmetrically. Guided by this intuition, we discover that the polynomial

$$\Phi(Y) = \big(Y - \varphi(r_1, r_2, r_3, r_4, r_5)\big) \cdot \big(Y - \varphi(r_1, r_3, r_2, r_4, r_5)\big) \\ \cdot \big(Y - \varphi(r_1, r_2, r_3, r_5, r_4)\big) \cdot \big(Y - \varphi(r_1, r_3, r_2, r_5, r_4)\big) \tag{14.6}$$

lies in $\mathbb{Q}[Y]$: computation yields

$$\Phi(Y) = Y^4 - 4Y^3 - 160Y^2 + 328Y + 4324.$$

This polynomial is irreducible in $\mathbb{Q}[Y]$ because no product of less than four factors in (14.6) yields a polynomial with rational coefficients. Therefore, according to the definition in Theorem 14.9 the Galois group of $P(X) = 0$ is

$$G = \left\{ \begin{array}{l} r_1 \ r_2 \ r_3 \ r_4 \ r_5 \\ r_1 \ r_3 \ r_2 \ r_4 \ r_5 \\ r_1 \ r_2 \ r_3 \ r_5 \ r_4 \\ r_1 \ r_3 \ r_2 \ r_5 \ r_4 \end{array} \right\}.$$

However, choosing a different irreducible factor of $\Omega(Y)$, for example the one with root $\varphi(r_2, r_4, r_1, r_5, r_3)$, yields a different group:

$$G' = \left\{ \begin{array}{l} r_2 \ r_4 \ r_1 \ r_5 \ r_3 \\ r_3 \ r_4 \ r_1 \ r_5 \ r_2 \\ r_2 \ r_5 \ r_1 \ r_4 \ r_3 \\ r_3 \ r_5 \ r_1 \ r_4 \ r_2 \end{array} \right\}.$$

These groups of arrangements have the same group of permutations, which is the group with four elements generated by the permutations $r_2 \leftrightarrow r_3$ and $r_4 \leftrightarrow r_5$. Note that $r_1$ is fixed under every permutation in this group; equivalently, the polynomial $x_1$ takes constant values on all the arrangements in $G$, or $x_3$ takes constant values on $G'$. According to Theorem 14.9, this reflects the fact that $r_1 \in \mathbb{Q}$.

As should be clear from this example, the main difficulty of the construction lies in finding an irreducible factor of $\Omega(Y)$. This is already challenging in the simple case above, and it turns out to be unworkable in more complex situations. However, for some special classes of equations the step-by-step construction can be bypassed: implicit in Galois' memoir is the fact that the group of permutations $\Pi(G)$ (defined in (14.2)) is characterized by Property (GP). (For a proof, see Proposition 14.17 below; see also Corollary 14.18.) This allows Galois to give examples without going through the arduous computation of the irreducible factor $\Phi$, as illustrated in the next two cases.

**Example 14.11.** *General polynomials.* Suppose $P$ is the general polynomial of degree $n$, with indeterminate roots $x_1$, ..., $x_n$ (see Definition 8.1, p. 94) and $F$ is the field of rational functions in the coefficients of $P$, i.e., $F = \mathbb{Q}(s_1, \ldots, s_n)$ where $s_1$, ..., $s_n$ are the elementary symmetric polynomials in $x_1$, ..., $x_n$. For any group $G$ of permutations of $x_1$, ..., $x_n$, one may find a polynomial with rational coefficients $f(x_1, \ldots, x_n)$ with isotropy group $G$, see Proposition 10.6. Thus, $f$ is constant on all the arrangements in the orbit $G\big((x_1, \ldots, x_n)\big)$; but according to the fundamental theorem of symmetric polynomials (Theorem 8.4, p. 95), $f(x_1, \ldots, x_n)$ belongs to $F$ if and only if $G = \mathrm{Sym}(\{x_1, \ldots, x_n\})$. Therefore, the only group that satisfies Property (GP) is the group of *all* arrangements of $x_1, \ldots, x_n$. Thus, the Galois group of the general polynomial $P$ is $\mathsf{A}(\{x_1, \ldots, x_n\})$; its associated permutation group is $\mathrm{Sym}(\{x_1, \ldots, x_n\})$.

**Example 14.12.** *Cyclotomic polynomials.* Suppose $P = \Phi_p$ is the cyclotomic polynomial of index some prime $p$, so $n = p - 1$, and let $F = \mathbb{Q}$. The roots of $P$ are the primitive $p$-th roots of unity, which can be ordered with the help of a primitive root $g$ of $p$ as in §12.3: starting with an arbitrary primitive $p$-th root of unity $\zeta \in \mathbb{C}$, let $\zeta_i = \zeta^{g^i}$ for $i = 0, \ldots, p - 2$. Then $R = \{\zeta_0, \zeta_1, \ldots, \zeta_{p-2}\}$, and we saw in Proposition 12.22 (p. 175) that for any polynomial $f \in \mathbb{Q}[x_1, \ldots, x_n]$, the equation

$$f(\zeta_0, \zeta_1, \ldots, \zeta_{p-3}, \zeta_{p-2}) = f(\zeta_1, \zeta_2, \ldots, \zeta_{p-2}, \zeta_0)$$

holds if and only if $f(\zeta_0, \zeta_1, \ldots, \zeta_{p-2})$ lies in $\mathbb{Q}$. Therefore, the smallest group $\mathsf{G}$ that contains the two arrangements

$$\zeta_0 \ \zeta_1 \ \zeta_2 \ \zeta_3 \ \cdots \ \zeta_{p-2} \qquad \text{and} \qquad \zeta_1 \ \zeta_2 \ \zeta_3 \ \cdots \ \zeta_{p-2} \ \zeta_0$$

satisfies Property (GP). The associated permutation group $\Pi(\mathsf{G})$ contains the following cyclic permutation:

$$\sigma: \quad \zeta_0 \mapsto \zeta_1 \mapsto \zeta_2 \mapsto \cdots \mapsto \zeta_{p-2} \mapsto \zeta_0.$$

As $\mathsf{G}$ is minimal, the group $\Pi(\mathsf{G})$ must be generated by $\sigma$, hence

$$\mathsf{G} = \left\{ \begin{array}{cccccc} \zeta_0 & \zeta_1 & \zeta_2 & \zeta_3 & \cdots & \zeta_{p-2} \\ \zeta_1 & \zeta_2 & \zeta_3 & \cdots & \zeta_{p-2} & \zeta_0 \\ \zeta_2 & \zeta_3 & \cdots & \zeta_{p-2} & \zeta_0 & \zeta_1 \\ \cdots & \cdots & \cdots & \cdots & & \\ \zeta_{p-2} & \zeta_0 & \zeta_1 & \zeta_2 & \cdots & \zeta_{p-3} \end{array} \right\}.$$

This group $\mathsf{G}$ must be a Galois group of $P$ since it satisfies Property (GP). Alternatively, we can also take for Galois group of $P$ the transforms under $\sigma, \sigma^2, \ldots$ of an arbitrary arrangement of $R$ instead of $(\zeta_0, \zeta_1, \ldots, \zeta_{p-2})$.

We now start our preparation for the proof of Theorem 14.9. A key technical tool is the observation that the roots of an irreducible polynomial all satisfy the same polynomial equations:

**Lemma 14.13.** *Let $v$, $w \in F(R)$ be roots of an irreducible polynomial $\Psi \in F[Y]$. For every polynomial $\Theta \in F[Y]$, the equation $\Theta(v) = 0$ implies $\Theta(w) = 0$.*

**Proof.** If $\Theta(v) = 0$, then $\Psi$ and $\Theta$ have a common root in $F(R)$. Since $\Psi$ is irreducible, it follows from Lemma 12.14 (p. 166) that $\Psi$ divides $\Theta$ in $F[Y]$, hence every root of $\Psi$ is a root of $\Theta$. $\qquad\square$

Lemma 14.13 is Lemma I in Galois' memoir [55, p. 111]. It will be used mainly for the roots of the irreducible polynomial $\Phi$ of (14.5).

For the proof of Theorem 14.9, we will use a different description of the Galois group of $P$, which is closer to Galois' own definition. The next lemma is the key to the alternative description of $\mathsf{G}$. It shows that the map carrying each arrangement $\alpha \in \mathsf{A}(R)$ to the value $\varphi(\alpha) \in F(R)$ has an inverse given by rational functions:

**Lemma 14.14.** *There exist rational functions* $\psi_1, \ldots, \psi_n \in F(Y)$ *such that for every arrangement* $\alpha \in \mathsf{A}(R)$

$$\alpha = \big(\psi_1(\varphi(\alpha)), \ldots, \psi_n(\varphi(\alpha))\big).$$

**Proof.** Because $\varphi$ is not invariant under any (non-trivial) permutation of the indeterminates we readily derive from Lagrange's theorem (Theorem 10.5, p. 135) that any rational function in $x_1, \ldots, x_n$ can be rationally expressed as a function of $\varphi$ and the elementary symmetric polynomials $s_1$, $\ldots, s_n$ in $x_1, \ldots, x_n$. This applies in particular to each indeterminate $x_i$, so there exist polynomials $f_1, g_1, \ldots, f_n, g_n$ with coefficients in $F$ such that

$$(x_1, \ldots, x_n) = \left( \frac{f_1(s_1, \ldots, s_n, \varphi)}{g_1(s_1, \ldots, s_n, \varphi)}, \ldots, \frac{f_n(s_1, \ldots, s_n, \varphi)}{g_n(s_1, \ldots, s_n, \varphi)} \right). \tag{14.7}$$

It remains to see that no denominator vanishes when the indeterminates $x_1, \ldots, x_n$ are evaluated on any arrangement $\alpha \in \mathsf{A}(R)$.

Inspection shows that the proof of Theorem 10.5 yields for $x_i$ an expression of the form $f_i(s_1, \ldots, s_n, \varphi)/g_i(s_1, \ldots, s_n, \varphi)$ where

$$g_i(s_1, \ldots, s_n, \varphi) = \prod_{\sigma \in S_n, \, \sigma \neq \mathrm{Id}} \big( \varphi(x_1, \ldots, x_n) - \varphi(x_{\sigma(1)}, \ldots, x_{\sigma(n)}) \big).$$

Because $\varphi(\alpha) \neq \varphi(\alpha')$ for $\alpha \neq \alpha'$ in $\mathsf{A}(R)$, it follows that $g_i(s_1, \ldots, s_n, \varphi)$ does not vanish when the entries of any arrangement of the roots of $P$ are substituted for $x_1, \ldots, x_n$. Therefore, we may define $\psi_i(Y) \in F(Y)$ to be the rational function obtained from $f_i(s_1, \ldots, s_n, Y)/g_i(s_1, \ldots, s_n, Y)$ by substituting the coefficients of $P$ for $s_1, \ldots, s_n$. For any arrangement $\alpha = (r_1, \ldots, r_n)$ of $R$, evaluation of each side of (14.7) at $(r_1, \ldots, r_n)$ yields

$$(r_1, \ldots, r_n) = \big(\psi_1(\varphi(\alpha)), \ldots, \psi_n(\varphi(\alpha))\big). \qquad\square$$

Lemma 14.14 is Lemma III in Galois' memoir [55, p. 111]. (Galois has a different proof, which was deemed insufficient by the referee; the proof above actually follows the referee's suggestion.) It shows that all the roots in $R$ can be rationally expressed in terms of the value $\varphi(\alpha)$ of $\varphi$ on any arrangement $\alpha \in A(R)$. Therefore, in order to determine the roots of $P$ it suffices to find $\varphi(\alpha)$, i.e., to solve $\Phi(Y) = 0$. Because of this observation, the element $\varphi(\alpha)$ is sometimes called a *Galois resolvent*[6] of $P(X) = 0$, and the equation $\Phi(Y) = 0$ is called a *Galois resolvent equation* of $P(X) = 0$.

Let $G = \{\alpha, \alpha', \alpha'', \ldots\}$, and let $v = \varphi(\alpha)$, $v' = \varphi(\alpha')$, $v'' = \varphi(\alpha''), \ldots$ be the roots of $\Phi$ corresponding to the various arrangements in $G$. Using this notation and Lemma 14.14, we may rewrite the elements of $G$ in the form

$$\big(\psi_1(v), \ldots, \psi_n(v)\big)$$
$$\big(\psi_1(v'), \ldots, \psi_n(v')\big)$$
$$\big(\psi_1(v''), \ldots, \psi_n(v'')\big)$$
$$\vdots$$

To avoid complications related to denominators, the following lemma is useful:

**Lemma 14.15.** *Let $\varpi \in G$ be a fixed arrangement of $R$ and let $v = \varphi(\varpi) \in F(R)$. For every $u \in F(R)$, there exists a polynomial $\theta \in F[Y]$ such that $u = \theta(v)$. Likewise, there are polynomials $\psi_1', \ldots, \psi_n' \in F[Y]$ such that $\psi_i'(w) = \psi_i(w)$ for every root $w$ of $\Phi$.*

**Proof.** Lemma 14.14 shows that every root of $P$ has a rational expression in $v$, so $F(R) = F(v)$. Since $v$ is a root of $\Phi$, it follows from Proposition 12.15 (p. 166) that every element in $F(v)$ has a polynomial expression in $v$, hence the first part of the lemma is proved.

To establish the second part, let $\psi_i = f_i/g_i \in F(Y)$ for some relatively prime polynomials $f_i$, $g_i \in F[Y]$. The first part of the lemma yields a polynomial $\psi_i' \in F[Y]$ such that $\psi_i(v) = \psi_i'(v)$, hence $f_i(v) = \psi_i'(v)g_i(v)$. This equation shows that $v$ is a root of the polynomial $f_i - \psi_i'g_i \in F[Y]$. Since Lemma 14.13 shows that all the roots of $\Phi$ satisfy the same polynomial equations it follows that $f_i(w) = \psi_i'(w)g_i(w)$ for every root $w$ of $\Phi$. If there is a root $w$ of $\Phi$ such that $g_i(w) = 0$, then another application of Lemma 14.13 shows that $g_i(v) = 0$. This is a contradiction because $\psi_i(v) =$

---

[6]The reader familiar with the modern version of Galois theory will recognize the property that $\varphi(\alpha)$ is a *primitive element* of $F(R)$.

$f_i(v)/g_i(v)$ is defined. Therefore, we have $g_i(w) \neq 0$, hence $\psi_i(w) = \psi_i'(w)$, for every root $w$ of $\Phi$. $\qquad\square$

From Lemmas 14.14 and 14.15, it follows that every element in $F(R)$ has a *polynomial* expression in the roots of $P$.

We may now prove that $\mathsf{G}$ is a group of arrangements: for $\alpha$, $\beta$, $\alpha' \in \mathsf{G}$, we have to show that the permutation $\binom{\alpha}{\beta}$ maps $\alpha'$ to an arrangement in $\mathsf{G}$. Let $v$, $v'$, $w \in F(R)$ be defined by

$$v = \varphi(\alpha), \qquad v' = \varphi(\alpha'), \qquad w = \varphi(\beta).$$

From the definition of $\mathsf{G}$, it follows that $v$, $v'$, and $w$ are roots of $\Phi$. By Lemma 14.15, there is a polynomial $\theta \in F[Y]$ such that $v' = \theta(v)$, and polynomials $\psi_1'$, ..., $\psi_n'$ such that

$$\varpi = \big(\psi_1'(\varphi(\varpi)), \ldots, \psi_n'(\varphi(\varpi))\big) \qquad \text{for every } \varpi \in \mathsf{G}.$$

In particular,

$$\begin{aligned}
\alpha &= \big(\psi_1'(v), \ldots, \psi_n'(v)\big), \\
\alpha' &= \big(\psi_1'(v'), \ldots, \psi_n'(v')\big), \\
\beta &= \big(\psi_1'(w), \ldots, \psi_n'(w)\big).
\end{aligned} \tag{14.8}$$

**Lemma 14.16.** $\theta(w)$ *is a root of* $\Phi$, *and the permutation* $\binom{\alpha}{\beta}$ *maps* $\alpha'$ *to the arrangement*

$$\big(\psi_1'(\theta(w)), \ldots, \psi_n'(\theta(w))\big).$$

*This arrangement lies in* $\mathsf{G}$.

**Proof.** By definition, $v$, $w$, and $v'$ are roots of $\Phi$. Since $v' = \theta(v)$, it follows that $v$ is a root of the polynomial $\Phi\big(\theta(Y)\big)$. Lemma 14.13 shows that all the roots of $\Phi$ satisfy the same polynomial equations, hence $\Phi\big(\theta(w)\big) = 0$. The first claim is thus proved.

We next determine the image of $\alpha'$ under the permutation $\binom{\alpha}{\beta}$, i.e., the arrangement $\beta'$ such that $\binom{\alpha}{\beta} = \binom{\alpha'}{\beta'}$. Since $\alpha$ and $\alpha'$ are arrangements of the $n$ roots of $P$, there exists a permutation $\sigma \in S_n$ such that $\alpha' = \alpha^\sigma$. Lemma 14.6 yields $\binom{\alpha}{\beta} = \binom{\alpha^\sigma}{\beta^\sigma}$, hence $\beta' = \beta^\sigma$.

In view of the description of $\alpha$ and $\alpha'$ in (14.8), the equation $\alpha' = \alpha^\sigma$ means that

$$\psi_i'(v') = \psi_{\sigma(i)}'(v) \qquad \text{for } i = 1, \ldots, n,$$

hence

$$\psi_i'\big(\theta(v)\big) - \psi_{\sigma(i)}'(v) = 0 \qquad \text{for } i = 1, \ldots, n.$$

As $w$ satisfies the same equations as $v$ by Lemma 14.13, it follows that

$$\psi_i'\big(\theta(w)\big) - \psi_{\sigma(i)}'(w) = 0 \qquad \text{for } i = 1, \ldots, n.$$

Because $\beta = \big(\psi_1'(w), \ldots, \psi_n'(w)\big)$, these equations yield

$$\big(\psi_1'(\theta(w)), \ldots, \psi_n'(\theta(w))\big) = \beta^\sigma.$$

We have thus proved the second claim, for $\beta^\sigma$ is the image of $\alpha'$ under $\binom{\alpha}{\beta}$. This arrangement lies in G because $\theta(w)$ is a root of $\Phi$. $\qquad\square$

**Proof of Theorem 14.9.** The theorem asserts that G is a group of arrangements (in the sense of Definition 14.4), and that it satisfies the following property:

**(GP)** for $f \in F[x_1, \ldots, x_n]$ and $\varpi \in$ G, we have $f(\varpi) \in F$ if and only if $f(\alpha) = f(\varpi)$ for all $\alpha \in$ G.

That G is a group of arrangements was shown in Lemma 14.16, so it only remains to prove Property (GP). As above, let $\psi_1', \ldots, \psi_n' \in F[Y]$ be polynomials as in Lemma 14.15. Let $v_1, \ldots, v_m$ be the roots of $\Phi$, so

$$\mathsf{G} = \big\{ \big(\psi_1'(v_i), \ldots, \psi_n'(v_i)\big) \mid i = 1, \ldots, m \big\}.$$

Let $\varpi \in$ G, say $\varpi = \big(\psi_1'(v_1), \ldots, \psi_n'(v_1)\big)$. For $f \in F[x_1, \ldots, x_n]$, consider the polynomial

$$g(Y) = f\big(\psi_1'(Y), \ldots, \psi_n'(Y)\big) \in F[Y].$$

If $f$ is constant on G, then

$$f(\varpi) = g(v_1) = \cdots = g(v_m) = \tfrac{1}{m}\big(g(v_1) + \cdots + g(v_m)\big).$$

Now, the polynomial $\tfrac{1}{m}\big(g(y_1) + \cdots + g(y_m)\big)$ in the indeterminates $y_1, \ldots, y_m$ is obviously symmetric, hence by the fundamental theorem of symmetric polynomials (Theorem 8.4, p. 95) it can be expressed as a polynomial in the elementary symmetric polynomials $s_1, \ldots, s_m$ in $y_1, \ldots, y_m$ with coefficients in $F$. Substituting the coefficients of $\Phi$ for $s_1, \ldots, s_m$ in this polynomial, we obtain for $f(\varpi)$ a value in $F$.

Conversely, suppose $f(\varpi) = a$ for some $a \in F$. Then $g(v_1) = a$, hence $v_1$ is a root of $g(Y) - a \in F[Y]$. But all the roots of $\Phi$ satisfy the same polynomial equations by Lemma 14.13, hence all the roots of $\Phi$ are roots of $g(Y) - a$, and therefore $f(\alpha) = a$ for all $\alpha \in$ G. $\qquad\square$

Our final result in this section shows that the group of permutations $\Pi(\mathsf{G})$ of the Galois group $\mathsf{G}$ (defined in (14.2)) is uniquely determined by Property (GP).

**Proposition 14.17.** *Let* $\mathsf{H}$ *be a group of arrangements of* $R$. *If* $\mathsf{H}$ *satisfies Property (GP), then* $\Pi(\mathsf{H}) = \Pi(\mathsf{G})$. *Moreover,* $\mathsf{G} = \mathsf{H}^\sigma$ *for any* $\sigma \in S_n$ *such that* $\mathsf{G}$ *and* $\mathsf{H}^\sigma$ *are not disjoint.*

**Proof.** Let $\varpi \in \mathsf{G}$, $\pi \in \mathsf{H}$, and let $\sigma \in S_n$ be such that $\pi^\sigma = \varpi$. For all arrangements $\alpha$, $\beta$ of $R$ we have $\binom{\alpha^\sigma}{\beta^\sigma} = \binom{\alpha}{\beta}$ by Lemma 14.6, hence the set

$$\mathsf{H}^\sigma = \{\alpha^\sigma \mid \alpha \in \mathsf{H}\}$$

is a group of arrangements with $\Pi(\mathsf{H}^\sigma) = \Pi(\mathsf{H})$. For $f \in F[x_1, \ldots, x_n]$, let[7]

$$\sigma(f)(x_1, \ldots, x_n) = f(x_{\sigma(1)}, \ldots, x_{\sigma(n)}) \in F[x_1, \ldots, x_n].$$

For any arrangement $\alpha$ of $R$ we have $\sigma(f)(\alpha) = f(\alpha^\sigma)$, hence $f$ is constant on $\mathsf{H}^\sigma$ if and only if $\sigma(f)$ is constant on $\mathsf{H}$. Therefore, $\mathsf{H}$ satisfies Property (GP) if and only if $\mathsf{H}^\sigma$ satisfies (GP). Substituting $\mathsf{H}^\sigma$ for $\mathsf{H}$, we may assume for the rest of the proof that $\mathsf{G}$ and $\mathsf{H}$ both contain $\varpi$. We will prove that $\mathsf{G} = \mathsf{H}$.

Consider the Galois resolvent equation $\Phi(Y) = 0$ used to define $\mathsf{G}$ in (14.5), and $\Phi\big(\varphi(x_1, \ldots, x_n)\big) \in F[x_1, \ldots, x_n]$. Because $\varpi \in \mathsf{G}$, the element $\varphi(\varpi)$ is a root of $\Phi$, hence $\Phi\big(\varphi(\varpi)\big) = 0$. Since $\mathsf{H}$ satisfies (GP), it follows that $\Phi\big(\varphi(\alpha)\big) = 0$ for every $\alpha \in \mathsf{H}$. But, by definition, $\mathsf{G}$ is the set of arrangements $\alpha$ such that $\varphi(\alpha)$ is a root of $\Phi$, hence $\mathsf{H} \subset \mathsf{G}$.

To prove the reverse inclusion, we will show that the polynomial

$$\Psi(Y) = \prod_{\alpha \in \mathsf{H}} (Y - \varphi(\alpha))$$

lies in $F[Y]$. Since it divides the irreducible polynomial $\Phi$, we must then have $\Psi = \Phi$, hence $\mathsf{H} = \mathsf{G}$. To see that the coefficients of $\Psi$ lie in $F$, it suffices to show that evaluating symmetric polynomials in the $\varphi(\alpha)$'s for $\alpha \in \mathsf{H}$ yields elements in $F$. For this, we use the subgroup $\Pi'(\mathsf{H}) \subset S_n$ defined at the end of §14.1. Recall from Proposition 14.7 that

$$\Pi'(\mathsf{H}) = \{\sigma \in S_n \mid \varpi^\sigma \in \mathsf{H}\} \quad \text{and} \quad \mathsf{H} = \varpi^{\Pi'(\mathsf{H})}.$$

For the ease of notation, fix a numbering of the elements:

$$\Pi'(\mathsf{H}) = \{\sigma_1, \ldots, \sigma_t\}, \quad \text{hence} \quad \mathsf{H} = \{\varpi^{\sigma_1}, \ldots, \varpi^{\sigma_t}\}.$$

---

[7]This notation is consistent with the action of $S_n$ on rational functions defined in §10.3: $\sigma(f)(x_1, \ldots, x_n) = \sigma\big(f(x_1, \ldots, x_n)\big)$.

Let $s \in F[y_1, \ldots, y_t]$ be a symmetric polynomial in $t$ variables. We want to show

$$s\big(\varphi(\varpi^{\sigma_1}), \ldots, \varphi(\varpi^{\sigma_t})\big) \in F.$$

Since H satisfies (GP), it suffices to prove

$$s\big(\varphi(\alpha^{\sigma_1}), \ldots, \varphi(\alpha^{\sigma_t})\big) = s\big(\varphi(\varpi^{\sigma_1}), \ldots, \varphi(\varpi^{\sigma_t})\big) \quad \text{for all } \alpha \in \mathsf{H}.$$

As $s$ is a symmetric polynomial, we only need to show

$$\{\alpha^{\sigma_1}, \ldots, \alpha^{\sigma_t}\} = \{\varpi^{\sigma_1}, \ldots, \varpi^{\sigma_t}\} \quad \text{for all } \alpha \in \mathsf{H}. \tag{14.9}$$

The right-hand side is H. But we saw in Proposition 14.7 that the definition of $\Pi'(\mathsf{H})$ does not depend on the choice of the reference arrangement, hence for all $\alpha \in \mathsf{H}$ we have

$$\{\sigma_1, \ldots, \sigma_t\} = \Pi'(\mathsf{H}) = \{\sigma \in S_n \mid \alpha^\sigma \in \mathsf{H}\}.$$

Therefore, the left side of (14.9) is also H, proving that equation (14.9) holds.                                                                          $\square$

**Corollary 14.18.** *Let $\Phi' \in F[Y]$ be a monic irreducible factor of the polynomial $\Omega$ of (14.4), and let $\mathsf{G}'$ be the corresponding group of arrangements of $R$, defined by*

$$\Phi'(Y) = \prod_{\alpha \in \mathsf{G}'} \big(Y - \varphi(\alpha)\big).$$

*If $\Phi' \neq \Phi$, then the groups $\mathsf{G}$, $\mathsf{G}'$ are disjoint, but $\Pi(\mathsf{G}) = \Pi(\mathsf{G}')$ and $\deg \Phi = \deg \Phi'$.*

**Proof.** Since they are different, the monic irreducible polynomials $\Phi$, $\Phi'$ are relatively prime, hence they cannot share a factor $Y - \varphi(\alpha)$. Therefore, $\mathsf{G}$ and $\mathsf{G}'$ are disjoint. By Theorem 14.9 each of these groups satisfies Property (GP), hence $\Pi(\mathsf{G}) = \Pi(\mathsf{G}')$ by Proposition 14.17. It follows that $\deg \Phi = \deg \Phi'$, since the degree of $\Phi$ (resp. $\Phi'$) is the number of elements in $\mathsf{G}$ (resp. $\mathsf{G}'$), which is the order of the group $\Pi(\mathsf{G})$ (resp. $\Pi(\mathsf{G}')$).                                    $\square$

## 14.3   The Galois group under base field extension

The main ingredient in Galois' characterization of the equations that are solvable by radicals is the analysis of the successive reductions in the Galois group when radicals are adjoined to the base field. These reductions are the focus of our discussion in this section.

We use the same notation as in the preceding section: $F$ is an arbitrary base field of characteristic zero and $R$ is the set of roots of the equation $P(X) = 0$ of degree $n$ that has to be solved. We let $\mathsf{A}(R)$ denote the set of arrangements of $R$, choose a polynomial $\varphi$ such that all the $\varphi(\alpha)$ for $\alpha \in \mathsf{A}(R)$ are pairwise distinct (as in Lemma 14.8) and let $\Phi \in F[Y]$ be an irreducible factor of $\Omega(Y) = \prod_{\alpha \in \mathsf{A}(R)} \bigl(Y - \varphi(\alpha)\bigr)$:

$$\Phi(Y) = \prod_{\alpha \in \mathsf{G}} \bigl(Y - \varphi(\alpha)\bigr).$$

Thus, $\Phi(Y) = 0$ is a Galois resolvent equation, and $\mathsf{G}$ is a Galois group of arrangements of $P$. We want to describe what happens to this construction when a root of an auxiliary equation of prime degree is adjoined to $F$.

Let $Q \in F[Z]$ be an irreducible polynomial of prime degree $p$, and let $u_1, \ldots, u_p$ be the roots of $Q$ in some field that contains $F$ and also the roots of $P$. We let $K = F(u_1)$ and $L = F(u_1, \ldots, u_p)$. Thus, $K$ is the new base field when one root of the equation $Q(Z) = 0$ is adjoined to $F$, whereas $L$ is the base field when *all* the roots of $Q(Z) = 0$ are adjoined. To determine the Galois group of $P(X) = 0$ over $K$, we have to pick an irreducible factor of $\Omega$ in $K[Y]$. If $\Phi$ remains irreducible when viewed in $K[Y]$, then we may still choose $\Phi$ hence the Galois group does not change. So, henceforth we assume that $\Phi$ factors in $K[Y]$: let

$$\Phi(Y) = \Lambda_1(Y) \ldots \Lambda_s(Y) \tag{14.10}$$

where $\Lambda_1, \ldots, \Lambda_s \in K[Y]$ are the monic irreducible factors of $\Phi$, and $s > 1$. (We will prove in Proposition 14.20 that $s = p$.) Corollary 14.18 shows that all the polynomials $\Lambda_i$ have the same degree, since they are irreducible factors of $\Omega$. Let

$$d = \deg \Lambda_1 = \cdots = \deg \Lambda_s, \qquad \text{hence} \quad \deg \Phi = ds.$$

For $i = 1, \ldots, s$, define $\mathsf{H}_i \subset \mathsf{G}$ by

$$\Lambda_i(Y) = \prod_{\alpha \in \mathsf{H}_i} \bigl(Y - \varphi(\alpha)\bigr).$$

The number of arrangements in each $\mathsf{H}_i$ is $d$, and the factorization (14.10) shows that $\mathsf{G}$ decomposes as a disjoint union

$$\mathsf{G} = \mathsf{H}_1 \cup \cdots \cup \mathsf{H}_s. \tag{14.11}$$

By definition of Galois group, each $\mathsf{H}_i$ is a Galois group of $P(X) = 0$ over $K$, and by Corollary 14.18 all these groups have the same group of permutations

$$\Pi(\mathsf{H}_1) = \cdots = \Pi(\mathsf{H}_s). \tag{14.12}$$

Thus, each $H_i$ is a subgroup of $d$ arrangements in $G$, and $\Pi(H_i)$ is a subgroup of $d$ permutations in $\Pi(G)$.

Following Galois, we will obtain another factorization of $\Phi$ in $L[Y]$. For this, we first observe that every element in $K = F(u_1)$ has a polynomial expression in $u_1$ by Proposition 12.15 (p. 166), hence we may find polynomials $a_{i,j} \in F[Z]$ for $i = 1, \ldots, s$ and $j = 0, \ldots, d-1$ such that

$$\Lambda_i(Y) = a_{i,0}(u_1) + a_{i,1}(u_1)Y + \cdots + a_{i,d-1}(u_1)Y^{d-1} + Y^d \quad \text{for } i = 1, \ldots, s.$$

For $i = 1, \ldots, s$, let

$$\lambda_i(Y, Z) = a_{i,0}(Z) + a_{i,1}(Z)Y + \cdots + a_{i,d-1}(Z)Y^{d-1} + Y^d \in F[Y, Z],$$

so $\Lambda_i(Y) = \lambda_i(Y, u_1)$ for $i = 1, \ldots, s$.

**Lemma 14.19.** *For all $i = 1, \ldots, s$ and $j = 1, \ldots, p$, we have* $\deg \lambda_i(Y, u_j) = d$ *and*

$$\Phi(Y) = \lambda_1(Y, u_j) \ldots \lambda_s(Y, u_j) \quad \text{in } F(u_j)[Y].$$

*Moreover, if $\lambda_i(Y, u_j)$ divides $\Theta(Y, u_j)$ for some polynomial $\Theta(Y, Z) \in F[Y, Z]$, then $\lambda_i(Y, u_k)$ divides $\Theta(Y, u_k)$ for all $k = 1, \ldots, p$.*

**Proof.** Clearly, $\lambda_i(Y, u_j)$ is monic of degree $d$ for all $i, j$. Let

$$\Phi(Y) - \lambda_1(Y, Z) \ldots \lambda_s(Y, Z) = c_0(Z) + c_1(Z)Y + \cdots + c_t(Z)Y^t$$

for some $t$ and some polynomials $c_0, \ldots, c_t \in F[Z]$. Equation (14.10) yields $\Phi(Y) = \lambda_1(Y, u_1) \ldots \lambda_s(Y, u_1)$, hence

$$c_0(u_1) = c_1(u_1) = \cdots = c_t(u_1) = 0.$$

Therefore, by Lemma 14.13, every root of the irreducible polynomial $Q$ is a root of $c_0, \ldots, c_t$, and it follows that

$$\Phi(Y) - \lambda_1(Y, u_j) \ldots \lambda_s(Y, u_j) = 0 \quad \text{for } j = 1, \ldots, p.$$

The last claim follows by similar arguments: let $\Theta(Y, Z) \in F[Y, Z]$, which we regard as a polynomial in $Y$ with coefficients in $F[Z]$. Since $\lambda_i(Y, Z)$ is monic in $Y$, we may use the Euclidean algorithm for polynomials in $Y$ with coefficients in $F[Z]$ (see Theorem 5.1, p. 43) to obtain polynomials $\Lambda, \Psi \in F[Y, Z]$ such that

$$\Theta(Y, Z) = \lambda_i(Y, Z)\Lambda(Y, Z) + \Psi(Y, Z)$$

and such that the degree $\deg_Y \Psi$ of $\Psi$ as a polynomial in $Y$ is strictly less than $d$ ($= \deg_Y \lambda_i$). If $\lambda_i(Y, u_j)$ divides $\Theta(Y, u_j)$, we have $\Psi(Y, u_j) = 0$ (because $\Psi(Y, u_j)$ is a remainder of the division of $\Theta(Y, u_j)$ by $\lambda_i(Y, u_j)$), which means that $u_j$ is a root of all the polynomials in $Z$ that are the coefficients of $\Psi$. As above, it follows that $\Psi(Y, u_k) = 0$ for all $k = 1, \ldots, p$, hence $\lambda_i(Y, u_k)$ divides $\Theta(Y, u_k)$. $\qquad\square$

We now obtain alternative factorizations of $\Phi$ in $L[Y]$:

**Proposition 14.20.** *With the notation above, we have $s = p$, and for all $i = 1, \ldots, s$*

$$\Phi(Y) = \lambda_i(Y, u_1)\lambda_i(Y, u_2) \ldots \lambda_i(Y, u_p) \quad in\ L[Y].$$

**Proof.** Let $\Phi_i(Y) = \prod_{j=1}^{p} \lambda_i(Y, u_j)$ for $i = 1, \ldots, s$. Since the product is symmetric in $u_1, \ldots, u_p$, it can be expressed in terms of the coefficients of $Q$, hence $\Phi_i \in F[Y]$. Multiplying the $p$ factorizations in Lemma 14.19 for $j = 1, \ldots, p$, we obtain

$$\Phi(Y)^p = \prod_{i=1}^{s}\prod_{j=1}^{p} \lambda_i(Y, u_j) = \Phi_1(Y) \ldots \Phi_s(Y).$$

Now, $\Phi$ is irreducible in $F[Y]$, hence the prime factorization of each $\Phi_i$ in $F[Y]$ only involves the irreducible factor $\Phi$. Let $\Phi_i = \Phi^{m_i}$ for some $m_i \geq 1$. We have $\deg \Phi = ds$ and, by Lemma 14.19, $\deg \Phi_i = dp$, hence

$$dp = dsm_i.$$

Because $p$ is prime and $s > 1$ we must have $s = p$ and $m_i = 1$. Therefore, $\Phi_i = \Phi$ for all $i = 1, \ldots, s$. $\qquad\square$

**Corollary 14.21.** *For all $i, j = 1, \ldots, p$ the polynomial $\lambda_i(Y, u_j)$ is irreducible in $F(u_j)[Y]$.*

**Proof.** The equation $s = p$ established in Proposition 14.20 shows that $\Phi$ decomposes in $F(u_1)[Y]$ into a product of $p$ irreducible polynomials of degree $d$. Since $u_1$ does not play any special role among the roots of $Q$, it follows that for any $j = 1, \ldots, p$ there is a factorization in $F(u_j)[Y]$

$$\Phi(Y) = \Lambda'_1(Y) \ldots \Lambda'_p(Y)$$

with $\Lambda'_1, \ldots, \Lambda'_p \in F(u_j)[Y]$ monic irreducible polynomials of degree $d$. On the other hand, from Lemma 14.19 we have

$$\Phi(Y) = \lambda_1(Y, u_j) \ldots \lambda_p(Y, u_j).$$

Therefore, each $\lambda_i(Y, u_j)$ must be divisible by some $\Lambda'_k(Y)$. But $\lambda_i(Y, u_j)$ and $\Lambda'_k(Y)$ are both monic of degree $d$, hence $\lambda_i(Y, u_j) = \Lambda'_k(Y)$, proving that $\lambda_i(Y, u_j)$ is irreducible. $\qquad\square$

As a result of the factorizations in Proposition 14.20, we obtain a new decomposition of the Galois group $G$. The following is Proposition II in Galois' memoir [55, p. 119]:

**Theorem 14.22.** *Assuming the Galois resolvent equation $\Phi(Y) = 0$ is not irreducible over $K$, the Galois group $G$ decomposes into a disjoint union of $p$ subgroups of $d$ arrangements*

$$G = G_1 \cup \cdots \cup G_p$$

*where each $G_i$ is a Galois group of $P$ over $F(u_i)$. Moreover, for any $i$, $j = 1, \ldots, p$ there is a permutation $\sigma \in \Pi(G)$ such that $G_j = \sigma(G_i)$, i.e., $G_j$ is obtained by letting $\sigma$ act on the arrangements of $G_i$.*

**Proof.** For $i = 1, \ldots, p$, let $G_i \subset G$ be defined by

$$\lambda_1(Y, u_i) = \prod_{\alpha \in G_i} (Y - \varphi(\alpha)).$$

Since $\lambda_1(Y, u_i)$ is an irreducible factor of $\Omega$ in $F(u_i)[Y]$ by Lemma 14.19 (or Proposition 14.20) and Corollary 14.21, it follows that $G_i$ is a Galois group of $P$ over $F(u_i)$ by definition. The number of arrangements in $G_i$ is $d$ because $\deg \lambda_1(Y, u_i) = d$ by Lemma 14.19. The following factorization of $\Phi$ from Proposition 14.20

$$\Phi(Y) = \lambda_1(Y, u_1)\lambda_1(Y, u_2) \ldots \lambda_1(Y, u_p)$$

shows that $G$ is the disjoint union of $G_1, \ldots, G_p$. To complete the proof, fix $i$ and $j = 1, \ldots, p$, and arrangements $\alpha \in G_i$ and $\beta \in G_j$. Consider the permutation $\sigma = \binom{\alpha}{\beta}$, such that $\sigma(\alpha) = \beta$. We will show that $\sigma(G_i) = G_j$. Since $G_i$ and $G_j$ have the same number $d$ of arrangements, it suffices to show that $\sigma(G_i) \subset G_j$, i.e., that $\sigma(\alpha') \in G_j$ for every $\alpha' \in G_i$.

Let $v = \varphi(\alpha)$, $w = \varphi(\beta)$, and $v' = \varphi(\alpha')$, and let $\theta \in F[Y]$ be such that $v' = \theta(v)$. Lemma 14.16 shows that $\varphi(\sigma(\alpha')) = \theta(w)$. Now, $v'$ is a root of $\lambda_1(Y, u_i)$ because $\alpha' \in G_i$, hence $v$ is a root of $\lambda_1(\theta(Y), u_i) \in F(u_i)[Y]$. But $v$ is also a root of the irreducible polynomial $\lambda_1(Y, u_i)$, hence by Lemma 12.14 (p. 166) $\lambda_1(Y, u_i)$ divides $\lambda_1(\theta(Y), u_i)$. By Lemma 14.19, it follows that $\lambda_1(Y, u_j)$ divides $\lambda_1(\theta(Y), u_j)$, hence every root of $\lambda_1(Y, u_j)$ is a root of $\lambda_1(\theta(Y), u_j)$. In particular, $w$ is a root of $\lambda_1(Y, u_j)$ because $\beta \in G_j$, hence $\theta(w)$ is a root of $\lambda_1(Y, u_j)$ and therefore $\sigma(\alpha') \in G_j$ since $\varphi(\sigma(\alpha')) = \theta(w)$. $\qquad\square$

Comparing Theorem 14.22 with (14.11), we have two decompositions of $G$ into disjoint unions of $p$ subgroups of $d$ arrangements (recall from Proposition 14.20 that $s = p$):

$$G = G_1 \cup \cdots \cup G_p = H_1 \cup \cdots \cup H_p.$$

By definition $G_i$ is the set of arrangements $\alpha$ such that $\varphi(\alpha)$ is a root of $\lambda_1(Y, u_i)$, whereas $H_i$ is the set of arrangements $\alpha$ such that $\varphi(\alpha)$ is a root of $\lambda_i(Y, u_1)$, hence $G_1 = H_1$. Therefore, if $p = 2$ we must also have $G_2 = H_2$ because $G_2 = G \setminus G_1$ and $H_2 = G \setminus H_1$. But if $p \geq 3$ the two decompositions may be different. Indeed, the subgroups $G_i$ and $H_i$ have different properties: we saw in (14.12) that

$$\Pi(H_1) = \Pi(H_2) = \cdots = \Pi(H_p),$$

whereas choosing $\sigma_2, \ldots, \sigma_p \in \Pi(G)$ such that $G_1 = \sigma_i(G_i)$ we have by Proposition 14.5

$$\Pi(G_1) = \sigma_2 \circ \Pi(G_2) \circ \sigma_2^{-1} = \cdots = \sigma_p \circ \Pi(G_p) \circ \sigma_p^{-1}.$$

When the two decompositions coincide (up to renumbering), we will say that the decomposition is *normal*.[8] More formally, we introduce the following definition:

**Definition 14.23.** Let $G \subset A(R)$ be a group of arrangements of a (finite) set $R$. A decomposition of $G$ into a disjoint union of subgroups

$$G = G_1 \cup \cdots \cup G_t$$

is said to be *normal* if $G_1, \ldots, G_t$ satisfy the following two conditions:

(i) $\Pi(G_1) = \cdots = \Pi(G_t)$ (hence $G_1, \ldots, G_t$ all have the same number of elements), and

(ii) for $i = 2, \ldots, t$ there exists $\sigma_i \in \Pi(G)$ such that $G_i = \sigma_i(G_1)$.

The number $t$ of parts is called the *index* of the decomposition.

Since $\Pi'(\sigma_i(G_1)) = \Pi'(G_1)$ by Lemma 14.6, it follows that for a normal decomposition we also have

(iii) $\Pi'(G_1) = \cdots = \Pi'(G_t)$.

Conditions (i) and (iii) actually imply (ii) (see Exercise 14.5), so a normal decomposition of index $t$ is also characterized by (i) and (iii).

We next translate this property in terms of permutation groups. Recall the following classical definitions from (modern) group theory: a subgroup $\Gamma'$ in a group $\Gamma$ is said to be *normal* if $\sigma \Gamma' \sigma^{-1} = \Gamma'$ for all $\sigma \in \Gamma$. The *index* $(\Gamma : \Gamma')$ of $\Gamma'$ in $\Gamma$ is the number of cosets; when $\Gamma$ is finite we have $(\Gamma : \Gamma') = |\Gamma|/|\Gamma'|$, see Theorem 10.4 (p. 134).

---

[8]Galois [55, p. 85] uses the term "proper decomposition," which could be confusing for modern readers, for whom "proper decomposition" would probably be a synonym for "decomposition into more than one part."

**Proposition 14.24.** *Let* $G \subset A(R)$ *be a group of arrangements.*

(i) *Suppose* $G = G_1 \cup \cdots \cup G_t$ *is a normal decomposition of index* $t$. *Then* $\Pi(G_1)$ *is a normal subgroup of index* $t$ *in* $\Pi(G)$. *Similarly,* $\Pi'(G_1)$ *is a normal subgroup of index* $t$ *in* $\Pi'(G)$.

(ii) *Suppose* $\Gamma \subset \Pi(G)$ *is a normal subgroup of index* $t$. *Then the orbits of* $G$ *under the action of* $\Gamma$ *yield a normal decomposition of index* $t$

$$G = G_1 \cup \cdots \cup G_t \quad \text{with } \Pi(G_1) = \cdots = \Pi(G_t) = \Gamma.$$

(iii) *Suppose* $\Gamma \subset \Pi'(G)$ *is a normal subgroup of index* $t$. *Then the orbits of* $G$ *under the action of* $\Gamma$ *yield a normal decomposition of index* $t$

$$G = G_1 \cup \cdots \cup G_t \quad \text{with } \Pi'(G_1) = \cdots = \Pi'(G_t) = \Gamma.$$

**Proof.** (i) In a normal decomposition, each part has the same number of elements, hence $|G| = t \cdot |G_1|$ and therefore $\big(\Pi(G) : \Pi(G_1)\big) = t$. Let $\sigma \in \Pi(G)$ and $\varpi \in G_1$, and suppose $\sigma(\varpi) \in G_i$. By Proposition 14.7 we have

$$G_i = \sigma(\varpi)^{\Pi'(G_i)}.$$

But $\sigma(\varpi)^\tau = \sigma(\varpi^\tau)$ for all $\tau$ (see Lemma 14.6), and $\Pi'(G_i) = \Pi'(G_1)$ because the decomposition of $G$ is normal, hence

$$G_i = \sigma\big(\varpi^{\Pi'(G_i)}\big) = \sigma\big(\varpi^{\Pi'(G_1)}\big) = \sigma(G_1).$$

By Proposition 14.5 it follows that $\Pi(G_i) = \sigma \circ \Pi(G_1) \circ \sigma^{-1}$. Since $\Pi(G_i) = \Pi(G_1)$, we thus have $\Pi(G_1) = \sigma \circ \Pi(G_1) \circ \sigma^{-1}$ for all $\sigma \in \Pi(G)$, proving that $\Pi(G_1)$ is a normal subgroup in $\Pi(G)$. The proof that $\Pi'(G_1)$ is a normal subgroup in $\Pi'(G)$ is similar.

(ii) Consider the decomposition of $G$ into orbits under the action of $\Gamma$:

$$G = \Gamma(\alpha_1) \cup \cdots \cup \Gamma(\alpha_s) \quad \text{for some } \alpha_1, \ldots, \alpha_s \in G. \tag{14.13}$$

For $i = 1, \ldots, s$ we have $|\Gamma(\alpha_i)| = |\Gamma|$ because $\sigma(\alpha_i) \neq \tau(\alpha_i)$ for $\sigma \neq \tau$, hence $s = t$ since $(\Pi(G) : \Gamma) = t$. Each $\Gamma(\alpha_i)$ is a group of arrangements with $\Pi\big(\Gamma(\alpha_i)\big) = \Gamma$ (see Exercise 14.1). Let $\sigma = \binom{\alpha_1}{\alpha_i} \in \Pi(G)$ be the permutation such that $\sigma(\alpha_1) = \alpha_i$. Since $\Gamma$ is a normal subgroup in $\Pi(G)$ we have $\sigma \circ \Gamma \circ \sigma^{-1} = \Gamma$, hence

$$\sigma\big(\Gamma(\alpha_1)\big) = \sigma \circ \Gamma \circ \sigma^{-1}\big(\sigma(\alpha_1)\big) = \Gamma(\alpha_i).$$

Therefore, (14.13) is a normal decomposition of index $t$ of $G$.

(iii) As in case (ii), since $\Gamma$ is a subgroup of index $t$ in $\Pi'(G)$ we have a decomposition of $G$ into $t$ orbits under the action of $\Gamma$:

$$G = \alpha_1^\Gamma \cup \cdots \cup \alpha_t^\Gamma \quad \text{for some } \alpha_1, \ldots, \alpha_t \in G. \tag{14.14}$$

For $i = 2, \ldots, t$, let $\sigma \in \Pi(\mathsf{G})$ and $\tau \in \Pi'(\mathsf{G})$ be such that $\alpha_i = \sigma(\alpha_1) = \alpha_1^\tau$. Then $\alpha_i^\Gamma = \sigma(\alpha_1^\Gamma)$ by Lemma 14.6. On the other hand, because $\Gamma$ is a normal subgroup in $\Pi'(\mathsf{G})$ we have

$$\alpha_i^\Gamma = (\alpha_1^\tau)^\Gamma = \alpha_1^{(\tau \circ \Gamma \circ \tau^{-1}) \circ \tau} = (\alpha_1^\Gamma)^\tau.$$

Since $\Pi\big((\alpha_1^\Gamma)^\tau\big) = \Pi(\alpha_1^\Gamma)$ by Lemma 14.6, it follows that $\Pi(\alpha_i^\Gamma) = \Pi(\alpha_1^\Gamma)$. Therefore, the decomposition (14.14) is normal of index $t$. $\qquad\square$

As pointed out by Galois in his Proposition III, normal decompositions occur in the Galois group of arrangements when all the roots $u_1, \ldots, u_p$ of $Q$ are rational functions of one of them, i.e., when $K = L$.

**Theorem 14.25.** *With the notation of this section, assume the Galois resolvent equation $\Phi(Y) = 0$ is not irreducible over $K$, and $K = L$. Then the decomposition of $\mathsf{G}$ in Theorem 14.22 is a normal decomposition of index $p$.*

**Proof.** Since $K = L$, Proposition 14.20 yields a factorization of $\Phi$ in $K[Y]$ which we can compare with (14.10):

$$\Phi(Y) = \Lambda_1(Y) \ldots \Lambda_p(Y) = \lambda_1(Y, u_1)\lambda_1(Y, u_2) \ldots \lambda_1(Y, u_p).$$

All the polynomials $\Lambda_i$ are irreducible in $K[Y]$ by hypothesis, hence each $\Lambda_i(Y)$ divides some $\lambda_1(Y, u_j)$ (and $\Lambda_1(Y) = \lambda_1(Y, u_1)$ by definition of $\lambda_1$). But if $\Lambda_i(Y)$ divides $\lambda_1(Y, u_j)$ we must have $\Lambda_i(Y) = \lambda_1(Y, u_j)$ because all the polynomials $\lambda_i(Y, u_j)$ are monic of degree $d$ by Lemma 14.19. Renumbering $u_2, \ldots, u_p$ (or $\Lambda_2, \ldots, \Lambda_p$) we may assume $\Lambda_i(Y) = \lambda_1(Y, u_i)$ for $i = 2, \ldots, p$, hence $\mathsf{H}_i = \mathsf{G}_i$ for all $i = 1, \ldots, p$. $\qquad\square$

Theorem 14.25 applies in particular when $K$ is a radical extension of height 1 of $F$, i.e., when $K = F(a^{1/p})$ for some $a \in F$ that is not a $p$-th power in $F$, provided that $F$ contains all the $p$-th roots of unity. For in this case $K$ is obtained from $F$ by adjoining a root of the polynomial $Q(Y) = Y^p - a$, which is irreducible over $F$ by Lemma 13.8, (p. 203), and has all its roots in $K$. We will see in §14.4 that this is indeed the main case to consider in order to establish Galois' solvability criterion (Theorem 14.29).

To complete this section, we discuss three examples illustrating the decomposition of Galois groups under base field extension. In each case, a direct computational approach would be rather clumsy, but we can infer the decompositions from their expected properties.

**Example 14.26.** Suppose $n = 3$, $R = \{r_1, r_2, r_3\}$, and the Galois group $\mathsf{G}$ is the full set $\mathsf{A}(R)$ of 6 arrangements of $R$. (This occurs for instance when

$P$ is the general polynomial of degree 3, see Example 14.11). Recall from §8.2 the polynomial

$$\Delta(x_1, x_2, x_3) = (x_1 - x_2)(x_1 - x_3)(x_2 - x_3).$$

Its square is a symmetric polynomial in $x_1$, $x_2$, $x_3$, hence $\Delta(r_1, r_2, r_3)^2$ can be expressed in terms of the coefficients of $P$ (see §8.2), and therefore $\Delta(r_1, r_2, r_3)^2 \in F$. Let $a = \Delta(r_1, r_2, r_3)^2$, and consider the polynomial $Q(Z) = Z^2 - a \in F[Z]$. If $Q$ is not irreducible, then $\Delta(r_1, r_2, r_3) \in F$, and therefore by Theorem 14.9 $\Delta$ must be constant on $G$. But we have

$$\Delta(r_1, r_2, r_3) = -\Delta(r_2, r_1, r_3),$$

hence if $\Delta$ is constant we must have $\Delta(r_1, r_2, r_3) = 0$, contradicting the hypothesis that the roots of $P$ are distinct. Therefore, $Q$ is irreducible, and the results of this section apply.

Let $K = F(\sqrt{a})$ be the field obtained by adjoining to $F$ a root of $Q$ (hence all the roots $\sqrt{a}$, $-\sqrt{a}$ of $Q$). Theorem 14.25 shows that $G$ has a normal decomposition of index 2; the two subgroups into which $G$ decomposes are Galois groups of arrangements of $P(X) = 0$ over $K$. Since $\Delta(r_1, r_2, r_3) = \pm\sqrt{a} \in K$, the polynomial $\Delta$ must be constant on each of these subgroups, hence the normal decomposition of $G$ is

$$G = \left\{ \begin{matrix} r_1\ r_2\ r_3 \\ r_2\ r_3\ r_1 \\ r_3\ r_1\ r_2 \end{matrix} \right\} \bigcup \left\{ \begin{matrix} r_1\ r_3\ r_2 \\ r_2\ r_1\ r_3 \\ r_3\ r_2\ r_1 \end{matrix} \right\}. \tag{14.15}$$

Note that the discussion above applies for any $n \geq 2$: the set of all arrangements $A(R)$ of any set $R$ of $n$ elements decomposes into a disjoint union of two subgroups $A_+(R)$, $A_-(R)$ of $n!/2$ arrangements according to the value taken by the polynomial $\Delta$ (defined for arbitrary $n$ in §8.2). This decomposition is necessarily normal since it has only two terms. The corresponding subgroup $\Pi(A_+(R)) = \Pi(A_-(R))$ of the symmetric group of permutations $\Pi(A(R)) = \mathrm{Sym}(R)$ is (by definition) the *alternating group* $\mathrm{Alt}(R)$. (See Exercise 13.7, p. 212.)

**Example 14.27.** Suppose $n = 4$, $R = \{r_1, r_2, r_3, r_4\}$, and the Galois group $G$ is the full set $A(R)$ of 24 arrangements of $R$. (Take for instance for $P$ the general polynomial of degree 4, see Example 14.11.) Consider the polynomial

$$\Theta(x_1, x_2, x_3, x_4) = x_1 x_2 + x_3 x_4.$$

As observed in Exercise 8.3 (p. 108), the elements

$$u_1 = r_1 r_2 + r_3 r_4, \qquad u_2 = r_1 r_3 + r_2 r_4, \qquad u_3 = r_1 r_4 + r_2 r_3$$

are the roots of a polynomial $Q \in F[Z]$ of degree 3, and the discriminant of $Q$ is the same as the discriminant of $P$. Since this discriminant is not 0 (because $r_1$, $r_2$, $r_3$, $r_4$ are pairwise distinct), it follows that $u_1$, $u_2$, and $u_3$ are pairwise distinct. If $Q$ is not irreducible in $F[Z]$, then it must have a factor of degree 1, hence one of $u_1$, $u_2$, $u_3$ lies in $F$. But then the polynomial $\Theta(x_1, x_2, x_3, x_4)$ must be constant on $\mathsf{G}$ by Theorem 14.9. This is a contradiction because the following values are distinct:

$$\Theta(r_1, r_2, r_3, r_4) = u_1, \qquad \Theta(r_1, r_3, r_4, r_2) = u_2, \qquad \Theta(r_1, r_4, r_2, r_3) = u_3.$$

Therefore, $Q$ is irreducible. Let $K = F(u_1)$. Because $\Theta(r_1, r_2, r_3, r_4) \in K$, the polynomial $\Theta(x_1, x_2, x_3, x_4)$ must be constant on the Galois group $\mathsf{H}_1$ of $P$ over $K$ containing the arrangement $(r_1, r_2, r_3, r_4)$, hence $\mathsf{H}_1 \subsetneq \mathsf{G}$. Since $\deg Q = 3$ we have $|\mathsf{H}_1| = |\mathsf{G}|/3 = 8$. We claim that the decomposition $\mathsf{G} = \mathsf{H}_1 \cup \mathsf{H}_2 \cup \mathsf{H}_3$ of (14.11) with $\Pi(\mathsf{H}_1) = \Pi(\mathsf{H}_2) = \Pi(\mathsf{H}_3)$ is the following:

$$\mathsf{G} = \left\{ \begin{matrix} r_1\ r_2\ r_3\ r_4 \\ r_2\ r_1\ r_3\ r_4 \\ r_1\ r_2\ r_4\ r_3 \\ r_2\ r_1\ r_4\ r_3 \\ r_3\ r_4\ r_1\ r_2 \\ r_3\ r_4\ r_2\ r_1 \\ r_4\ r_3\ r_1\ r_2 \\ r_4\ r_3\ r_2\ r_1 \end{matrix} \right\} \cup \left\{ \begin{matrix} r_1\ r_3\ r_4\ r_2 \\ r_2\ r_3\ r_4\ r_1 \\ r_1\ r_4\ r_3\ r_2 \\ r_2\ r_4\ r_3\ r_1 \\ r_3\ r_1\ r_2\ r_4 \\ r_3\ r_2\ r_1\ r_4 \\ r_4\ r_1\ r_2\ r_3 \\ r_4\ r_2\ r_1\ r_3 \end{matrix} \right\} \cup \left\{ \begin{matrix} r_1\ r_3\ r_2\ r_4 \\ r_2\ r_3\ r_1\ r_4 \\ r_1\ r_4\ r_2\ r_3 \\ r_2\ r_4\ r_1\ r_3 \\ r_3\ r_1\ r_4\ r_2 \\ r_3\ r_2\ r_4\ r_1 \\ r_4\ r_1\ r_3\ r_2 \\ r_4\ r_2\ r_3\ r_1 \end{matrix} \right\}.$$

This follows because on all the arrangements of the second group $\Theta(x_1, x_2, x_3, x_4)$ takes the value $u_2$, and it takes the value $u_3$ on all the arrangements of the third group, so $\mathsf{H}_1$ must be contained in the first group. It coincides with this first group because $|\mathsf{H}_1| = 8$. The groups $\mathsf{H}_2$ and $\mathsf{H}_3$ are then obtained as $\mathsf{H}_2 = \Pi(\mathsf{H}_1)(r_1, r_3, r_4, r_2)$ and $\mathsf{H}_3 = \Pi(\mathsf{H}_1)(r_1, r_3, r_2, r_4)$.

Note that the decomposition $\mathsf{G} = \mathsf{H}_1 \cup \mathsf{H}_2 \cup \mathsf{H}_3$ is *not* normal. The decomposition of $\mathsf{G}$ following Theorem 14.25 is $\mathsf{G} = \mathsf{G}_1 \cup \mathsf{G}_2 \cup \mathsf{G}_3$ with $\mathsf{G}_i$ a Galois group of $P(X) = 0$ over $F(u_i)$ (hence $\mathsf{G}_1 = \mathsf{H}_1$); this decomposition is as follows:

$$\mathsf{G} = \left\{ \begin{matrix} r_1\ r_2\ r_3\ r_4 \\ r_2\ r_1\ r_3\ r_4 \\ r_1\ r_2\ r_4\ r_3 \\ r_2\ r_1\ r_4\ r_3 \\ r_3\ r_4\ r_1\ r_2 \\ r_3\ r_4\ r_2\ r_1 \\ r_4\ r_3\ r_1\ r_2 \\ r_4\ r_3\ r_2\ r_1 \end{matrix} \right\} \cup \left\{ \begin{matrix} r_1\ r_3\ r_2\ r_4 \\ r_3\ r_1\ r_2\ r_4 \\ r_1\ r_3\ r_4\ r_2 \\ r_3\ r_1\ r_4\ r_2 \\ r_2\ r_4\ r_1\ r_3 \\ r_2\ r_4\ r_3\ r_1 \\ r_4\ r_2\ r_1\ r_3 \\ r_4\ r_2\ r_3\ r_1 \end{matrix} \right\} \cup \left\{ \begin{matrix} r_1\ r_4\ r_2\ r_3 \\ r_4\ r_1\ r_2\ r_3 \\ r_1\ r_4\ r_3\ r_2 \\ r_4\ r_1\ r_3\ r_2 \\ r_2\ r_3\ r_1\ r_4 \\ r_2\ r_3\ r_4\ r_1 \\ r_3\ r_2\ r_1\ r_4 \\ r_3\ r_2\ r_4\ r_1 \end{matrix} \right\}.$$

In this last decomposition, the second group is $\sigma_2(\mathsf{H}_1)$ where $\sigma_2 \colon r_2 \leftrightarrow r_3$, and the third is $\sigma_3(\mathsf{H}_1)$ where $\sigma_3 \colon r_2 \mapsto r_4 \mapsto r_3 \mapsto r_2$.

**Example 14.28.** Suppose again $n = 4$ and $R = \{r_1, r_2, r_3, r_4\}$, but now assume the Galois group $\mathsf{G}$ is a set $\mathsf{A}_+(R)$ of 12 arrangements as defined at the end of Example 14.26. For instance, we may start with a general polynomial of degree 4 and adjoin to the base field a square root of $\Delta(r_1, r_2, r_3, r_4)^2$. Consider the same polynomial $Q \in F[Z]$ of degree 3 as in Example 14.27. Extending the base field to $K = F(u_1)$ now yields a *normal* decomposition of $\mathsf{G}$ of index 3:

$$
\mathsf{G} = \left\{\begin{matrix} r_1 \ r_2 \ r_3 \ r_4 \\ r_2 \ r_1 \ r_4 \ r_3 \\ r_3 \ r_4 \ r_1 \ r_2 \\ r_4 \ r_3 \ r_2 \ r_1 \end{matrix}\right\} \quad \bigcup \quad \left\{\begin{matrix} r_1 \ r_3 \ r_4 \ r_2 \\ r_2 \ r_4 \ r_3 \ r_1 \\ r_3 \ r_1 \ r_2 \ r_4 \\ r_4 \ r_2 \ r_1 \ r_3 \end{matrix}\right\} \quad \bigcup \quad \left\{\begin{matrix} r_1 \ r_4 \ r_2 \ r_3 \\ r_2 \ r_3 \ r_1 \ r_4 \\ r_3 \ r_2 \ r_4 \ r_1 \\ r_4 \ r_1 \ r_3 \ r_2 \end{matrix}\right\}.
$$

Details are left as exercises.

## 14.4 Solvability by radicals

Galois' characterization of the equations that are solvable by radicals follows in a surprisingly easy way from his normal decomposition theorem (Theorem 14.25), as we show in this section. We first note however that the notion of solvability by radicals in Galois' memoir is slightly different from that of §13.1, in that Galois requires *all* the roots of the equation (instead of just one of them) to have an expression by radicals. To distinguish this condition from that of §13.1, we say that a polynomial equation with coefficients in a field $F$ is *completely solvable (by radicals) over $F$* if there is a radical extension of $F$ containing all the roots of the equation.

This distinction is significant when dealing with arbitrary equations, and more specifically with equations $P(X) = 0$ in which $P$ is a reducible polynomial. In this case, solving the equation amounts to finding a root of one of the factors of $P$, and the difficulty of finding such a root can be completely different from factor to factor. For instance, over $\mathbb{C}(s_1, \ldots, s_n)$ the equation

$$
(X - 1)(X^n - s_1 X^{n-1} + s_2 X^{n-2} - \cdots + (-1)^n s_n) = 0
$$

is solvable by radicals, since $X - 1 = 0$ is solvable, but it is not completely solvable by radicals if $n \geq 5$ because general equations of degree at least 5 are not solvable by radicals (Theorem 13.15, p. 211).

We will see however that if the polynomial $P$ is irreducible over $F$, then the equation $P(X) = 0$ is solvable by radicals over $F$ if and only if it is completely solvable by radicals over $F$ (see Corollary 14.36). Since the crucial case is the solution of irreducible equations (to which one is led by factoring the given polynomial) and since in this case both notions are equivalent, Galois' results are actually sufficient to investigate the more general notion of solvability of equations by radicals.

As in the preceding sections, we consider an equation $P(X) = 0$ where $P$ is a polynomial of degree $n$ with coefficients in a base field $F$ of characteristic zero, without multiple roots in any extension of $F$. All our computations will take place in some universal field containing $F$, all the roots of $P$, and the roots of any auxiliary equation we might need. (If $F$ is the field $\mathbb{Q}$ of rational numbers, or indeed any subfield of $\mathbb{C}$, the universal field can be assumed to be $\mathbb{C}$.) Let $R = \{r_1, \ldots, r_n\}$ be the set of roots of $P$. To make a specific choice of Galois group, we will only consider Galois groups containing the arrangement $\varpi = (r_1, \ldots, r_n)$. For any extension $K$ of $F$, we write $\mathsf{G}_K$ for the Galois group of $P$ over the base field $K$. Our goal is to translate in terms of $\mathsf{G}_F$ the condition that $P$ is completely solvable by radicals.

Suppose first this condition holds. Following the discussion in §13.1, this means that there is a radical extension $E$ of $F$ containing all the roots of $P$. Galois' first observation is that $\mathsf{G}_E$ is then reduced to the single arrangement $\varpi$, because each of the polynomials $x_1, \ldots, x_n$ takes on $\varpi$ a value in $E$: if $\mathsf{G}_E$ contains another arrangement $\alpha \neq \varpi$, then one of $x_1$, $\ldots$, $x_n$ is not constant on $\mathsf{G}_E$, contradicting Property (GP). Therefore, as we go up the tower of radical extensions of height 1 from $F$ to $E$

$$F \subset E_1 \subset E_2 \subset \cdots \subset E,$$

the Galois group of $P$ must undergo a series of reductions, although not necessarily at each step. In any case, we may consider the first occurrence of a reduction, and look for a radical of the least degree that achieves this reduction: there is a field $F_0$ and a radical extension of height 1 of the form $K = F_0(a^{1/p})$, for some prime $p$ and some $a \in F_0$ that is not a $p$-th power in $F_0$, such that

$$\mathsf{G}_{F_0} = \mathsf{G}_F, \qquad \mathsf{G}_K \subsetneq \mathsf{G}_F,$$

and such that $p$ is minimal for these properties. Thus, $K$ is obtained from $F_0$ by adjoining a root of $Z^p - a$, which is irreducible in $F_0[Z]$ by Lemma 13.8 (p. 203), and we may apply the results of §14.3: $\mathsf{G}_F$ decomposes into $p$

parts (one of which is $\mathsf{G}_K$) under base field extension from $F_0$ to $K$. Let $\zeta$ be a primitive $p$-th root of unity and let $F_0' = F_0(\zeta)$ and $K' = K(\zeta)$. Proposition 13.3 (p. 199) shows that $F_0'$ lies in a radical extension of $F_0$ obtained by adjoining roots of equations $Z^q - b$ with $q < p$, hence the minimality condition on $p$ yields $\mathsf{G}_{F_0'} = \mathsf{G}_{F_0} = \mathsf{G}_F$. On the other hand, since $K \subset K'$ we have $\mathsf{G}_K \supset \mathsf{G}_{K'}$, hence $\mathsf{G}_{K'} \subsetneq \mathsf{G}_{F_0'}$. Note that $K' = F_0'(a^{1/p})$; if $a$ is a $p$-th power in $F_0'$, then $K' = F_0'$ because all the roots of $Z^p - a$ then belong to $F_0'$. This is a contradiction since $\mathsf{G}_{K'} \neq \mathsf{G}_{F_0'}$. Thus, $K'$ is a radical extension of height 1 of $F_0'$; substituting $F_0'$ for $F_0$ and $K'$ for $K$, we may assume that $F_0$ contains a primitive $p$-th root of unity. Then $K$ contains all the roots of $Z^p - a$, hence by Theorem 14.25 the decomposition of $\mathsf{G}_F$ under the base field extension from $F_0$ to $K$ is normal.

We may argue similarly for the next reduction of the Galois group (or use induction on the number of elements in $\mathsf{G}_F$); we have thus proved the "only if" part of the main theorem:

**Theorem 14.29.** *The equation $P(X) = 0$ is completely solvable by radicals over $F$ if and only if its Galois group $\mathsf{G}_F$ can be reduced to a single arrangement by a series of normal decompositions of prime index.*

To prove the converse, suppose $\mathsf{G}_F$ has a normal decomposition of prime index

$$\mathsf{G}_F = \mathsf{G}_0 \cup \cdots \cup \mathsf{G}_{p-1}, \tag{14.16}$$

with $p$ minimal among all such decompositions. Say $\varpi \in \mathsf{G}_0$. We will find a radical extension $L$ of $F$ such that $\mathsf{G}_L = \mathsf{G}_0$. Arguing similarly for the subsequent decompositions, we finally obtain a radical extension $E$ of $F$ such that $\mathsf{G}_E = \{\varpi\}$. Then each of the polynomials $x_1, \ldots, x_n$ is constant on $\mathsf{G}_E$, hence $r_1, \ldots, r_n \in E$ and it follows that $P(X) = 0$ is completely solvable by radicals.

It remains to find a radical extension $L$ of $F$ such that $\mathsf{G}_L = \mathsf{G}_0$. The construction of $L$ is directly inspired by the Lagrange resolvents of §10.2. Let $\zeta$ be a primitive $p$-th root of unity. As pointed out in the proof of the "only if" part, Proposition 13.3 shows that $\zeta$ lies in a radical extension of $F$ obtained by adjoining roots of equations $Z^q - b = 0$ with $q < p$. Since $p$ was chosen minimal, none of these adjunctions produces a decomposition of the Galois group. Therefore, we may assume $\zeta \in F$.

Let $\rho \in S_n$ be such that $\varpi^\rho \in \mathsf{G}_1$. Thus, the groups of arrangements $\mathsf{G}_0^\rho$ and $\mathsf{G}_1$ both contain $\varpi^\rho$. We have $\Pi(\mathsf{G}_0^\rho) = \Pi(\mathsf{G}_0)$ by Lemma 14.6, but also $\Pi(\mathsf{G}_1) = \Pi(\mathsf{G}_0)$ because the decomposition (14.16) is normal, hence

$$\mathsf{G}_0^\rho = \{\sigma(\varpi^\rho) \mid \sigma \in \Pi(\mathsf{G}_0^\rho)\} = \{\sigma(\varpi^\rho) \mid \sigma \in \Pi(\mathsf{G}_1)\} = \mathsf{G}_1.$$

Similarly, $\mathsf{G}_i^\rho \in \{\mathsf{G}_0, \ldots, \mathsf{G}_{p-1}\}$ for all $i$, and the action of $\rho$ defines a partition of $\{\mathsf{G}_0, \ldots, \mathsf{G}_{p-1}\}$ into cycles $\{\mathsf{G}_i, \mathsf{G}_i^\rho, \mathsf{G}_i^{\rho^2}, \ldots\}$. These cycles all have the same length: if $\mathsf{G}_i^{\rho^t} = \mathsf{G}_i$ for some $i$ and $t$, then $\rho^t \in \Pi'(\mathsf{G}_i)$; but since the decomposition (14.16) is normal we have $\Pi'(\mathsf{G}_0) = \cdots = \Pi'(\mathsf{G}_{p-1})$ (see (iii) in Definition 14.23), hence $\rho^t \in \Pi'(\mathsf{G}_j)$ for all $j$, and therefore $\mathsf{G}_j^{\rho^t} = \mathsf{G}_j$ for all $j = 0, \ldots, p-1$. Thus, the length of the cycles divides $p$. Since $p$ is prime, there must be just one cycle, hence the least integer $t > 0$ such that $\rho^t \in \Pi'(\mathsf{G}_0)$ is $p$. Renumbering $\mathsf{G}_2, \ldots, \mathsf{G}_{p-1}$, we may assume

$$\mathsf{G}_i = \mathsf{G}_0^{\rho^i} \qquad \text{for } i = 0, \ldots, p-1.$$

Recall from the proof of Proposition 14.17 the action of $S_n$ on polynomials in $n$ indeterminates: for $\sigma \in S_n$ and $f \in F[x_1, \ldots, x_n]$, we let

$$\sigma(f)(x_1, \ldots, x_n) = f(x_{\sigma(1)}, \ldots, x_{\sigma(n)}).$$

Thus, $\sigma(f)(\alpha) = f(\alpha^\sigma)$ for every arrangement $\alpha \in \mathsf{A}(R)$. We will need the following result:

**Lemma 14.30.** *There exists a polynomial $\theta \in F[x_1, \ldots, x_n]$ such that*

(i) $\sigma(\theta) = \theta$ for all $\sigma \in \Pi'(\mathsf{G}_0)$, and
(ii) *for the polynomial $f = \theta + \zeta \rho(\theta) + \cdots + \zeta^{p-1} \rho^{p-1}(\theta) \in F[x_1, \ldots, x_n]$,* we have $f(\varpi) \neq 0$.

Note that since $\rho^p \in \Pi'(\mathsf{G}_0)$ condition (i) implies $\rho^i(\theta) = \rho^j(\theta)$ if $i \equiv j \bmod p$.

Assuming the lemma holds, and picking $\theta$ satisfying (i) and (ii), we complete the proof of Theorem 14.29 as follows: observe that for $\alpha \in \mathsf{G}_i = \mathsf{G}_0^{\rho^i}$ there exists $\sigma \in \Pi'(\mathsf{G}_0)$ such that $\alpha = \varpi^{\sigma \rho^i}$, hence for $j = 0, \ldots, p-1$

$$\rho^j(\theta)(\alpha) = \rho^j(\theta)(\varpi^{\sigma \rho^i}) = \sigma \rho^{i+j}(\theta)(\varpi). \tag{14.17}$$

Since the decomposition (14.16) is normal, $\Pi'(\mathsf{G}_0)$ is a normal subgroup of $\Pi'(\mathsf{G}_F)$, hence $\sigma \rho^{i+j} = \rho^{i+j} \sigma'$ for some $\sigma' \in \Pi'(\mathsf{G}_0)$. Therefore, using condition (i), we have

$$\sigma \rho^{i+j}(\theta)(\varpi) = \rho^{i+j} \sigma'(\theta)(\varpi) = \rho^{i+j}(\theta)(\varpi). \tag{14.18}$$

Comparing (14.17) and (14.18), we see that

$$\rho^j(\theta)(\alpha) = \rho^{i+j}(\theta)(\varpi) \quad \text{for every } \alpha \in \mathsf{G}_i \text{ and } j = 0, \ldots, p-1.$$

For the polynomial $f$ in condition (ii), and for $\alpha \in \mathsf{G}_i$, it follows that

$$\begin{aligned}
f(\alpha) &= \theta(\alpha) + \zeta \rho(\theta)(\alpha) + \cdots + \zeta^{p-1} \rho^{p-1}(\theta)(\alpha) \\
&= \rho^i(\theta)(\varpi) + \zeta \rho^{i+1}(\theta)(\varpi) + \cdots + \zeta^{p-1} \rho^{i+p-1}(\theta)(\varpi) \\
&= \zeta^{-i} f(\varpi). \tag{14.19}
\end{aligned}$$

Since $\zeta^p = 1$, raising each side to the $p$-th power yields

$$f(\alpha)^p = f(\varpi)^p \qquad \text{for all } \alpha \in \mathsf{G}_i \text{ and } i = 0, \ldots, p-1.$$

Therefore, the polynomial $f^p$ is constant on $\mathsf{G}_F$, and Property (GP) implies that $f(\varpi)^p \in F$. On the other hand, since $f(\varpi) \neq 0$ by condition (ii), equation (14.19) shows that $f$ is not constant on $\mathsf{G}_F$. Therefore, $f(\varpi) \notin F$, and similarly $\zeta^i f(\varpi) \notin F$ for $i = 1, \ldots, p-1$, hence $f(\varpi)^p$ is not a $p$-th power in $F$. The field $L = F\big(f(\varpi)\big)$ is thus a radical extension of height 1 of $F$. Since $f(\varpi) \in L$, the polynomial $f$ must be constant on $\mathsf{G}_L$, hence $f(\alpha) = f(\varpi)$ for $\alpha \in \mathsf{G}_L$. Since

$$\mathsf{G}_L \subset \mathsf{G}_F = \mathsf{G}_0 \cup \cdots \cup \mathsf{G}_{p-1},$$

it follows from (14.19) that $\mathsf{G}_L \subset \mathsf{G}_0$. But under base field extension from $F$ to $L$ the Galois group decomposes into exactly $p$ parts, hence the number of elements in $\mathsf{G}_L$ is $|\mathsf{G}_F|/p$, just like $\mathsf{G}_0$. Therefore, $\mathsf{G}_L = \mathsf{G}_0$. We have thus found a radical extension that achieves the reduction of the Galois group to $\mathsf{G}_0$, and Theorem 14.29 is proved. $\qquad\qquad\qquad\square$

**Proof of Lemma 14.30.** Let $\varphi \in F[x_1, \ldots, x_n]$ be a polynomial that takes $n!$ different values when evaluated on the arrangements of $R$ (see Lemma 14.8). Consider the following polynomial in $n + 1$ indeterminates:

$$\Theta(x_1, \ldots, x_n, Y) = \prod_{\sigma \in \Pi'(\mathsf{G}_0)} \big(Y - \sigma(\varphi)(x_1, \ldots, x_n)\big) \in F[x_1, \ldots, x_n, Y].$$

Evaluating $x_1, \ldots, x_n$ in $\varpi$ yields a polynomial $\Theta(\varpi, Y) \in F(R)[Y]$ with roots $\varphi(\varpi^\sigma)$ for $\sigma \in \Pi'(\mathsf{G}_0)$. The roots are thus the evaluations $\varphi(\alpha)$ for $\alpha \in \mathsf{G}_0$. Similarly, the roots of the polynomial $\Theta(\varpi^{\rho^i}, Y)$, for $i = 1, \ldots,$ $p-1$, are the evaluations $\varphi(\alpha)$ for $\alpha \in \mathsf{G}_0^{\rho^i} = \mathsf{G}_i$. Thus, the polynomials $\Theta(\varpi, Y)$, $\Theta(\varpi^\rho, Y)$, $\ldots$, $\Theta(\varpi^{\rho^{p-1}}, Y)$ have different roots and are therefore distinct. It follows that

$$\prod_{i=1}^{p-1} \big(\Theta(\varpi, Y) - \Theta(\varpi^{\rho^i}, Y)\big) \neq 0 \qquad \text{in } F(R)[Y].$$

Since $F$ is infinite (because its characteristic is zero), we may find $t \in F$ that is not a root of this polynomial (see Theorem 5.15, p. 52). Fix such a $t \in F$ and consider the following elements[9] in $F(R)$, for $d = 0, 1, \ldots, p-1$:

$$u_d = \Theta(\varpi, t)^d + \zeta \, \Theta(\varpi^\rho, t)^d + \cdots + \zeta^{p-1} \Theta(\varpi^{\rho^{p-1}}, t)^d.$$

---

[9]In fact $u_0 = 0$ because $\zeta$ is a root of the cyclotomic polynomial $1 + X + \cdots + X^{p-1}$, but this observation is not relevant for the argument.

For every polynomial $g(Z) = a_0 + a_1 Z + \cdots + a_{p-1} Z^{p-1} \in F(R)[Z]$ we have

$$g\big(\Theta(\varpi,t)\big) + \zeta g\big(\Theta(\varpi^\rho,t)\big) + \cdots + \zeta^{p-1} g\big(\Theta(\varpi^{\rho^{p-1}},t)\big)$$
$$= a_0 u_0 + a_1 u_1 + \cdots + a_{p-1} u_{p-1}.$$

In particular, for $g(Z) = \big(Z - \Theta(\varpi^\rho,t)\big) \ldots \big(Z - \Theta(\varpi^{\rho^{p-1}},t)\big)$ we have

$$g\big(\Theta(\varpi,t)\big) + \zeta\, g\big(\Theta(\varpi^\rho,t)\big) + \cdots + \zeta^{p-1} g\big(\Theta(\varpi^{\rho^{p-1}},t)\big) = g\big(\Theta(\varpi,t)\big)$$
$$= \big(\Theta(\varpi,t) - \Theta(\varpi^\rho,t)\big) \ldots \big(\Theta(\varpi,t) - \Theta(\varpi^{\rho^{p-1}},t)\big) \neq 0.$$

Therefore, we cannot have $u_0 = u_1 = \cdots = u_{p-1} = 0$: we can pick $d \in \{0, 1, \ldots, p-1\}$ such that $u_d \neq 0$. Let

$$\theta(x_1, \ldots, x_n) = \Theta(x_1, \ldots, x_n, t)^d$$
$$= \prod_{\sigma \in \Pi'(\mathsf{G}_0)} \big(t - \sigma(\varphi)(x_1, \ldots, x_n)\big)^d \in F[x_1, \ldots, x_n].$$

It is clear from the definition that $\theta$ satisfies (i). Moreover, for the polynomial $f$ from (ii) we have

$$f(\varpi) = \Theta(\varpi,t)^d + \zeta\,\Theta(\varpi^\rho,t)^d + \cdots + \zeta^{p-1}\Theta(\varpi^{\rho^{p-1}},t)^d = u_d \neq 0.$$

Therefore, condition (ii) also holds. □

The translation of Theorem 14.29 in terms of permutation groups is easy to achieve, since normal decompositions of groups of arrangements correspond to normal subgroups of permutation groups by Proposition 14.24:

**Corollary 14.31.** *The equation $P(X) = 0$ is completely solvable by radicals over $F$ if and only if the permutation group $\Pi(\mathsf{G}_F)$ contains a sequence of subgroups*

$$\Pi(\mathsf{G}_F) = \Gamma_0 \supset \Gamma_1 \supset \cdots \supset \Gamma_t = \{\mathrm{Id}_R\}$$

*such that each $\Gamma_i$ is normal of prime index in $\Gamma_{i-1}$ for $i = 1, \ldots, t$.*

As a consequence, (permutation) groups that satisfy the condition of the corollary are said to be *solvable*.

To illustrate Theorem 14.29 and Corollary 14.31, we consider general equations of degree 3 and 4 (or, more generally, any equation of degree 3 or 4 with Galois group the full set of arrangements of the roots). For degree 3, we have the following sequence of normal decompositions: $\mathsf{G} = \mathsf{A}_+(R) \cup \mathsf{A}_-(R)$, where $\mathsf{A}_+(R)$ and $\mathsf{A}_-(R)$ are defined in Example 14.26 (see (14.15)), and

$$\left\{ \begin{array}{l} r_1\, r_2\, r_3 \\ r_2\, r_3\, r_1 \\ r_3\, r_1\, r_2 \end{array} \right\} = \big\{ r_1\, r_2\, r_3 \big\} \ \cup \ \big\{ r_2\, r_3\, r_1 \big\} \ \cup \ \big\{ r_3\, r_1\, r_2 \big\}.$$

The corresponding sequence of subgroups in $\Pi(\mathsf{G}_F) = \mathrm{Sym}(R)$ is

$$\mathrm{Sym}(R) \supset \mathrm{Alt}(R) \supset \{\mathrm{Id}_R\}$$

where $\mathrm{Alt}(R)$ is the alternating group (see Example 14.26).

For degree 4, the group $\mathsf{G}_F = \mathsf{A}(R)$ first decomposes into $\mathsf{A}_+(R) \cup \mathsf{A}_-(R)$, then $\mathsf{A}_+(R)$ has the normal decomposition into three groups of 4 elements displayed in Example 14.28. Finally, we may decompose

$$
\left\{
\begin{array}{llll}
r_1 & r_2 & r_3 & r_4 \\
r_2 & r_1 & r_4 & r_3 \\
r_3 & r_4 & r_1 & r_2 \\
r_4 & r_3 & r_2 & r_1
\end{array}
\right\}
=
\left\{
\begin{array}{llll}
r_1 & r_2 & r_3 & r_4 \\
r_2 & r_1 & r_4 & r_3
\end{array}
\right\}
\ \bigcup\
\left\{
\begin{array}{llll}
r_3 & r_4 & r_1 & r_2 \\
r_4 & r_3 & r_2 & r_1
\end{array}
\right\}
\tag{14.20}
$$

and

$$
\left\{
\begin{array}{llll}
r_1 & r_2 & r_3 & r_4 \\
r_2 & r_1 & r_4 & r_3
\end{array}
\right\}
=
\left\{
\begin{array}{llll}
r_1 & r_2 & r_3 & r_4
\end{array}
\right\}
\ \bigcup\
\left\{
\begin{array}{llll}
r_2 & r_1 & r_4 & r_3
\end{array}
\right\}.
$$

To describe the corresponding sequence of subgroups in $\Pi(\mathsf{G}_F) = \mathrm{Sym}(R)$, let $V(R)$ denote the group of permutations of the group of 4 arrangements on the left-hand side of (14.20). (This group is known as Klein's *Vierergruppe*.) We have

$$V(R) = \{\mathrm{Id}_R, \sigma, \tau, \nu\}$$

where

$$\sigma: r_1 \leftrightarrow r_2,\ r_3 \leftrightarrow r_4, \quad \tau: r_1 \leftrightarrow r_3,\ r_2 \leftrightarrow r_4, \quad \nu: r_1 \leftrightarrow r_4,\ r_2 \leftrightarrow r_3.$$

The sequence of subgroups of $\Pi(\mathsf{G}_F) = \mathrm{Sym}(R)$ reflecting the complete solvability of general equations of degree 4 is

$$\mathrm{Sym}(R) \supset \mathrm{Alt}(R) \supset V(R) \supset \{\mathrm{Id}_R, \sigma\} \supset \{\mathrm{Id}_R\}.$$

The first Galois group reduction is obtained by adjoining a square root of the discriminant (which is the same as the discriminant of a cubic resolvent, see Exercise 8.3, p. 108); the second follows under the adjunction of a root of a cubic resolvent (see Example 14.28). The complete solution is then obtained by the resolution of two quadratic equations, as in §3.2 or §6.2.

For general equations of degree $n \geq 5$, we still have the decomposition $\mathsf{A}(R) = \mathsf{A}_+(R) \cup \mathsf{A}_-(R)$, but there is no further normal decomposition, as the following lemma shows:

**Lemma 14.32.** *Let $X$ be a finite set of $n$ elements, with $n \geq 5$, and let $\Gamma \subset \mathrm{Sym}(X)$ be a subgroup containing all the cyclic permutations of three elements (i.e., permutations of the type $a \mapsto b \mapsto c \mapsto a$). If $\Gamma' \subset \Gamma$ is a normal subgroup of prime index, then $\Gamma'$ also contains all the cyclic permutations of three elements.*

When applied to $\Gamma = \mathrm{Sym}(R)$, Lemma 14.32 shows that any normal subgroup of prime index in $\mathrm{Sym}(R)$ contains the subgroup generated by the cyclic permutations of three elements. But this subgroup is easily seen to be the alternating group $\mathrm{Alt}(R)$, because $\mathrm{Alt}(R)$ consists of permutations that decompose into products of an even number of transpositions (i.e., cyclic permutations of two elements; see Exercise 13.7, p. 212), and every product of two transpositions is a product of cyclic permutations of three elements: the composition of two transpositions with non-disjoint cycles is a cyclic permutation of three elements, and the composition of two transpositions $a \leftrightarrow b$ and $c \leftrightarrow d$ with disjoint cycles is the product $\sigma \circ \tau$ where $\sigma \colon a \mapsto b \mapsto c \mapsto a$ and $\tau \colon b \mapsto c \mapsto d \mapsto b$. Therefore, the only normal subgroup of prime index in $\mathrm{Sym}(R)$ is $\mathrm{Alt}(R)$. When applied to $\Gamma = \mathrm{Alt}(R)$, Lemma 14.32 shows that $\mathrm{Alt}(R)$ does not contain any normal subgroup of prime index,[10] because such a subgroup would contain all the cyclic permutations of three elements, hence it would contain $\mathrm{Alt}(R)$.

**Proof of Lemma 14.32.** Let $p$ be the index of $\Gamma'$ in $\Gamma$, so $|\Gamma| = p \cdot |\Gamma'|$, and let $\rho \in \Gamma \setminus \Gamma'$. Let $e$ be the least strictly positive integer such that $\rho^e \in \Gamma'$; thus, $e > 1$. Because $\Gamma'$ is assumed to be a normal subgroup in $\Gamma$, we have $\rho \circ \Gamma' = \Gamma' \circ \rho$, hence the following set is a subgroup of $\Gamma$:

$$\Gamma' \cup (\rho \circ \Gamma') \cup \cdots \cup (\rho^{e-1} \circ \Gamma') = \Gamma' \cup (\Gamma' \circ \rho) \cup \cdots \cup (\Gamma' \circ \rho^{e-1}) \subset \Gamma. \quad (14.21)$$

This subgroup has $e \cdot |\Gamma'|$ elements, and the order of every subgroup of $\Gamma$ divides $|\Gamma| = p \cdot |\Gamma'|$ by Theorem 10.4, hence $e = p$ and the inclusion in (14.21) is an equality.

Now, let $\sigma, \tau \in \Gamma$. We may find integers $i, j \in \{0, \ldots, p-1\}$ such that $\sigma \in \rho^i \circ \Gamma'$ and $\tau \in \rho^j \circ \Gamma'$, hence $\sigma \circ \tau \in \rho^i \circ \Gamma' \circ \rho^j \circ \Gamma' = \rho^{i+j} \circ \Gamma'$ and, similarly, $\tau \circ \sigma \in \rho^{i+j} \circ \Gamma'$. If $\theta, \theta' \in \Gamma'$ are such that $\sigma \circ \tau = \rho^{i+j} \circ \theta$ and $\tau \circ \sigma = \rho^{i+j} \circ \theta'$, then $\tau^{-1} \circ \sigma^{-1} \circ \tau \circ \sigma = \theta^{-1} \circ \theta' \in \Gamma'$. To see that $\Gamma'$ contains every cyclic permutation of three elements, it suffices to apply this computation to suitably selected $\sigma, \tau \in \Gamma$: in order to obtain the permutation $\pi \colon a \mapsto b \mapsto c \mapsto a$, pick two other elements $c, d \in R$ (this is where we need to assume $n \geq 5$) and let $\sigma \colon b \mapsto c \mapsto d \mapsto b$ and $\tau \colon a \mapsto c \mapsto e \mapsto a$. We have $\sigma, \tau \in \Gamma$ because $\Gamma$ is assumed to contain all the cyclic permutations of three elements, and it is readily verified that $\pi = \tau^{-1} \circ \sigma^{-1} \circ \tau \circ \sigma$. $\qquad\Box$

---

[10]In fact, it does not contain *any* non-trivial normal subgroup: see for instance [41, Th. 4.11].

**Corollary 14.33.** *The general equation of degree $n \geq 5$ is not completely solvable by radicals.*

**Proof.** This readily follows from the solvability criterion in Corollary 14.31, since Lemma 14.32 shows that the group $\mathrm{Sym}(R)$ is not solvable when $|R| \geq 5$. □

This corollary is not stated explicitly by Galois. As pointed out by Neumann [55, p. 384], Galois may not have given much thought to alternating groups. He gives instead a solvability criterion for equations of prime degree, which we discuss in the next section. From this criterion, it readily follows that general equations of prime degree $p \geq 5$ are not completely solvable by radicals.

To conclude this section, we turn to the (not necessarily complete) solvability of equations. We use the notation set up at the beginning of this section. A field $E$ contains a root $r_1$ of $P$ if and only if every permutation $\sigma \in \Pi(\mathsf{G}_E)$ leaves $r_1$ fixed, because Property (GP) shows that $r_1 \in E$ if and only if the polynomial $x_1$ is constant on $\mathsf{G}_E$. Therefore, a straightforward modification of the arguments in the proof of Theorem 14.29 (substituting the condition that $r_1$ be fixed under the permutations of the Galois group for the condition that the Galois group be reduced to a single arrangement) yields the following:

**Proposition 14.34.** *The root $r_1$ of $P(X) = 0$ has an expression by radicals if and only if the Galois group of arrangements $\mathsf{G}_F$ can be reduced by a series of normal decompositions of prime index to a group $\mathsf{G}'$ such that all the permutations in $\Pi(\mathsf{G}')$ fix $r_1$. This condition holds if and only if $\Pi(\mathsf{G}_F)$ contains a series of subgroups*

$$\Pi(\mathsf{G}_F) = \Gamma_0 \supset \Gamma_1 \supset \cdots \supset \Gamma_t$$

*such that each $\Gamma_i$ is normal of prime index in $\Gamma_{i-1}$ for $i = 1, \ldots, t$, and all the permutations in $\Gamma_t$ fix $r_1$.*

We next consider the particular case where $P$ is irreducible. Note that all the roots of an irreducible polynomial (over a field of characteristic zero) are simple by Corollary 5.22 (p. 55), hence the Galois group of $P$ is defined.

**Proposition 14.35.** *The polynomial $P$ is irreducible in $F[X]$ if and only if the group of permutations $\Pi(\mathsf{G}_F)$ of its Galois group is transitive on $R$; i.e., for any two roots $r$, $r'$ of $P$, there exists a permutation $\sigma \in \Pi(\mathsf{G}_F)$ such that $\sigma(r) = r'$.*

**Proof.** Suppose $P = P_1 P_2$ for some non-constant polynomials $P_1$, $P_2 \in F[X]$, and let $r$ be a root of $P_1$ and $r'$ a root of $P_2$. Renumbering the roots, we may assume $r = r_1$, $r' = r_2$, and $\mathsf{G}_F$ contains the arrangement $\varpi = (r_1, r_2, \ldots, r_n)$. Since $P$ does not have multiple roots, we have $P_1(r_2) \neq 0$. But, viewed as a polynomial in $F[x_1, \ldots, x_n]$, the polynomial $P_1(x_1)$ satisfies

$$P_1(\varpi) = P_1(r_1) = 0 \in F,$$

hence Property (GP) shows that $P_1(x_1)$ must be constant on all the arrangements of $\mathsf{G}_F$. Therefore, $\mathsf{G}_F$ does not contain any arrangement with $r_2$ in the first slot, and it follows that there is no permutation $\sigma \in \Pi(\mathsf{G}_F)$ such that $\sigma(r_1) = r_2$.

Conversely, suppose $\Pi(\mathsf{G}_F)$ is not transitive on $R$. Renumbering the elements in $R$, we may assume that the images of $r_1$ under the action of $\Pi(\mathsf{G}_F)$ are $r_2$, ..., $r_m$ with $m < n$, and $\mathsf{G}_F$ contains $\varpi = (r_1, \ldots, r_n)$. Let $P_1(X) = (X - r_1) \ldots (X - r_m)$. The coefficients of $P_1$ are the elementary symmetric polynomials in $x_1$, ..., $x_m$ evaluated in $r_1$, ..., $r_m$. Because all the arrangements in $\mathsf{G}_F$ have $r_1$, ..., $r_m$ in the first $m$ slots, these polynomials are constant on $\mathsf{G}_F$, hence $P_1 \in F[X]$ by Property (GP). Since $P_1$ divides $P$, it follows that $P$ is not irreducible in $F[X]$. $\qquad\square$

Proposition 14.35 shows in particular that general polynomials are irreducible, since we saw in Example 14.11 that the group of permutations of their Galois group is the full symmetric group. Therefore, the following corollary can be used to strengthen Corollary 14.33: general equations of degree at least 5 are not solvable by radicals in the (weaker) sense of §13.1.

**Corollary 14.36.** *If the polynomial $P$ is irreducible, the equation $P(X) = 0$ is solvable by radicals if and only if it is completely solvable by radicals.*

**Proof.** It obviously suffices to prove the "only if" part. Suppose the root $r$ of $P$ has an expression by radicals. By Proposition 14.34 we can find a series of subgroups

$$\Pi(\mathsf{G}_F) = \Gamma_0 \supset \Gamma_1 \supset \cdots \supset \Gamma_t$$

such that each $\Gamma_i$ is normal of prime index in $\Gamma_{i-1}$ for $i = 1$, ..., $t$, and all the permutations in $\Gamma_t$ fix $r$. By Proposition 14.35, for any other root $r' \in R$ we may find $\sigma \in \Pi(\mathsf{G}_F)$ such that $\sigma(r) = r'$. For $i = 0$, ..., $t$, let $\Gamma_i' = \sigma \circ \Gamma_i \circ \sigma^{-1}$. Then

$$\Pi(\mathsf{G}_F) = \Gamma_0' \supset \Gamma_1' \supset \cdots \supset \Gamma_t',$$

each $\Gamma'_i$ is normal of prime index in $\Gamma'_{i-1}$ for $i = 1, \ldots, t$, and all the permutations in $\Gamma'_t$ fix $r'$. By Proposition 14.34 again, it follows that $r'$ has an expression by radicals. Thus, each root of $P$ lies in a radical extension of $F$, and Corollary 13.5 shows that there is a radical extension of $F$ that contains all the roots of $P$. Therefore, $P(X) = 0$ is completely solvable by radicals. $\qquad\qquad\qquad\qquad\qquad\qquad\qquad\qquad\qquad\qquad\qquad\qquad\qquad$ $\square$

## 14.5 Applications

We present two applications of Galois' theory: the first one is due to Galois himself; the second one establishes the theorem of Abel stated in the introduction (Theorem 14.1), proving that equations that satisfy Abel's condition are solvable by radicals.

Throughout the section, we use the same notation as in the preceding sections: the equation to be solved is $P(X) = 0$, where $P$ is a polynomial with coefficients in an arbitrary field $F$ of characteristic zero, without multiple roots in any extension of $F$. We fix some universal field containing $F$, the set $R$ of all roots of $P$, and the roots of any auxiliary equation we may need to solve.

### 14.5.1 *Irreducible equations of prime degree*

In the last part of his memoir, Galois considers the case where the polynomial $P$ is irreducible of prime degree. First, he gives a necessary and sufficient condition for the solvability[11] of $P(X) = 0$ in terms of the Galois group, see Theorem 14.41. He then rephrases the condition in a statement where the Galois group does not appear explicitly, see Theorem 14.42.

A major role in Galois' results is played by the subgroup $GA(p)$ of the symmetric group $S_p$ (for $p$ prime), which arose in relation with Lagrange resolvents in §10.3 (see Theorem 10.8, p. 140): in effect, Galois proves that every transitive solvable subgroup of $S_p$ is conjugate to a subgroup of $GA(p)$, see Exercise 14.10. In preparation for the proofs of Galois' theorems, we first discuss a few properties of $GA(p)$ which are used by Galois.

Throughout this subsection, we fix a prime $p$ and consider $S_p$ as the group of permutations of $\{0, 1, \ldots, p-1\}$. Let $\tau \in S_p$ be the following

---

[11]For $P(X) = 0$, solvability and complete solvability are equivalent properties by Corollary 14.36 because $P$ is irreducible.

cyclic permutation:

$$\tau : 0 \mapsto 1 \mapsto 2 \mapsto \cdots \mapsto p-1 \mapsto 0.$$

Using addition modulo $p$ (in fact, identifying $\{0, 1, \ldots, p-1\}$ with the field $\mathbb{F}_p$ of $p$ elements, see Remark 12.4, p. 158), we may write

$$\tau(x) = x + 1 \qquad \text{for } x \in \mathbb{F}_p.$$

The group $\mathrm{GA}(p) \subset S_p$ is the set of permutations of the form

$$x \mapsto ax + b \qquad \text{for } x \in \mathbb{F}_p,$$

with $a$, $b \in \mathbb{F}_p$ and $a \neq 0$. It is clear that $a$ and $b$ are uniquely determined by such a map: if $ax + b = a'x + b'$ for all $x \in \mathbb{F}_p$, then $a = a'$ and $b = b'$. Therefore, $|\mathrm{GA}(p)| = p(p-1)$.

**Lemma 14.37.** *Let $\sigma \in S_p$. If $\sigma \circ \tau \circ \sigma^{-1} = \tau^k$ for some $k \in \{1, \ldots, p-1\}$, then*

$$\sigma(x) = kx + \sigma(0) \qquad \text{for all } x \in \mathbb{F}_p.$$

*In particular, $\sigma \in \mathrm{GA}(p)$.*

**Proof.** From $\sigma \circ \tau = \tau^k \circ \sigma$, it follows that

$$\sigma(y+1) = \sigma(y) + k \qquad \text{for all } y \in \mathbb{F}_p.$$

Then $\sigma(y+2) = \sigma(y+1) + k = \sigma(y) + 2k$, $\sigma(y+3) = \sigma(y) + 3k$, etc., and in general

$$\sigma(y+x) = \sigma(y) + kx \qquad \text{for all } x,\ y \in \mathbb{F}_p.$$

The lemma follows by letting $y = 0$. $\qquad\qquad\qquad\qquad\qquad\qquad\square$

Note that the order of a permutation $\sigma$ that decomposes into disjoint cycles of length $a$, $b$, $c$, $\ldots$ (i.e., the smallest strictly positive integer $e$ such that $\sigma^e = \mathrm{Id}$) is the least common multiple of $a$, $b$, $c$, $\ldots$. Since $\tau^2, \tau^3, \ldots,$ $\tau^{p-1}$ all have order $p$, they are all cyclic permutations of all the elements of $\mathbb{F}_p$.

**Lemma 14.38.** *Suppose $\Gamma \subset S_p$ is a subgroup containing $\tau$. If $|\Gamma| < p^2$, then $\tau$, $\tau^2$, $\ldots$, $\tau^{p-1}$ are the only permutations in $\Gamma$ consisting of a single cycle of all the elements of $\mathbb{F}_p$.*

**Proof.** Suppose $\tau' \in \Gamma$ is a cyclic permutation of all the elements of $\mathbb{F}_p$, hence $\tau'^p = \mathrm{Id}_{\mathbb{F}_p}$. Because $|\Gamma| < p^2$, the products $\tau^i \tau'^j$ for $i$, $j = 0, \ldots,$ $p-1$ cannot be all distinct: we must have $\tau^i \tau'^j = \tau^k \tau'^\ell$ for some $i$, $j$, $k$, $\ell \in \{0, 1, \ldots, p-1\}$ with $i \neq k$ and $j \neq \ell$. Because $\mathbb{F}_p$ is a field, we may then find an integer $m$ such that $(j - \ell)m \equiv 1 \bmod p$, and then

$$\tau' = \tau'^{(j-\ell)m} = \tau^{(k-i)m} \in \{\tau, \tau^2, \ldots, \tau^{p-1}\}. \qquad\qquad\square$$

The main property of $GA(p)$ for the purposes of Galois' results is the following:

**Proposition 14.39.** *Let* $\Gamma \subset S_p$ *be a subgroup. If* $\Gamma$ *contains a normal subgroup* $\Gamma'$ *with* $\tau \in \Gamma'$ *and* $|\Gamma'| < p^2$, *then* $\Gamma \subset GA(p)$.

**Proof.** Let $\sigma \in \Gamma$. Since $\Gamma'$ is a normal subgroup of $\Gamma$ containing $\tau$, we have $\sigma \circ \tau \circ \sigma^{-1} \in \Gamma'$. But $\sigma \circ \tau \circ \sigma^{-1}$ is a cyclic permutation of all the elements of $\mathbb{F}_p$:

$$\sigma \circ \tau \circ \sigma^{-1} \colon \sigma(0) \mapsto \sigma(1) \mapsto \sigma(2) \mapsto \cdots \mapsto \sigma(p-1) \mapsto \sigma(0).$$

Since $|\Gamma'| < p^2$, Lemma 14.38 yields

$$\sigma \circ \tau \circ \sigma^{-1} \in \{\tau, \tau^2, \ldots, \tau^{p-1}\}.$$

Lemma 14.37 then shows that $\sigma \in GA(p)$. Therefore, $\Gamma \subset GA(p)$.  $\square$

Note that if $|\Gamma| < p^2$ and $\tau \in \Gamma$ we may apply Proposition 14.39 with $\Gamma' = \Gamma$ to conclude that $\Gamma \subset GA(p)$.

We now turn to the solution of $P(X) = 0$. We write $\mathsf{G}$ for a Galois group of $P$, and $p$ (instead of $n$) for the degree of $P$, which is assumed to be prime.

First, suppose $P(X) = 0$ is solvable by radicals. By Corollary 14.31, we know there is a sequence of subgroups in the permutation group $\Pi(\mathsf{G})$

$$\Pi(\mathsf{G}) = \Gamma_0 \supset \Gamma_1 \supset \cdots \supset \Gamma_t = \{\mathrm{Id}_R\}, \tag{14.22}$$

where each $\Gamma_i$ is normal of prime index in $\Gamma_{i-1}$ for $i = 1, \ldots, t$. The first step in Galois' analysis is to show that $\Gamma_{t-1}$ is the group generated by a cyclic permutation of the $p$ roots of $P$. For this, he argues along the lines of the Galois resolvent factorization in §14.3. We substitute an easy group-theoretic result for his somewhat delicate argument:

**Lemma 14.40.** *Let* $\Gamma \subset \mathrm{Sym}(R)$ *be a group of permutations of* $R$, *where* $|R| = p$ *is prime, and let* $\Gamma' \subset \Gamma$ *be a normal subgroup. If* $\Gamma$ *is transitive on* $R$, *then either* $\Gamma'$ *is transitive on* $R$, *or* $\Gamma' = \{\mathrm{Id}_R\}$.

**Proof.** Consider the decomposition of $R$ into disjoint orbits under the action of $\Gamma'$:

$$R = \Gamma'(r_1) \cup \cdots \cup \Gamma'(r_s) \qquad \text{for some } r_1, \ldots, r_s \in R. \tag{14.23}$$

Because $\Gamma$ is transitive on $R$ we may find for each $r_i$ a permutation $\sigma \in \Gamma$ such that $\sigma(r_1) = r_i$. Now, $\sigma \circ \Gamma' \circ \sigma^{-1} = \Gamma'$ because $\Gamma'$ is a normal subgroup in $\Gamma$, hence

$$\Gamma'(r_i) = \sigma \circ \Gamma' \circ \sigma^{-1}\big(\sigma(r_1)\big) = \sigma\big(\Gamma'(r_1)\big).$$

Therefore, all the subsets in the disjoint union (14.23) have the same number of elements, and it follows that $|R| = s \cdot |\Gamma'(r_1)|$. Since $|R| = p$ is prime, either $s = 1$ or $s = p$. If $s = 1$, then $R = \Gamma'(r_1)$, hence $\Gamma'$ is transitive on $R$. If $s = p$, then $\{r_1, \ldots, r_s\} = R$ and each $\Gamma'(r)$ has only one element, which means that $\Gamma' = \{\mathrm{Id}_R\}$. $\qquad\square$

The lemma applies recursively to all the groups in the sequence (14.22) because the irreducibility of $P$ implies by Proposition 14.35 that $\Pi(\mathsf{G})$ is transitive on $R$. It follows that $\Gamma_{t-1}$ also is transitive on $R$. Its order is prime because $\{\mathrm{Id}_R\}$ has prime index in $\Gamma_{t-1}$. Let $\gamma \in \Gamma_{t-1}$ be a non-identity element. The set of powers of $\gamma$ is a subgroup in $\Gamma_{t-1}$, but the order of any subgroup of $\Gamma_{t-1}$ divides $|\Gamma_{t-1}|$ (see Theorem 10.4, p. 134), hence $\Gamma_{t-1}$ is the collection of powers of $\gamma$. Since $\Gamma_{t-1}$ is transitive on $R$, the cycle decomposition of $\gamma$ must consist of a single cycle involving all the elements of $R$. We may number the roots of $P$ so that $\gamma$ is the following cyclic permutation:

$$\gamma\colon r_0 \mapsto r_1 \mapsto r_2 \mapsto \cdots \mapsto r_{p-1} \mapsto r_0.$$

It follows in particular that $|\Gamma_{t-1}| = p$. Consider the following arrangement of $R$:

$$\varpi = (r_0, r_1, \ldots, r_{p-1}).$$

For $i = 0, \ldots, t$, let $\mathsf{G}_i = \Gamma_i(\varpi)$ be the orbit of $\varpi$ under $\Gamma_i$, and suppose $\mathsf{G}$ is the Galois group of $P$ containing $\varpi$. Thus, $\mathsf{G}_0 = \mathsf{G}$, and in the sequence of successive normal decompositions of $\mathsf{G}$ corresponding to the resolution sequence (14.22), $\mathsf{G}_i$ is the group of arrangements containing $\varpi$ that appears at the $i$-th stage (and $\mathsf{G}_t = \{\varpi\}$). Since $\Gamma_{t-1} = \{\mathrm{Id}_R, \gamma, \gamma^2, \ldots, \gamma^{p-1}\}$, we have

$$\mathsf{G}_{t-1} = \left\{ \begin{array}{cccccc} r_0 & r_1 & r_2 & \cdots & r_{p-3} & r_{p-2} & r_{p-1} \\ r_1 & r_2 & r_3 & \cdots & r_{p-2} & r_{p-1} & r_0 \\ r_2 & r_3 & r_4 & \cdots & r_{p-1} & r_0 & r_1 \\ & \cdots & \cdots & \cdots & & \\ r_{p-1} & r_0 & r_1 & \cdots & r_{p-4} & r_{p-3} & r_{p-2} \end{array} \right\}.$$

The second line is $\varpi^\tau$ where $\tau \in S_p$ is (as above) the cyclic permutation

$$\tau\colon 0 \mapsto 1 \mapsto 2 \mapsto \cdots \mapsto p-1 \mapsto 0.$$

We thus have

$$\Pi'(\mathsf{G}_{t-1}) = \{\mathrm{Id}, \tau, \ldots, \tau^{p-1}\}.$$

Because $G_{t-1}$ is part of a normal decomposition of $G_{t-2}$, we know by Proposition 14.24 that $\Pi'(G_{t-1})$ is a normal subgroup in $\Pi'(G_{t-2})$. We may therefore apply Proposition 14.39 to see that $\Pi'(G_{t-2}) \subset GA(p)$, hence $|\Pi'(G_{t-2})| < p^2$ since $|GA(p)| = p(p-1)$. Now $\Pi'(G_{t-2})$ is a normal subgroup in $\Pi'(G_{t-3})$, and we may apply Proposition 14.39 again to obtain $\Pi'(G_{t-3}) \subset GA(p)$, etc. By induction, we finally see that $\Gamma'_0 \subset GA(p)$. We have thus proved the "only if" part of the following theorem:

**Theorem 14.41.** *The irreducible equation $P(X) = 0$ of prime degree $p$ is solvable by radicals if and only if its Galois group $G$ satisfies $\Pi'(G) \subset GA(p)$ for a suitable numbering of the roots.*

To prove the "if" part, suppose $R = \{r_0, r_1, \ldots, r_{p-1}\}$ is a numbering of the roots of $P$ such that $\Pi'(G) \subset GA(p)$. We may assume $G$ contains the arrangement $\varpi = (r_0, r_1, \ldots, r_{p-1})$, so

$$G = \varpi^{\Pi'(G)} \subset \varpi^{GA(p)}.$$

Because all the roots of unity have expressions by radicals (see Proposition 13.3, p. 199), we may assume the base field $F$ contains all the roots of unity of exponent $p$ and of exponent $p - 1$. (Adjoining roots of unity to $F$ may reduce the Galois group, but the hypothesis that $\Pi'(G) \subset GA(p)$ is preserved.)

Let $g \in \mathbb{N}$ be a primitive root of $p$ (see Theorem 12.1, p. 156), and let $\sigma_g, \tau \in GA(p)$ be defined by

$$\sigma_g(x) = gx \quad \text{and} \quad \tau(x) = x + 1 \quad \text{for } x \in \mathbb{F}_p.$$

Let also $k \in \mathbb{N}$ be such that $gk \equiv 1 \bmod p$. Since $g$ is a primitive root of $p$, the group $GA(p)$ is generated by $\sigma_g$ and $\tau$, and since $\tau \circ \sigma_g(x) = gx + 1 = g(x + k)$ for $x \in \mathbb{F}_p$ we have

$$\tau \circ \sigma_g = \sigma_g \circ \tau^k. \tag{14.24}$$

For any $p$-th root of unity $\omega$, consider the following Lagrange resolvent in the indeterminates $x_0, \ldots, x_{p-1}$:

$$t_\omega(x_0, \ldots, x_{p-1}) = x_0 + \omega x_1 + \omega^2 x_2 + \cdots + \omega^{p-1} x_{p-1}.$$

We have

$$\tau(t_\omega)(x_0, \ldots, x_{p-1}) = t_\omega(x_1, \ldots, x_{p-1}, x_0) = \omega^{-1} t_\omega(x_0, \ldots, x_{p-1}),$$

hence

$$\tau(t_\omega^p) = t_\omega^p. \tag{14.25}$$

From (14.24) it follows that

$$\tau \circ \sigma_g(t_\omega^p) = \sigma_g \circ \tau^k(t_\omega^p) = \sigma_g(t_\omega^p). \tag{14.26}$$

Now, for any $(p-1)$-st root of unity $\zeta$, let

$$\Theta_{\omega,\zeta} = t_\omega^p + \zeta\sigma_g(t_\omega^p) + \zeta^2\sigma_g^2(t_\omega^p) + \cdots + \zeta^{p-2}\sigma_g^{p-2}(t_\omega^p).$$

By (14.25) and (14.26) we have $\tau(\Theta_{\omega,\zeta}) = \Theta_{\omega,\zeta}$; moreover,

$$\sigma_g(\Theta_{\omega,\zeta}) = \zeta^{-1}\Theta_{\omega,\zeta},$$

hence

$$\sigma_g(\Theta_{\omega,\zeta}^{p-1}) = \tau(\Theta_{\omega,\zeta}^{p-1}) = \Theta_{\omega,\zeta}^{p-1}.$$

Since $\mathrm{GA}(p)$ is generated by $\sigma_g$ and $\tau$, it follows that

$$\gamma(\Theta_{\omega,\zeta}^{p-1}) = \Theta_{\omega,\zeta}^{p-1} \quad \text{for all } \gamma \in \mathrm{GA}(p),$$

hence

$$\Theta_{\omega,\zeta}(\varpi^\gamma)^{p-1} = \Theta_{\omega,\zeta}(\varpi)^{p-1} \quad \text{for all } \gamma \in \mathrm{GA}(p).$$

As $\mathsf{G} \subset \varpi^{\mathrm{GA}(p)}$, the last equation shows that the polynomial $\Theta_{\omega,\zeta}(x_0,\ldots,x_{p-1})^{p-1}$ is constant on $\mathsf{G}$. By Property (GP), it follows that $\Theta_{\omega,\zeta}(\varpi)^{p-1} \in F$. Proposition 13.2 then shows that $\Theta_{\omega,\zeta}(\varpi)$ lies in a radical extension of $F$. By Corollary 13.5, we may find a radical extension $E$ of $F$ containing $\Theta_{\omega,\zeta}(\varpi)$ for all $p$-th roots of unity $\omega$ and all $(p-1)$-st roots of unity $\zeta$. Lagrange's formula (Proposition 10.2, p. 130) yields

$$t_\omega(\varpi)^p = \frac{1}{p-1}\sum_\zeta \Theta_{\omega,\zeta}(\varpi) \in E,$$

hence Proposition 13.2 shows that $t_\omega(\varpi)$ lies in a radical extension of $E$ (hence of $F$). By Corollary 13.5 again, there is a radical extension $E'$ of $F$ which contains all the $t_\omega(\varpi)$'s. Lagrange's formula (Proposition 10.2) now yields

$$r_i = \frac{1}{p}\sum_\omega \omega^{-i}t_\omega(\varpi) \in E' \quad \text{for } i = 0,\ldots,p-1.$$

Therefore, $P(X) = 0$ is solvable by radicals. $\qquad\square$

In order to justify the introduction of groups and to demonstrate the power and the usefulness of this new tool, it is necessary to produce new results that do not refer to groups in their statement but require some group theory in their proof. The final result in Galois' memoir recasts the necessary and sufficient condition for the solvability of irreducible equations of prime degree in a form that does not involve groups:

**Theorem 14.42.** *The irreducible equation of prime degree $P(X) = 0$ is solvable by radicals if and only if all the roots of $P$ can be rationally expressed from any[12] two of them; i.e., $R \subset F(r, r')$ for any $r$, $r' \in R$ with $r \neq r'$.*

**Proof.** Suppose first $P(X) = 0$ is solvable. By Theorem 14.41, we may number the roots $R = \{r_0, r_1, \ldots, r_{p-1}\}$ so that the Galois group $\mathsf{G}$ satisfies $\Pi'(\mathsf{G}) \subset \mathrm{GA}(p)$. We may assume $\mathsf{G}$ contains the arrangement $\varpi = (r_0, r_1, \ldots, r_{p-1})$. Pick $i$, $j \in \{0, 1, \ldots, p-1\}$ with $i \neq j$, and let $\mathsf{G}_{F(r_i, r_j)} \subset \mathsf{G}$ be the Galois group of $P$ containing $\varpi$ over $F(r_i, r_j)$. We have $\Pi'(\mathsf{G}_{F(r_i, r_j)}) \subset \Pi'(\mathsf{G}) \subset \mathrm{GA}(p)$, but over $F(r_i, r_j)$ the polynomials $x_i$ and $x_j$ evaluated in $\varpi$ take rational values (namely, $r_i$ and $r_j$). By Property (GP), these polynomials must be constant on $\mathsf{G}_{F(r_i, r_j)}$, which means that all the permutations in $\Pi'(\mathsf{G}_{F(r_i, r_j)})$ must leave $i$ and $j$ fixed. A permutation in $\Pi'(\mathsf{G}_{F(r_i, r_j)})$ has the form $x \mapsto ax + b$ for some $a$, $b \in \mathbb{F}_p$ with $a \neq 0$ because $\Pi'(\mathsf{G}_{F(r_i, r_j)}) \subset \mathrm{GA}(p)$, and the conditions

$$\begin{cases} ai + b = i \\ aj + b = j \end{cases}$$

imply $a = 1$ and $b = 0$. Therefore, $\Pi'(\mathsf{G}_{F(r_i, r_j)}) = \{\mathrm{Id}\}$, and $\mathsf{G}_{F(r_i, r_j)} = \{\varpi\}$. It follows that every polynomial $x_k$ is constant on $\mathsf{G}_{F(r_i, r_j)}$, hence $r_k \in F(r_i, r_j)$ for all $k$.

For the converse, we first translate the condition that $R \subset F(r, r')$ into a property of the Galois group $\mathsf{G}$:

**Lemma 14.43.** *Suppose $R \subset F(r, r')$ for some $r$, $r' \in R$ with $r \neq r'$. If $\sigma \in \Pi(\mathsf{G})$ fixes $r$ and $r'$, then $\sigma = \mathrm{Id}_R$. Moreover, $|\Pi(\mathsf{G})| \leq p(p-1)$.*

**Proof of Lemma 14.43.** As a first step in the proof, we show that every element in $F(r, r')$ (hence every root of $P$) has a polynomial expression in

---

[12]The condition that there exist $r$, $r' \in R$ with $r \neq r'$ such that $R \subset F(r, r')$ is formally weaker, but can be shown to be equivalent with the help of more advanced group theory: see Exercise 14.10.

$r$ and $r'$. The proof follows by a double application of Proposition 12.15 (p. 166): viewing $F(r, r')$ as $F(r)(r')$, we see from Proposition 12.15 that every $u \in F(r, r')$ has a polynomial expression in $r'$ of the form

$$u = a_0(r) + a_1(r)r' + \cdots + a_{d-1}(r)r'^{d-1}$$

where $a_0(r), \ldots, a_{d-1}(r) \in F(r)$ (and $d$ is the degree of the irreducible factor of $P$ in $F(r)[X]$ which has $r'$ as a root). Proposition 12.15 also yields polynomials $b_0, \ldots, b_{d-1} \in F[X]$ such that $a_i(r) = b_i(r)$ for $i = 0, \ldots, d-1$, hence we obtain a polynomial expression for $u$:

$$u = b_0(r) + b_1(r)r' + \cdots + b_{d-1}(r)r'^{d-1}.$$

Now, number the elements of $R$ so that $r = r_0$ and $r' = r_1$, and assume $\mathsf{G}$ contains the arrangement $\varpi = (r_0, r_1, \ldots, r_{p-1})$. Since $R \subset F(r, r')$, we may find for each $k = 2, \ldots, p-1$ a polynomial $f_k(x_0, x_1) \in F[x_0, x_1]$ such that

$$r_k = f_k(r_0, r_1).$$

Evaluation of the polynomial $x_k - f_k(x_0, x_1)$ on $\varpi$ yields $0 \in F$, hence by Property (GP) this polynomial must be constant on $\mathsf{G}$. Therefore, if $\sigma \in \Pi(\mathsf{G})$, so

$$\sigma(\varpi) = \bigl(\sigma(r_0), \ldots, \sigma(r_{p-1})\bigr) \in \mathsf{G},$$

we must have $\sigma(r_k) - f_k\bigl(\sigma(r_0), \sigma(r_1)\bigr) = 0$. In particular, if $\sigma$ fixes $r_0$ and $r_1$, then

$$\sigma(r_k) = f_k(r_0, r_1) = r_k \qquad \text{for } k = 2, \ldots, p-1,$$

hence $\sigma = \mathrm{Id}_R$. This proves the first part of the lemma.

From this first part, it follows that every $\sigma \in \Pi(\mathsf{G})$ is determined by $\sigma(r)$ and $\sigma(r')$, because if $\sigma, \rho \in \Pi(\mathsf{G})$ satisfy $\sigma(r) = \rho(r)$ and $\sigma(r') = \rho(r')$, then $\rho^{-1} \circ \sigma \in \Pi(\mathsf{G})$ leaves $r$ and $r'$ fixed, hence $\rho^{-1} \circ \sigma = \mathrm{Id}_R$. Therefore, the map $\Pi(\mathsf{G}) \to R \times R$ that carries $\sigma$ to $\bigl(\sigma(r), \sigma(r')\bigr)$ is injective. Since $\sigma(r) \neq \sigma(r')$, its image does not contain any point on the diagonal of $R \times R$, hence $|\Pi(\mathsf{G})| \leq p(p-1)$. $\qquad\square$

**Proof of Theorem 14.42, completed.** To prove the "if" part of Theorem 14.42, suppose $R \subset F(r, r')$ for any $r, r' \in R$ with $r \neq r'$, so by Lemma 14.43 the only permutation in $\Pi(\mathsf{G})$ that leaves any two elements in $R$ fixed is the identity. Our first goal is to show that $\Pi(\mathsf{G})$ contains a cyclic permutation of all the elements of $R$. For $r \in R$, let $I(r) \subset \Pi(\mathsf{G})$ be the set of permutations that leave $r$ fixed. Since $P$ is irreducible, $\Pi(\mathsf{G})$ is transitive on $R$ by Proposition 14.35, hence Lagrange's theorem (Proposition 14.3) yields $|\Pi(\mathsf{G})| = p \cdot |I(r)|$ for all $r \in R$. If each $\sigma \in \Pi(\mathsf{G})$ leaves an element of $R$ fixed, then

$$\Pi(\mathsf{G}) = \bigcup_{r \in R} I(r).$$

But all the $I(r)$'s contain $\mathrm{Id}_R$, so we may rewrite the preceding equation as

$$\Pi(\mathsf{G}) = \{\mathrm{Id}_R\} \cup \bigcup_{r \in R} (I(r) \setminus \{\mathrm{Id}_R\}). \tag{14.27}$$

Since $|I(r)| = \frac{1}{p}|\Pi(\mathsf{G})|$ for all $r \in R$, the number of elements in the set on the right-hand side is at most

$$1 + p(\tfrac{1}{p}|\Pi(\mathsf{G})| - 1) = |\Pi(\mathsf{G})| - p + 1 < |\Pi(\mathsf{G})|.$$

Therefore, (14.27) is impossible, which means that $\Pi(\mathsf{G})$ contains permutations that move all the elements of $R$. Let $\theta \in \Pi(\mathsf{G})$ be such a permutation. We will prove that $\theta$ is a cyclic permutation. For this, consider the subgroup $\Gamma \subset \Pi(\mathsf{G})$ generated by $\theta$:

$$\Gamma = \{\mathrm{Id}_R, \theta, \theta^2, \dots\}.$$

The set $R$ decomposes into a disjoint union of orbits under the action of $\Gamma$:

$$R = \bigcup_{i=1}^{s} \Gamma(r_i) \qquad \text{for some } r_1, \dots, r_s \in R.$$

Because $\theta$ moves all the elements of $R$, each $\Gamma(r_i)$ has at least two elements. In fact, they all have the same number of elements: if $\theta^k(r_i) = r_i$ for some $i$, then applying $\theta$ to each side we obtain $\theta^k(\theta(r_i)) = \theta(r_i)$, hence $\theta^k$ leaves $r_i$ and $\theta(r_i)$ fixed, and therefore $\theta^k = \mathrm{Id}_R$ because $\theta^k$ fixes two elements in $R$. Since $|R| = p$ is prime, there can be only one orbit of $\Gamma$, which means that $\theta$ is a cyclic permutation of all the elements of $R$. (The same kind of argument has been used in the proof of the "if" part of Theorem 14.29.) Renumbering the elements of $R$, we may assume $\theta$ is the following permutation:

$$\theta \colon r_0 \mapsto r_1 \mapsto r_2 \mapsto \cdots \mapsto r_{p-1} \mapsto r_0,$$

and assume $\mathsf{G}$ contains the arrangement $\varpi = (r_0, r_1, \dots, r_{p-1})$. Then $\mathsf{G}$ also contains

$$\theta(\varpi) = (r_1, r_2, \dots, r_{p-1}, r_0),$$

hence $\Pi'(\mathsf{G})$ contains the permutation $\tau \colon 0 \mapsto 1 \mapsto \cdots \mapsto p - 1 \mapsto 0$, which is such that $\theta(\varpi) = \varpi^\tau$. Note that the condition $R \subset F(r, r')$ also yields $|\Pi'(\mathsf{G})| \le p(p - 1) < p^2$ by Lemma 14.43, hence we may apply Proposition 14.39 (with $\Gamma = \Gamma' = \Pi'(\mathsf{G})$) to see that $\Pi'(\mathsf{G}) \subset \mathrm{GA}(p)$. It then follows from Theorem 14.41 that $P(X) = 0$ is solvable by radicals. $\qquad \square$

Theorem 14.42 can be effectively used to produce examples of nonsolvable equations over the field $\mathbb{Q}$ of rational numbers, as we now show.

**Corollary 14.44.** *Let $P$ be an irreducible polynomial of prime degree over $\mathbb{Q}$. If at least two roots of $P$, but not all, are real, then $P(X) = 0$ is not solvable by radicals over $\mathbb{Q}$.*

**Proof.** Let $R = \{r_0, \ldots, r_{p-1}\}$, and assume $r_0$, $r_1 \in \mathbb{R}$ and $r_{p-1} \notin \mathbb{R}$. Then $\mathbb{Q}(r_0, r_1) \subset \mathbb{R}$, hence $r_{p-1} \notin \mathbb{Q}(r_0, r_1)$ and Theorem 14.42 shows that $P(X) = 0$ is not solvable over $\mathbb{Q}$. □

The condition on the roots of $P$ in Corollary 14.44 is not hard to check in specific examples. If the degree of $P$ is a prime congruent to 1 mod 4, it can even be verified purely arithmetically with the aid of the discriminant, as the next corollary shows:

**Corollary 14.45.** *Let $P$ be a monic irreducible polynomial of prime degree $p$ over $\mathbb{Q}$. Assume $p \equiv 1$ mod 4. If the discriminant of $P$ is negative, then $P(X) = 0$ is not solvable by radicals over $\mathbb{Q}$.*

**Proof.** By Exercise 8.4 (p. 108), the condition on the discriminant readily implies that the number of real roots of $P$ is not 1 nor $p$. □

As a specific example, equations of the form

$$X^5 - pqX + p = 0$$

where $p$ is prime and $q$ is an integer, $q \geq 2$ (or $q \geq 1$ and $p \geq 13$) are not solvable by radicals over $\mathbb{Q}$, since $X^5 - pqX + p$ is irreducible over $\mathbb{Q}$, as Eisenstein's criterion (Proposition 12.12, p. 164) readily shows, and its discriminant is negative (see Exercise 8.1 (p. 107) for the calculation of this discriminant).

### 14.5.2   Abelian equations

The goal of this subsection is to show how Galois' solvability criterion (Theorem 14.29) can be used to prove that equations satisfying Abel's condition are solvable by radicals, thus establishing Abel's theorem (Theorem 14.1). This application of Theorem 14.29 was not mentioned by Galois.

Throughout this subsection, $P$ is a polynomial of degree $n$ over a field $F$ of characteristic zero. We let $R = \{r_1, \ldots, r_n\}$ be the set of all roots of

$P$ in some field containing $F$, and assume there are rational functions $\theta_2$, ..., $\theta_n \in F(X)$ such that

$$r_i = \theta_i(r_1) \qquad \text{for } i = 2, \ldots, n. \tag{14.28}$$

In other words, $R \subset F(r_1)$, hence $P$ is completely solvable by radicals if and only if there is a radical extension of $F$ containing $r_1$. To study the complete solvability of $P$, we may therefore substitute for $P$ its irreducible factor that has $r_1$ as a root, and thus assume henceforth that $P$ is irreducible. In particular, it follows by Corollary 5.22 (p. 55) that all the roots of $P$ are simple, hence its Galois groups of arrangements are defined. Let $\mathsf{G}$ be the Galois group of $P$ containing the arrangement $\varpi = (r_1, \ldots, r_n)$.

**Lemma 14.46.** *For every* $\sigma \in \Pi(\mathsf{G})$,

$$\sigma(\varpi) = \big(\sigma(r_1), \theta_2(\sigma(r_1)), \ldots, \theta_n(\sigma(r_1))\big).$$

**Proof.** Let $\theta_i = f_i/g_i$ for some relatively prime polynomials $f_i, g_i \in F[X]$, and consider the following polynomial:

$$x_i g_i(x_1) - f_i(x_1) \in F[x_1, \ldots, x_n].$$

By (14.28), this polynomial vanishes on $\varpi$, hence by Property (GP) it must be constant on $\mathsf{G}$. In particular, it vanishes on $\sigma(\varpi) = \big(\sigma(r_1), \ldots, \sigma(r_n)\big)$, and therefore

$$\sigma(r_i)g\big(\sigma(r_1)\big) - f_i\big(\sigma(r_1)\big) = 0. \tag{14.29}$$

If $g_i\big(\sigma(r_1)\big) = 0$, then $g_i$ has a root in common with $P$, which is irreducible. By Lemma 12.14 (p. 166), it follows that $P$ divides $g_i$; but then $g_i(r_1) = 0$ and $\theta_i(r_1)$ is not defined. This contradiction shows that $g_i\big(\sigma(r_1)\big) \neq 0$, hence (14.29) yields $\sigma(r_i) = \theta_i\big(\sigma(r_1)\big)$. $\qquad\square$

Lemma 14.46 shows that all the arrangements in $\mathsf{G}$ have the form

$$\big(r, \theta_2(r), \ldots, \theta_n(r)\big) \qquad \text{for some } r \in R. \tag{14.30}$$

They are thus completely determined by their first slot (and the rational functions $\theta_2, \ldots, \theta_n$).

**Proposition 14.47.** *Suppose the rational functions* $\theta_2, \ldots, \theta_n$ *satisfy Abel's condition:*

$$\theta_i\big(\theta_j(r_1)\big) = \theta_j\big(\theta_i(r_1)\big) \qquad \text{for all } i, j = 2, \ldots, n. \tag{14.31}$$

*Then* $\sigma \circ \tau = \tau \circ \sigma$ *for all* $\sigma, \tau \in \Pi(\mathsf{G})$.

**Proof.** Let $\sigma, \tau \in \Pi(\mathsf{G})$. Since $\sigma(\varpi)$ and $\tau(\varpi)$ lie in $\mathsf{G}$, these arrangements have the form (14.30), hence there exist $j$, $k \in \{2, \ldots, n\}$ such that

$$\sigma(\varpi) = \big(r_j, \theta_2(r_j), \ldots, \theta_n(r_j)\big) \quad \text{and} \quad \tau(\varpi) = \big(r_k, \theta_2(r_k), \ldots, \theta_n(r_k)\big).$$

The second equation yields $\tau(r_1) = r_k$, and the first equation shows that $\sigma(r_k) = \theta_k(r_j)$, hence $(\sigma \circ \tau)(r_1) = \theta_k(r_j)$. Therefore, Lemma 14.46 yields

$$(\sigma \circ \tau)(\varpi) = \big(\theta_k(r_j), \theta_2(\theta_k(r_j)), \ldots, \theta_n(\theta_k(r_j))\big).$$

Likewise, we have $\sigma(r_1) = r_j$ and $\tau(r_j) = \theta_j(r_k)$, hence $(\tau \circ \sigma)(r_1) = \theta_j(r_k)$, and therefore

$$(\tau \circ \sigma)(\varpi) = \big(\theta_j(r_k), \theta_2(\theta_j(r_k)), \ldots, \theta_n(\theta_j(r_k))\big).$$

As $\theta_j(r_1) = r_j$ and $\theta_k(r_1) = r_k$, Abel's condition (14.31) yields $\theta_k(r_j) = \theta_j(r_k)$, hence $(\sigma \circ \tau)(\varpi) = (\tau \circ \sigma)(\varpi)$, and it follows that $\sigma \circ \tau = \tau \circ \sigma$. $\square$

Groups in which all the elements pairwise commute are said to be *abelian*, in reference to Abel's condition (14.31). As a consequence of Proposition 14.47, we see that this condition implies that $\Pi(\mathsf{G})$ is abelian. To derive Abel's theorem, it therefore suffices to show that equations with abelian Galois group are solvable. This is a direct consequence of Galois' criterion (as formulated in Corollary 14.31) and the following group-theoretic statement:

**Proposition 14.48.** *Every finite abelian group is solvable.*

**Proof.** Since every subgroup of an abelian group is normal, it suffices to prove that every finite abelian group $\Gamma \neq \{1\}$ contains a subgroup $\Gamma_1$ of prime index. Arguing by induction on the order of $\Gamma$, we then construct a sequence of subgroups

$$\Gamma \supset \Gamma_1 \supset \Gamma_2 \supset \cdots \supset \Gamma_t = \{1\}$$

each of which is normal of prime index in the preceding one. This sequence shows that $\Gamma$ is solvable.

The existence of a subgroup of prime index is a special case ($\Gamma' = \{1\}$) of the following result:

**Lemma 14.49.** *Let $\Gamma'$ be a subgroup of a finite abelian group $\Gamma$. If $\Gamma' \neq \Gamma$, then there exists in $\Gamma$ a subgroup $\Gamma_1$ of prime index which contains $\Gamma'$.*

**Proof.** We argue by induction on the index $(\Gamma : \Gamma')$ $(= \frac{|\Gamma|}{|\Gamma'|})$, which is assumed to be at least 2. If $(\Gamma : \Gamma') = 2, 3$ or any other prime number, then $\Gamma'$ satisfies the required conditions, hence we may take $\Gamma_1 = \Gamma'$. Assume then $(\Gamma : \Gamma')$ is not prime. Pick $\sigma$ in $\Gamma$ but not in $\Gamma'$ and consider the minimal exponent $e > 0$ for which $\sigma^e \in \Gamma'$. Let also $p$ be a prime factor of $e$ and

$$\rho = \sigma^{e/p}.$$

Then $\rho \notin \Gamma'$ (otherwise $e$ would not be minimal), and $\rho^p \in \Gamma'$. Consider then

$$\Gamma'' = \{\rho^i \mu \mid i = 0, \ldots, p - 1;\ \mu \in \Gamma'\}.$$

It is easily checked that $\Gamma''$ is a subgroup of $\Gamma$ containing $\Gamma'$, and that the cosets of $\Gamma'$ in $\Gamma''$ are $\Gamma'$, $\rho\Gamma'$, $\rho^2\Gamma'$, ..., $\rho^{p-1}\Gamma'$, so that $(\Gamma'' : \Gamma') = p$. Therefore, $\Gamma'' \neq \Gamma$, since by hypothesis $(\Gamma : \Gamma')$ is not prime, and

$$(\Gamma : \Gamma'') < (\Gamma : \Gamma').$$

From the induction hypothesis, it follows that there exists in $\Gamma$ a subgroup $\Gamma_1$ of prime index containing $\Gamma''$, hence also $\Gamma'$. $\qquad\qquad\square$

## Exercises

In the first five exercises, $R$ is an arbitrary finite set of $n$ elements, with $n \geq 2$.

*Exercise 14.1.* Fix an arrangement $\varpi \in A(R)$. Show that for any group of permutations $\Gamma \subset \mathrm{Sym}(R)$ the set $G = \Gamma(\varpi) \subset A(R)$ is a group of arrangements with $\Pi(G) = \Gamma$. Similarly, show that for any subgroup $\Gamma' \subset S_n$ the set $G' = \varpi^{\Gamma'} \subset A(R)$ is a group of arrangements with $\Pi'(G') = \Gamma'$.

*Exercise 14.2.* Let $G \subset A(R)$ be a group of arrangements. Show that every arrangement $\alpha \in G$ defines an isomorphism $f_\alpha \colon \Pi(G) \to \Pi'(G)$ by mapping $\sigma \in \Pi(G)$ to $\tau \in \Pi'(G)$ if $\sigma(\alpha) = \alpha^\tau$. For $\alpha, \beta \in G$ and $\sigma \in \Pi(G)$, show that $f_\beta^{-1} \circ f_\alpha(\sigma) = \binom{\alpha}{\beta} \circ \sigma \circ \binom{\beta}{\alpha}$.

*Exercise 14.3.* Let $G_1$, $G_2 \subset A(R)$ be groups of arrangements and let $\alpha_1 \in G_1$, $\alpha_2 \in G_2$. Show that if $\Pi(G_1) = \Pi(G_2)$ and $\sigma \in S_n$ is such that $\alpha_2 = \alpha_1^\sigma$, then $G_2 = G_1^\sigma$. Show that if $\Pi'(G_1) = \Pi'(G_2)$ and $\tau \in \mathrm{Sym}(R)$ is such that $\alpha_2 = \tau(\alpha_1)$, then $G_2 = \tau(G_1)$.

*Exercise 14.4.* Let $\mathsf{G} \subset \mathsf{A}(R)$ be a group of arrangements. Show that the following conditions are equivalent:

(a) $\Pi(\mathsf{G})$ is a normal subgroup of $\mathrm{Sym}(R)$;
(b) $\Pi'(\mathsf{G})$ is a normal subgroup of $S_n$;
(c) for every $\sigma \in \mathrm{Sym}(R)$ and $\tau \in S_n$ such that $\sigma(\mathsf{G}) \cap \mathsf{G}^\tau \neq \varnothing$, we have $\sigma(\mathsf{G}) = \mathsf{G}^\tau$.

*Exercise 14.5.* Let $\mathsf{G} \subset \mathsf{A}(R)$ be a group of arrangements, which decomposes into a disjoint union of subgroups $\mathsf{G} = \mathsf{G}_1 \cup \cdots \cup \mathsf{G}_t$. Show that this decomposition is normal if and only if $\Pi(\mathsf{G}_1) = \cdots = \Pi(\mathsf{G}_t)$ and $\Pi'(\mathsf{G}_1) = \cdots = \Pi'(\mathsf{G}_t)$.

*Exercise 14.6.* Let $\mathsf{G}$ be a Galois group of an equation $P(X) = 0$ of degree $n$. Show that $|\mathsf{G}|$ divides $n!$, and that $n$ divides $|\mathsf{G}|$ if $P$ is irreducible.

*Exercise 14.7.* Let $\mathsf{G}$ be a Galois group of an equation $P(X) = 0$ over a field $F$. Show that if $Q \in F[Z]$ is an irreducible polynomial of prime degree $p$ and if $p$ does not divide $|\mathsf{G}|$, then the Galois group does not change when a root of $Q$ is adjoined to the base field.

*Exercise 14.8.* Prove the statements in Example 14.28.

*Exercise 14.9.* Show that an equation $P(X) = 0$ is completely solvable by radicals over a field $F$ if and only if for each irreducible factor $P'$ of $P$, the equation $P'(X) = 0$ is solvable by radicals over $F$.

*Exercise 14.10.* Let $p$ be a prime integer, and let $\Gamma \subset S_p$ be a subgroup that is transitive on $\{0, \ldots, p-1\}$. Show that the following conditions are equivalent:

(a) $\Gamma$ is solvable;
(b) $\Gamma$ is conjugate to a subgroup of $\mathrm{GA}(p)$, i.e., there exists $\sigma \in S_p$ such that $\sigma \circ \Gamma \circ \sigma^{-1} \subset \mathrm{GA}(p)$;
(c) for any $i, j \in \{0, \ldots, p-1\}$ with $i \neq j$, the only element $\sigma \in \Gamma$ that fixes $i$ and $j$ is the identity;
(d) there exist $i, j \in \{0, \ldots, p-1\}$ with $i \neq j$ such that the only element $\sigma \in \Gamma$ that fixes $i$ and $j$ is the identity.

[Hint: The equivalence of (a) and (b) (resp. (b) and (c)) is essentially established in the proof of Theorem 14.41 (resp. 14.42); note that conjugation in $S_p$ amounts to renumbering $\{0, \ldots, p-1\}$. To prove (d) $\Rightarrow$ (b), use a theorem of Cauchy (published in 1845, see [53]) to see that $\Gamma$ contains a cyclic permutation of order $p$.]

# Chapter 15

# Epilogue

Although Galois' memoir is nowadays regarded as the climax of several decades of research on algebraic equations, the first reactions to Galois' theory were negative. It was rejected by the referees, because the arguments were "neither clear enough nor well enough developed" [55, p. 148], but also for another, deeper motive: it did not yield any workable criterion to determine whether an equation is solvable by radicals. In that respect, even the application to equations of prime degree indicated by Galois (see Theorem 14.42, p. 262) is hardly useful, as the referees pointed out:

> It must be noted however that [the memoir] does not contain, as [its] title promised, the condition for solubility of equations by radicals; for, even accepting Mr Galois' proposition as true, one is hardly further forward in knowing whether or not an equation of prime degree is soluble by radicals, because it would first be necessary to convince oneself whether the equation is irreducible, and then whether any one of its roots may be expressed as a rational function of two others. The condition for solubility, if it exists, should be an external character which one might verify by inspection of the coefficients of a given equation, or at the worst, by solving other equations of degree lower than the one given [55, p. 148].

Galois' criterion (see Theorem 14.29, p. 248) is very far from being external; indeed, Galois always worked with the roots of the proposed equation, never with its coefficients.[1] Thus, Galois' theory did not match expectations, it was too novel to be readily accepted.

After the publication of Galois' memoir by Liouville, its importance dawned upon the mathematical world, and it was eventually realized that Galois had discovered a mathematical gem much more valuable than any hypothetical external characterization of solvable equations. After all, the

---

[1] It is telling that the proposed equation is nowhere displayed in Galois' memoir.

problem of solving equations by radicals was utterly artificial. It had focused the efforts of several generations of brilliant mathematicians because it displayed some strange, puzzling phenomena. It contained something mysterious, profoundly appealing. Galois had taken the pith out of the problem, by showing that the difficulty of an equation was related to the ambiguity of its roots and pointing out how this ambiguity could be measured by means of a group. He had thus set the theory of equations and, indeed, the whole subject of algebra, on a completely different track.

> Thus I believe that the simplifications produced by elegance of calculations (intellectual simplifications, of course; there are no material ones) have their limits; I believe that the time will come when the algebraic transformations foreseen by the speculations of analysts will find neither the time nor the place for their realisation; at which point one will have to be content with having foreseen them. [...]

> Jump with both feet on calculations, put operations into groups, class them according to their difficulty and not according to their form; that is, according to me, the mission of future geometers, that is the path that I have entered in this work [55, p. 253].

Thereafter, research on the theory of equations slowly came to an end, while new subjects emerged, such as the theory of groups and of various algebraic structures. This final stage in the evolution of a mathematical theory has been beautifully described by Weil:

> Nothing is more fruitful, as all mathematicians know, than these dim analogies, these foggy glimpses from one theory to the other, these stealthy caresses, these inexplicable jumbles; nothing also gives more pleasure to the researcher. A day comes when the illusion dissipates; the vagueness changes into certainty; the twin theories disclose their common fount before vanishing; as the Gītā teaches, one reaches knowledge and indifference at the same time. Metaphysics has become mathematics, ready to make the substance of a treatise whose cold beauty could not move us any more [86, p. 52].

The subsequent developments arising from Galois theory do not fall within the scope of this work, so we refer to the papers by Kiernan [44] and by van der Waerden [78] and to the monographs by Ehrhardt [27] and by Nový [60] for detailed accounts. There is however one major trend in this evolution that we want to point out: the gradual elimination of polynomials and equations from the foundations of Galois theory. Indeed, it is revealing of the profoundness of Galois' ideas to see, through the various textbook expositions, how this theory initially designed to answer a question about equations progressively outgrew its original context.

The first step in this direction is the emergence of the notion of field, through the works of Kronecker and Dedekind. Their approaches were quite different but complementary. Kronecker's point of view was constructivist. To define a field according to this point of view is to describe a process by which the elements of the field can be constructed. By contrast, Dedekind's approach was set-theoretic. He did not hesitate to define the field generated by a set $P$ of complex numbers as the intersection of all the fields which contain $P$. This definition is hardly useful for determining whether a given complex number belongs to the field thus defined. Although Dedekind's approach has become the usual point of view nowadays, Kronecker's constructivism also led to important results, such as the algebraic construction of fields in which polynomials split into linear factors, see §9.1. The next step is the observation by Dedekind, around the end of the nineteenth century, that the permutations in the Galois group of an equation can be considered as automorphisms of the field of rational functions of the roots. The Galois group is thus related to a field extension instead of a polynomial; it can be defined without reference to any specific equation. Moreover, the newly developed linear algebra was brought to bear on the theory of fields, as the larger field in an extension can be regarded as a vector space over the smaller field.

These ideas came to fruition in the first decades of the twentieth century, as witnessed by van der Waerden's famous treatise "Moderne Algebra" (1930) (of which [77] is the seventh edition). The exposition of Galois theory in this book is based on lectures by Emil Artin (1898–1962). It states as its "fundamental theorem" a one-to-one correspondence between subfields of certain extensions (those that are obtained by adjoining all the roots of a polynomial without multiple roots), nowadays called *Galois* extensions, and the subgroups of the associated Galois group. This correspondence is nowhere to be found in Galois' memoir, which achieves its goal of characterizing equations that are solvable by radicals by working with a shockingly elementary notion of field extension. The only traces of a correspondence between groups and fields can be observed in the discussion of the Galois group reductions that can be obtained by the adjunction of a root of an auxiliary equation, see §14.3.

In van der Waerden's book, the treatment of Galois theory clearly emphasizes fields and groups, while polynomials and equations play a secondary role. They are used as tools in the proofs, but the main theorems do not involve polynomials in their statement.

A few years later, the exposition of Galois theory further evolved under the influence of Emil Artin, who once wrote:

> Since my mathematical youth, I have been under the spell of the classical theory of Galois. This charm has forced me to return to it again and again, and to try to find new ways to prove these fundamental theorems [3, p. 380].

In his book "Galois theory" [2] of 1942, Artin proposes a new, highly original, definition of Galois extension. The extension is looked at from the point of view of the larger field instead of the smaller. An extension of fields is then said to be *Galois* if the smaller field is the field of invariants under a (finite) group of automorphisms of the larger. This definition and some improvements in the proofs enabled Artin to further reduce the role of polynomials in the basic results of Galois theory, so that the fundamental theorem can now be proved without ever mentioning polynomials. (See Appendix 1.)

Artin's exposition has nowadays become the classical treatment of Galois theory from an elementary point of view. However, several other expositions have been proposed in more recent times, inspired by the applications of Galois theory in related areas. For instance, the Jacobson–Bourbaki correspondence [40, p. 22] yields a uniform treatment of both the classical Galois theory and the Galois theory for purely inseparable field extensions of height 1, where restricted $p$-Lie algebras are substituted for groups. In another direction, the Galois theory of commutative rings due to Chase, Harrison and Rosenberg [15], and Grothendieck's new foundation of algebraic geometry have inspired new expositions which stress the analogy between extensions of fields and coverings of locally compact topological spaces, see Douady [25] (compare also the new version of Bourbaki's treatise [8]) and Szamuely [72].

Through its applications in various areas and as a source of inspiration for new investigations, Galois theory is far from being a closed issue.

## Appendix 1: The fundamental theorem of Galois theory

In this section, we give an account of the one-to-one correspondence that is now regarded as the fundamental theorem of Galois theory, after Artin's classical exposition in [2].

**Definitions 15.1.** Any field $K$ containing a field $F$ as a subfield is said to be an *extension* (or a *field extension*) of $F$. The dimension of $K$, regarded as a vector space over $F$, is called the *degree* of $K$ over $F$, and is denoted by $[K : F]$, so

$$[K : F] = \dim_F K.$$

The extension $K$ of $F$ is called a *Galois* extension if $F$ is the field of all elements that are invariant under some finite group of (field-) automorphisms of $K$. In other words, letting $K^G$ denote the field of invariants under a group $G$ of automorphisms of $K$, i.e.,

$$K^G = \{x \in K \mid \sigma(x) = x \text{ for all } \sigma \in G\},$$

the extension $K$ of $F$ is Galois if and only if there exists a finite group $G$ of automorphisms of $K$ such that $F = K^G$.

As part of Theorem 15.2 below, it will be shown that when the condition holds, the group $G$ such that $F = K^G$ is the group $\mathrm{Aut}_F K$ of all automorphisms of $K$ that leave $F$ elementwise invariant (i.e., the $F$-linear automorphisms of $K$); it is therefore uniquely determined by $K$ and $F$. We will write $\mathrm{Gal}(K/F)$ for $\mathrm{Aut}_F K$ when the extension $K$ of $F$ is Galois, and call $\mathrm{Gal}(K/F)$ the *Galois group of $K$ over $F$*.

For example, we can take for $K$ a field of rational functions in $n$ indeterminates over an arbitrary field $k$,

$$K = k(x_1, \ldots, x_n),$$

and take for $G$ the symmetric group $S_n$, viewed as a group of automorphisms of $K$ by letting (as in §10.3)

$$\sigma\big(f(x_1, \ldots, x_n)\big) = f(x_{\sigma(1)}, \ldots, x_{\sigma(n)}) \quad \text{for } f \in k(x_1, \ldots, x_n) \text{ and } \sigma \in S_n.$$

The fundamental theorem of symmetric functions (Theorem 8.3, p. 95) shows that $K^G = k(s_1, \ldots, s_n)$ where $s_1$, ..., $s_n$ are the elementary symmetric polynomials in $x_1$, ..., $x_n$. Therefore, $k(x_1, \ldots, x_n)$ is a Galois extension of $k(s_1, \ldots, s_n)$ with Galois group $S_n$.

Other examples of Galois extensions are obtained as follows: let $F$ be an arbitrary field and let $P \in F[X]$ be a monic non-constant polynomial. Let $L$ be a field extension of $F$ such that $P$ splits into a product of linear factors in $L[X]$ (see Theorem 9.3, p. 110):

$$P(X) = (X - r_1) \ldots (X - r_n) \in L[X].$$

Assume $r_1$, ..., $r_n$ are pairwise distinct. Then the subfield $F(r_1, \ldots, r_n) \subset L$ is a Galois extension of $F$ with Galois group isomorphic to $\Pi(\mathsf{G})$ for $\mathsf{G}$ a Galois group (of arrangements) of $P(X) = 0$: see Remark 15.16.

**Theorem 15.2 (Fundamental theorem of Galois theory).** *Let $K$ be an extension of a field $F$. If $F = K^G$ for some finite group $G$ of automorphisms of $K$, then*

$$[K : F] = |G| \quad \text{and} \quad G = \text{Aut}_F K.$$

*The field $K$ is then a Galois extension of every subfield containing $F$. Moreover, there is a one-to-one correspondence between the subfields of $K$ containing $F$ and the subgroups of $G$, which associates to any subfield $L$ the Galois group $\text{Gal}(K/L) \subset G$ and to any subgroup $H \subset G$ its field of invariants $K^H$.*

*Under this correspondence, the degree over $F$ of a subfield of $K$ corresponds to the index in $G$ of the associated subgroup,*

$$[L : F] = \big(G : \text{Gal}(K/L)\big) \quad \text{and} \quad (G : H) = [K^H : F].$$

*Furthermore, a subfield $L$ of $K$ containing $F$ is a Galois extension of $F$ if and only if the corresponding subgroup $\text{Gal}(K/L)$ is normal in $G$. When the condition holds, the Galois group $\text{Gal}(L/F)$ is obtained by restricting to $L$ the automorphisms in $G$, and the restriction homomorphism induces an isomorphism $G/\text{Gal}(K/L) \xrightarrow{\sim} \text{Gal}(L/F)$.*

The proof of this theorem requires some preparation. We start with an easy observation on degrees of extensions:

**Lemma 15.3.** *Let $K \supset L \supset F$ be a tower of fields. Then*

$$[K : F] = [K : L] \cdot [L : F].$$

**Proof.** Let $(k_i)_{i \in I}$ be a basis of $K$ over $L$ and $(\ell_j)_{j \in J}$ be a basis of $L$ over $F$. If we prove $(k_i \ell_j)_{(i,j) \in I \times J}$ is a basis of $K$ over $F$, the lemma readily follows.

The family $(k_i \ell_j)_{(i,j) \in I \times J}$ spans $K$ since every element $x \in K$ can be written $x = \sum_{i \in I} k_i x_i$ for some $x_i \in L$, and decomposing $x_i = \sum_{j \in J} \ell_j y_{ij}$ with $y_{ij} \in F$ yields

$$x = \sum_{\substack{i \in I \\ j \in J}} (k_i \ell_j) y_{ij}.$$

To show that the family $(k_i \ell_j)_{(i,j) \in I \times J}$ is linearly independent over $F$, consider

$$\sum_{\substack{i \in I \\ j \in J}} k_i \ell_j y_{ij} = 0 \quad \text{for some } y_{ij} \in F.$$

Collecting terms which have the same index $i$, we get

$$\sum_{i \in I} k_i \Big( \sum_{j \in J} \ell_j y_{ij} \Big) = 0.$$

It follows that $\sum_{j \in J} \ell_j y_{ij} = 0$ for all $i \in I$, since $(k_i)_{i \in I}$ is linearly independent over $L$. Therefore,

$$y_{ij} = 0 \qquad \text{for all } i \in I, \, j \in J,$$

because $(\ell_j)_{j \in J}$ is linearly independent over $F$. $\qquad\qquad\square$

The basic observation which lies at the heart of the proof of the fundamental theorem is known as the lemma *of linear independence of homomorphisms*. It is due to Artin, and generalizes an earlier result of Dedekind.

**Lemma 15.4.** *Consider distinct homomorphisms $\sigma_1, \ldots, \sigma_n$ of a field $L$ into a field $K$. Then $\sigma_1, \ldots, \sigma_n$, viewed as elements of the $K$-vector space of all maps from $L$ to $K$, are linearly independent over $K$. In other words, if $a_1, \ldots, a_n \in K$ are such that*

$$a_1 \sigma_1(x) + \cdots + a_n \sigma_n(x) = 0 \qquad \text{for all } x \in L,$$

*then $a_1 = \cdots = a_n = 0$.*

**Proof.** Assume by way of contradiction that $\sigma_1, \ldots, \sigma_n$ are not independent, and choose $a_1, \ldots, a_n \in K$ such that

$$a_1 \sigma_1(x) + \cdots + a_n \sigma_n(x) = 0 \qquad \text{for all } x \in L \qquad (15.1)$$

with $a_1, \ldots, a_n$ not all zero, but such that the number of $a_i \neq 0$ be minimal. This number is at least equal to 2, otherwise one of the $\sigma_i$ would map $L$ to $\{0\}$. This is impossible since, by definition of homomorphisms of fields, $\sigma_i(1) = 1$ for all $i$. Renumbering $\sigma_1, \ldots, \sigma_n$ if necessary, we may thus assume without loss of generality that $a_1 \neq 0$ and $a_2 \neq 0$.

Choose $\ell \in L$ such that $\sigma_1(\ell) \neq \sigma_2(\ell)$. (This is possible since $\sigma_1 \neq \sigma_2$.) Multiplying each side of (15.1) by $\sigma_1(\ell)$, we obtain

$$a_1 \sigma_1(\ell) \sigma_1(x) + \cdots + a_n \sigma_1(\ell) \sigma_n(x) = 0 \qquad \text{for all } x \in L. \qquad (15.2)$$

On the other hand, substituting $\ell x$ for $x$ in equation (15.1), and using the multiplicative property of $\sigma_i$, we obtain

$$a_1 \sigma_1(\ell) \sigma_1(x) + \cdots + a_n \sigma_n(\ell) \sigma_n(x) = 0 \qquad \text{for all } x \in L. \qquad (15.3)$$

Subtracting (15.3) from (15.2), the first terms of each equation cancel out and we obtain

$$a_2 \big( \sigma_1(\ell) - \sigma_2(\ell) \big) \sigma_2(x) + \cdots + a_n \big( \sigma_1(\ell) - \sigma_n(\ell) \big) \sigma_n(x) = 0 \quad \text{for all } x \in L.$$

The coefficients are not all zero since $\sigma_1(\ell) \neq \sigma_2(\ell)$, but this linear combination has fewer non-zero terms than (15.1). This is a contradiction. $\quad\square$

**Remark.** Only the multiplicative property of $\sigma_1, \ldots, \sigma_n$ has been used. The same proof thus establishes the linear independence of distinct homomorphisms from any group to the multiplicative group of a field.

**Corollary 15.5.** *Let* $\sigma_1, \ldots, \sigma_n$ *be as in Lemma 15.4 and let*

$$F = \{x \in L \mid \sigma_1(x) = \cdots = \sigma_n(x)\}.$$

*Then* $[L : F] \geq n$.

**Proof.** Suppose $[L : F] < n$, and let $[L : F] = m$. Choose a basis $\ell_1, \ldots, \ell_m$ of $L$ over $F$ and consider the matrix $(\sigma_i(\ell_j))_{\substack{1 \leq i \leq n \\ 1 \leq j \leq m}}$ with entries in $K$.

The rank of this matrix is at most $m$, since the number of its columns is $m$, hence its rows are linearly dependent over $K$. We can therefore find elements $a_1, \ldots, a_n \in K$, not all zero, such that

$$a_1\sigma_1(\ell_j) + \cdots + a_n\sigma_n(\ell_j) = 0 \qquad \text{for } j = 1, \ldots, m. \tag{15.4}$$

Now, any $x \in L$ can be expressed as $x = \sum_{j=1}^{m} \ell_j x_j$ for some $x_j \in F$. Multiplying equations (15.4) by $\sigma_1(x_j)$ (which is equal to $\sigma_i(x_j)$ for all $i$, since $x_j \in F$) and adding the equations thus obtained, we get

$$a_1\sigma_1(x) + \cdots + a_n\sigma_n(x) = 0.$$

This contradicts Lemma 15.4. $\qquad\qquad\qquad\qquad\qquad\qquad\qquad\qquad\square$

**Corollary 15.6.** *Let* $\sigma_1, \ldots, \sigma_n$ *be as in Lemma 15.4. There exists* $u \in L$ *such that* $\sigma_1(u) + \cdots + \sigma_n(u) \neq 0$.

**Proof.** If no such $u$ exists, then $\sigma_1(x) + \cdots + \sigma_n(x) = 0$ for all $x \in L$. This contradicts Lemma 15.4. $\qquad\qquad\qquad\qquad\qquad\qquad\qquad\qquad\qquad\qquad\square$

We are now ready for the proof of Theorem 15.2. Assume $G$ is a finite group of automorphisms of the field $K$ such that $F = K^G$. Since all the automorphisms in $G$ fix $F$ we have $G \subset \text{Aut}_F K$.

*Step 1:* $[K : F] = |G|$.

Applying Corollary 15.5 with $L = K$ and $\{\sigma_1, \ldots, \sigma_n\} = G$, we already have

$$[K : F] \geq |G|.$$

If $[K : F] > |G|$, then for some $m > n \ (= |G|)$ we can find a sequence $k_1, \ldots, k_m$ of elements of $K$ which are linearly independent over $F$. The matrix

$$(\sigma_i(k_j))_{\substack{1 \leq i \leq n \\ 1 \leq j \leq m}}$$

has rank at most $n$, hence its columns are linearly dependent over $K$. Let $a_1, \ldots, a_m \in K$, not all zero, such that

$$\sigma_i(k_1)a_1 + \cdots + \sigma_i(k_m)a_m = 0 \qquad \text{for } i = 1, \ldots, n. \tag{15.5}$$

Renumbering $k_1, \ldots, k_m$ if necessary, we may assume $a_1 \neq 0$. By Corollary 15.6, we may pick $u \in K$ such that $\sigma_1(u) + \cdots + \sigma_n(u) \neq 0$. Multiplying all the equations (15.5) by $a_1^{-1}u$, we obtain another set of equations like (15.5), with the additional property that the coefficient of $\sigma_i(k_1)$ is $u$. Therefore, for the rest of the argument we may assume $\sigma_1(a_1) + \cdots + \sigma_n(a_1) \neq 0$.

Applying $\sigma_i^{-1}$ to (15.5) and adding up the equations thus obtained for $i = 1, \ldots, n$, we obtain

$$k_1\big(\sigma_1^{-1}(a_1) + \cdots + \sigma_n^{-1}(a_1)\big) + \cdots + k_m\big(\sigma_1^{-1}(a_m) + \cdots + \sigma_n^{-1}(a_m)\big) = 0$$

i.e., since $G = \{\sigma_1, \ldots, \sigma_n\} = \{\sigma_1^{-1}, \ldots, \sigma_n^{-1}\}$,

$$k_1\Big(\sum_{\sigma \in G} \sigma(a_1)\Big) + \cdots + k_m\Big(\sum_{\sigma \in G} \sigma(a_m)\Big) = 0.$$

The coefficients of $k_1, \ldots, k_m$ are invariant under $G$ and are not all zero, by the hypothesis on $a_1$. Therefore, this equation is in contradiction with the hypothesis that $k_1, \ldots, k_m$ are linearly independent over $K$. Thus,

$$[K : F] = |G|.$$

*Step 2:* $G = \operatorname{Aut}_F K$.

Let $G = \{\sigma_1, \ldots, \sigma_n\}$. We already have $G \subseteq \operatorname{Aut}_F K$. Suppose, by way of contradiction, that $\operatorname{Aut}_F K$ contains an element $\tau$ that is not in $G$. Clearly,

$$\{x \in K \mid \sigma_1(x) = \cdots = \sigma_n(x)\} \supseteq \{x \in K \mid \sigma_1(x) = \cdots = \sigma_n(x) = \tau(x)\}.$$

But $F = \{x \in K \mid \sigma_1(x) = \cdots = \sigma_n(x)\}$ since $F = K^G$, and

$$\{x \in K \mid \sigma_1(x) = \cdots = \sigma_n(x) = \tau(x)\} \supseteq F$$

since $\sigma_1, \ldots, \sigma_n, \tau \in \operatorname{Aut}_F K$. Therefore, the last inclusion is an equality, and Corollary 15.5 yields

$$[K : F] \geq n + 1.$$

This is a contradiction, since it was seen in Step 1 that $[K : F] = n$.

*Step 3:* $\text{Aut}_{K^H} K = H$ and $(G : H) = [K^H : F]$ for any subgroup $H$ of $G$.

The first equation follows from Step 2, with $H$ instead of $G$. To obtain the second equation, we compare the following equations which are derived from Step 1:

$$[K : F] = |G| \quad \text{and} \quad [K : K^H] = |H|.$$

Since the degrees of field extensions are multiplicative (by Lemma 15.3), we have

$$[K : F] = [K : K^H] \cdot [K^H : F].$$

Therefore the preceding equations yield

$$[K^H : F] = \frac{|G|}{|H|} = (G : H).$$

*Step 4:* $K^{\text{Aut}_L K} = L$ and $[L : F] = (G : \text{Aut}_L K)$ for any subfield $L$ of $K$ containing $F$.

By restriction to $L$, each $\sigma \in G$ induces a homomorphism of fields

$$\sigma|_L \colon L \to K.$$

If two such homomorphisms coincide, say

$$\sigma|_L = \tau|_L \quad \text{for some } \sigma, \tau \in G,$$

then $\sigma(\ell) = \tau(\ell)$ for all $\ell \in L$, hence $\tau^{-1} \circ \sigma(\ell) = \ell$ for all $\ell \in L$. Therefore $\tau^{-1} \circ \sigma \in \text{Aut}_L K$, and it follows that

$$\sigma \circ \text{Aut}_L K = \tau \circ \text{Aut}_L K.$$

Therefore, if $(G : \text{Aut}_L K) = r$ and if $\sigma_1, \ldots, \sigma_r$ are elements of $G$ in pairwise different cosets of $\text{Aut}_L K$, then the homomorphisms $\sigma_i|_L$ are pairwise different. Since $\sigma_1(x) = \cdots = \sigma_r(x)$ for any $x \in F$, Corollary 15.5 implies that

$$[L : F] \geq r \quad (= (G : \text{Aut}_L K)).$$

On the other hand, the inclusion $L \subseteq K^{\text{Aut}_L K}$ and Step 3 yield

$$[L : F] \leq [K^{\text{Aut}_L K} : F] = (G : \text{Aut}_L K),$$

hence, comparing the preceding inequalities,

$$[L : F] = (G : \text{Aut}_L K).$$

Moreover, since $L \subseteq K^{\text{Aut}_L K}$ and since these fields both have the same (finite) dimension over $F$, we have

$$L = K^{\text{Aut}_L K}.$$

We have thus proved that $K$ is a Galois extension of every subfield $L$ containing $F$. Henceforth, we will use the notation $\mathrm{Gal}(K/L)$ for $\mathrm{Aut}_L K$.

Steps 3 and 4 show that the maps $H \mapsto K^H$ and $L \mapsto \mathrm{Gal}(K/L)$ are reciprocal bijections between the set of subgroups of $G$ and the set of subfields of $K$ containing $F$. These maps clearly reverse the inclusions,

$$H \subseteq J \Longrightarrow K^H \supseteq K^J \qquad \text{and} \qquad L \subseteq M \Longrightarrow \mathrm{Gal}(K/L) \supseteq \mathrm{Gal}(K/M).$$

To complete the proof of the fundamental theorem, it now suffices to show that normal subgroups correspond to fields that are Galois over $F$.

*Step 5:* $\mathrm{Gal}\big(K/\sigma(L)\big) = \sigma \circ \mathrm{Gal}(K/L) \circ \sigma^{-1}$ for any $\sigma \in G$ and any subfield $L$ of $K$ containing $F$.

This readily follows from the inclusions

$$\sigma \circ \mathrm{Gal}(K/L) \circ \sigma^{-1} \subseteq \mathrm{Gal}\big(K/\sigma(L)\big) \text{ and } \sigma^{-1} \circ \mathrm{Gal}\big(K/\sigma(L)\big) \circ \sigma \subseteq \mathrm{Gal}(K/L),$$

which are both obvious.

*Step 6:* If $L$ is a subfield of $K$ containing $F$ such that $\mathrm{Gal}(K/L)$ is normal in $G$, then $L$ is Galois over $F$, and there is an isomorphism $G/\mathrm{Gal}(K/L) \to \mathrm{Gal}(L/F)$ obtained by restricting to $L$ the automorphisms in $G$.

By Step 5, the hypothesis that $\mathrm{Gal}(K/L)$ is normal in $G$ implies that any $\sigma \in G$ induces by restriction to $L$ an automorphism

$$\sigma|_L \colon L \to L$$

which leaves $F$ elementwise invariant. We thus have a restriction map

$$\mathrm{res} \colon G \to \mathrm{Gal}(L/F).$$

Since the kernel of this map is $\mathrm{Gal}(K/L)$, there is an induced injective map

$$\overline{\mathrm{res}} \colon G/\mathrm{Gal}(K/L) \to \mathrm{Gal}(L/F).$$

Now, Steps 4 and 1 yield

$$\big(G : \mathrm{Gal}(K/L)\big) = [L : F] = |\mathrm{Gal}(L/F)|,$$

hence the groups $G/\mathrm{Gal}(K/L)$ and $\mathrm{Gal}(L/F)$ have the same finite order, and the injective map $\overline{\mathrm{res}}$ is therefore an isomorphism.

*Step 7:* If $L$ is a Galois extension of $F$ contained in $K$, then $\mathrm{Gal}(K/L)$ is a normal subgroup of $G$.

Let $\big(G : \mathrm{Gal}(K/L)\big) = r$ $(= [L : F]$, by Step 4$)$, and let $\sigma_1, \ldots, \sigma_r$ be elements of $G$ in the $r$ different cosets of $\mathrm{Gal}(K/L)$. We have shown in the proof of Step 4 that the restrictions $\sigma_i|_L \colon L \to K$ yield pairwise different homomorphisms of $L$ in $K$ that restrict to the identity on $F$. On the other hand, it follows from Corollary 15.5 that there are at most $r$ homomorphisms of $L$ in $K$ that restrict to the identity on $F$. Therefore, any such homomorphism has the form $\sigma_i|_L$ for some $i = 1, \ldots, r$.

Now, since $L$ is assumed to be Galois over $F$, we can find $r$ automorphisms of $L$ leaving $F$ elementwise invariant. These automorphisms can be regarded as homomorphisms from $L$ into $K$, since $L \subseteq K$, hence they have the form $\sigma_i|_L$. Thus,

$$\mathrm{Gal}(L/F) = \{\sigma_1|_L, \ldots, \sigma_r|_L\},$$

and consequently $\sigma_1, \ldots, \sigma_r$ map $L$ into $L$,

$$\sigma_i(L) = L \qquad \text{for } i = 1, \ldots, r.$$

By Step 5, it follows that

$$\sigma_i \circ \mathrm{Gal}(K/L) \circ \sigma_i^{-1} = \mathrm{Gal}(K/L) \qquad \text{for } i = 1, \ldots, r.$$

Since every element $\sigma \in G$ has the form $\sigma_i \circ \tau$ for some $i = 1, \ldots, r$ and some $\tau \in \mathrm{Gal}(K/L)$ we also have

$$\sigma \circ \mathrm{Gal}(K/L) \circ \sigma^{-1} = \mathrm{Gal}(K/L) \qquad \text{for all } \sigma \in G,$$

hence $\mathrm{Gal}(K/L)$ is normal in $G$. $\qquad\qquad\square$

**Corollary 15.7.** *Let $K$ be a finite-degree extension of a field $F$. The following conditions are equivalent:*

(a) $K$ *is a Galois extension of $F$;*
(b) $|\mathrm{Aut}_F K| \geq [K : F]$;
(c) $|\mathrm{Aut}_F K| = [K : F]$.

**Proof.** Steps 1 and 2 in the proof of Theorem 15.2 yield (a) $\Rightarrow$ (c), and (c) $\Rightarrow$ (b) is obvious. To prove (b) $\Rightarrow$ (a), let $[K : F] = n$ and pick $\sigma_1, \ldots,$ $\sigma_m \in \mathrm{Aut}_F K$, with $m \geq n$ and $\sigma_1 = \mathrm{Id}$. Let

$$F' = \{x \in K \mid \sigma_1(x) = \cdots = \sigma_m(x)\}.$$

By Corollary 15.5, we have $[K : F'] \geq m$. But since $\sigma_1 = \mathrm{Id}$ and $\sigma_2, \ldots,$ $\sigma_m$ fix $F$ we have $F \subset F'$, hence

$$[K : F] \geq [K : F'].$$

As $[K : F'] \geq m \geq n = [K : F]$, it follows that $m = n$ and $F' = F$. Therefore, $\mathrm{Aut}_F K$ has exactly $n$ elements, and the equation $F' = F$ shows that $F = K^{\mathrm{Aut}_F K}$. This proves (a). $\qquad\qquad\square$

## Appendix 2: Galois theory à *la* Grothendieck

In this appendix, we survey the Grothendieck approach to Galois theory. Our purpose is to highlight the striking convergence between the latest re-interpretations of Galois theory and Galois' original ideas.

### *Étale algebras*

The basic objects in Grothendieck's approach are the étale algebras, which are defined next:

**Definitions 15.8.** Let $F$ be an arbitrary field. A non-constant polynomial $P \in F[X]$ is said to be *separable* if $P$ does not have multiple roots in any field extension of $F$. By Proposition 5.19 (p. 54), this condition holds if and only if $P$ is relatively prime to its derivative $\partial(P)$. An *étale algebra* over $F$ is a finite-dimensional $F$-algebra that is isomorphic to a direct product $L_1 \times \cdots \times L_t$ of fields of the form $L_i = F[X]/(P_i)$, where $P_i$ is a separable irreducible polynomial. In other words, an étale $F$-algebra is a finite direct product of finite separable field extensions of $F$. Other equivalent characterizations can be found in Knus *et al.* [47, (18.3)].

For example, for any separable polynomial $P \in F[X]$ the factor ring $F[X]/(P)$ is an étale algebra because in the prime factorization $P = P_1 \ldots P_t$ the irreducible factors are separable and pairwise distinct, and the Chinese Remainder Theorem (see for instance [8, Proposition 4, p. VII.3]) yields an isomorphism

$$F[X]/(P) \simeq F[X]/(P_1) \times \cdots \times F[X]/(P_t).$$

Conversely, if the field $F$ is infinite, every étale $F$-algebra is isomorphic to $F[X]/(P)$ for some separable polynomial $P \in F[X]$; see Bourbaki [8, Prop. 7, p. V.40].

Using the construction of splitting fields of polynomials (Theorem 9.3, p. 110) and transfinite induction, or by a Zorn's lemma argument (see Kaplansky [42, Th. 66] or Bourbaki [8, Ch. V, §4, §7]), one can establish that every field $F$ has a field extension $F_s$ with the following properties:

(i) every $x \in F_s$ is a root of some separable polynomial in $F[X]$;
(ii) every separable polynomial in $F_s[X]$ splits into a product of linear factors in $F_s[X]$.

Such a field $F_s$ is uniquely determined up to isomorphism (see Bourbaki [8, Corollaire, p. V.44]); it is called a *separable closure* of $F$.

For the rest of the appendix, fix a separable closure $F_s$ of $F$, and let $\Gamma$ be the group of all automorphisms of $F_s$ that restrict to the identity on $F$. (The group $\Gamma$ is called the *absolute Galois group* of $F$.) To each étale $F$-algebra $E$, we associate the set $R(E)$ of $F$-algebra homomorphisms $\rho\colon E \to F_s$:

$$R(E) = \mathrm{Hom}_{F\text{-alg}}(E, F_s).$$

This set is finite (see Bourbaki [8, Corollaire, p. V.29]), hence there is a finite-dimensional extension $M$ of $F$ in $F_s$ that contains $\rho(E)$ for all $\rho \in R(E)$. Note that $R(E)$ carries an action of $\Gamma$, because for $\rho \in R(E)$ and $\gamma \in \Gamma$ we have $\gamma \circ \rho \in R(E)$. Of course, if $\gamma$ is the identity on $M$, then $\gamma \circ \rho = \rho$ for all $\rho \in R(E)$. We summarize these properties by declaring $R(E)$ a $\Gamma$-*set*, as per the following definition:

**Definition 15.9.** A $\Gamma$-*set* is a finite set $R$ with an action of $\Gamma$ (i.e., a group homomorphism $\Gamma \to \mathrm{Sym}(R)$, see §14.1) whose kernel contains all the automorphisms of $F_s$ that restrict to the identity on some finite-dimensional extension of $F$ in $F_s$. (This property means that the action is continuous for the discrete topology on $\mathrm{Sym}(R)$ and the natural (Krull) topology on $\Gamma$.)

**Example 15.10.** If $E = F[X]/(P)$ for some separable polynomial $P \in F[X]$, the set $R(E)$ can be identified with the set of roots of $P$ in $F_s$, by mapping each homomorphism $\rho\colon E \to F_s$ to $\rho\big(X + (P)\big) \in F_s$ (which must be a root of $P$ because $P\big(X + (P)\big) = 0$). This follows because for every root $r \in F_s$ of $P$ there is a unique homomorphism $\rho\colon E \to F_s$ that maps $X + (P)$ to $r$, namely the map $\rho$ induced by the evaluation of polynomials in $F[X]$ at $r$. Under this identification, the action of $\Gamma$ on $R(E)$ is just the restriction of the action of $\Gamma$ on $F_s$. This example suggests the following (non-standard) terminology: for any étale $F$-algebra $E$, the $\Gamma$-set $R(E)$ is the set of *roots* of $E$.

Étale $F$-algebras and $F$-algebra homomorphisms form a category $\mathsf{Ét}_F$. Likewise, $\Gamma$-sets and $\Gamma$-equivariant maps form a category $\mathsf{Set}_\Gamma$, and the assignment $E \mapsto R(E)$ is functorial (but contravariant) because every $F$-algebra homomorphism $\varphi\colon E \to E'$ induces a $\Gamma$-equivariant map $R(\varphi)\colon R(E') \to R(E)$ carrying $\rho \in R(E')$ to $\rho \circ \varphi \in R(E)$. The following theorem is crucial for Grothendieck's interpretation of Galois theory (see Deligne [21, Prop. 2.4.3]):

**Theorem 15.11.** *The contravariant functor "roots": $E \mapsto R(E)$ defines* an anti-equivalence of categories

$$\text{Ét}_F \equiv \text{Set}_\Gamma.$$

*Under this anti-equivalence, the dimension for étale algebras corresponds to the cardinality for $\Gamma$-sets:* $\dim E = |R(E)|$ *for every étale $F$-algebra $E$. Moreover, the direct product of étale algebras corresponds to the disjoint union of $\Gamma$-sets: for $E_1, \ldots, E_t$ étale $F$-algebras,*

$$R(E_1 \times \cdots \times E_t) = R(E_1) \cup \cdots \cup R(E_t).$$

The last equation is to be understood as follows: if

$$E = E_1 \times \cdots \times E_t,$$

then for each $i = 1, \ldots, t$ projection on the $i$-th component is an algebra homomorphism $p_i \colon E \to E_i$; the corresponding $\Gamma$-equivariant map $R(p_i) \colon R(E_i) \to R(E)$ identifies $R(E_i)$ with a sub-$\Gamma$-set of $R(E)$.

For a proof of Theorem 15.11, see Bourbaki [8, Ch. V, §10] or Knus *et al.* [47, (18.4)].

From the definition, it is clear that étale algebras that have no (nontrivial) direct product decomposition are fields. On the other hand, every $\Gamma$-set decomposes into a disjoint union of orbits (see §14.1), each of which is a sub-$\Gamma$-set. Therefore, the last statement in Theorem 15.11 shows that the sets of roots of fields are the $\Gamma$-sets with a transitive action of $\Gamma$, since these are the indecomposable $\Gamma$-sets. (Compare Proposition 14.35, p. 254.) At the other extreme, $\Gamma$-sets with a trivial action of $\Gamma$ are the sets of roots of algebras of the form $F \times \cdots \times F$.

If $E = F[X]/(P)$ where $P$ is a separable polynomial with prime factorization $P = P_1 \ldots P_t$, we obtain with $E_i = F[X]/(P_i)$

$$E \simeq E_1 \times \cdots \times E_t \quad \text{and} \quad R(E) \simeq R(E_1) \cup \cdots \cup R(E_t).$$

The latter decomposition can be viewed as the partition of the roots of $P$ into the sets of roots of $P_1, \ldots, P_t$.

### Galois algebras

We next consider actions of finite groups on étale $F$-algebras, which correspond under the anti-equivalence of Theorem 15.11 to actions on $\Gamma$-sets. Let $G$ be a finite group acting by $F$-algebra automorphisms on an étale $F$-algebra $E$. We write $E^G$ for the subalgebra of fixed elements:

$$E^G = \{x \in E \mid g(x) = x \text{ for all } g \in G\}.$$

For $g \in G$ and $\rho \in R(E)$, we define $\rho^g \in R(E)$ by

$$\rho^g(x) = \rho(g(x)) \qquad \text{for } x \in E.$$

Thus, $\rho \mapsto \rho^g$ is the action of $g$ on $R(E)$ functorially associated to the action of $g$ on $E$. Since $E^G$ is the equalizer of the automorphisms $x \mapsto g(x)$ for $g \in G$, it follows that $R(E^G)$ is the co-equalizer of the automorphisms $\rho \mapsto \rho^g$, which means that $R(E^G)$ is the set of orbits of $R(E)$ under the action of $G$:

$$R(E^G) = R(E)/G. \tag{15.6}$$

**Definitions 15.12.** An étale $F$-algebra $E$ with an action of a finite group $G$ by algebra automorphisms is said to be *G-Galois* if $|G| = \dim E$ and $E^G = F$. By Theorem 15.11 and (15.6), these conditions hold if and only if $|G| = |R(E)|$ and $G$ acts transitively on $R(E)$. It then follows by Lagrange's theorem (Proposition 14.3, p. 221) that the action of $G$ is *simply transitive*: for any $\rho$, $\rho' \in R(E)$ there is a unique $g \in G$ such that $\rho' = \rho^g$. In particular, the action of $G$ on $E$ and $R(E)$ is faithful: only the identity of $G$ fixes all the elements in $E$ or all the elements in $R(E)$.

A $\Gamma$-set $R$ with a simply transitive action of a finite group $G$ by $\Gamma$-equivariant permutations (i.e., such that

$$\gamma(\rho^g) = \gamma(\rho)^g \qquad \text{for all } \gamma \in \Gamma, \ \rho \in R, \text{ and } g \in G)$$

is called a *G-torsor* (or a *principal homogeneous $\Gamma$-set under $G$*). Thus, under the anti-equivalence of Theorem 15.11, $G$-Galois étale algebras correspond to $G$-torsors.

For example, let $L$ be a finite-dimensional field extension of $F$. If $L$ is a Galois extension of $F$, i.e., if the group of algebra automorphisms $\text{Aut}_F L$ has $[L : F]$ elements (see Corollary 15.7), then $L$ is $G$-Galois for any group $G$ isomorphic to $\text{Aut}_F L$. But there are $G$-Galois étale algebras that are not fields; their structure is described in the next theorem. Its statement uses the following notion: an *idempotent* in an étale $F$-algebra $E$ is an element $e \in E$ such that $e^2 = e$; a non-zero idempotent is *primitive* if it does not decompose into a sum of non-zero idempotents. In a product of fields $L_1 \times \cdots \times L_t$ the idempotents are the $t$-uples $(\varepsilon_1, \ldots, \varepsilon_t)$ where $\varepsilon_i = 0$ or 1 for each $i$, and the primitive idempotents are

$$e_1 = (1, 0, \ldots, 0), \quad e_2 = (0, 1, 0, \ldots, 0), \quad \ldots, \quad e_t = (0, \ldots, 0, 1).$$

**Theorem 15.13.** *Let $E$ be a $G$-Galois étale $F$-algebra with primitive idempotents $e_1, \ldots, e_t$. For $i = 1, \ldots, t$, let $L_i = e_i E \subset E$; then each $F$-algebra*

$L_i$ is a field (with unit element $e_i$), and the map $x \mapsto (e_1 x, \ldots, e_t x)$ defines an $F$-algebra isomorphism

$$E \simeq L_1 \times \cdots \times L_t.$$

If $E$ is $G$-Galois, let

$$G_i = \{ g \in G \mid g(e_i) = e_i \} \quad \text{for } i = 1, \ldots, t.$$

Then each $L_i$ is a Galois extension of $F$ with Galois group isomorphic to $G_i$ under the map $G_i \to \operatorname{Aut}_F(L_i)$ given by the action of $G$. Moreover, $G$ acts transitively on $e_1, \ldots, e_t$, so $G_1, \ldots, G_t$ are conjugate subgroups of $G$, and

$$L_1 \simeq L_2 \simeq \cdots \simeq L_t.$$

For a proof, see Knus *et al.* [47, (18.18)] or Bourbaki [9, §16, n° 7].

In terms of torsors, Theorem 15.13 can be translated as follows: every $G$-torsor $R$ decomposes into a disjoint union of $\Gamma$-orbits

$$R = \Gamma(\rho_1) \cup \cdots \cup \Gamma(\rho_t),$$

which are permuted under the action of $G$. Each orbit $\Gamma(\rho_i)$ is a $G_i$-torsor with $G_i = \{ g \in G \mid \rho_i^g \in \Gamma(\rho_i) \} \subset G$, and the subgroups $G_1, \ldots, G_t$ are conjugate in $G$. If the $G$-torsor $R$ is the $\Gamma$-set of roots of a $G$-Galois étale $F$-algebra $E$, and if $E \simeq L_1 \times \cdots \times L_t$ is the decomposition of $E$ into fields with $L_i = e_i E$ for $e_1, \ldots, e_t$ the primitive idempotents of $E$ (as in Theorem 15.13), then the $\Gamma$-orbit decomposition of $R(E)$ is

$$R(E) = R(L_1) \cup \cdots \cup R(L_t)$$

(see Theorem 15.11), and for $i = 1, \ldots, t$ the group $G_i$ is

$$\{ g \in G \mid g(e_i) = e_i \} = \{ g \in G \mid R(L_i)^g = R(L_i) \}.$$

## Galois groups

To each $\Gamma$-set $R$ of $n$ elements, we attach an $S_n$-torsor by the following construction: $\mathsf{A}(R)$ is (as in §14.1) the set of arrangements of $R$, i.e., the set of bijections $\{1, \ldots, n\} \to R$, which we provide with an action of $\Gamma$ by setting

$$\gamma(r_1, \ldots, r_n) = \big( \gamma(r_1), \ldots, \gamma(r_n) \big) \quad \text{for } \gamma \in \Gamma \text{ and } (r_1, \ldots, r_n) \in \mathsf{A}(R),$$

and an action of $S_n$ by

$$(r_1, \ldots, r_n)^\sigma = (r_{\sigma(1)}, \ldots, r_{\sigma(n)}) \quad \text{for } \sigma \in S_n \text{ and } (r_1, \ldots, r_n) \in \mathsf{A}(R).$$

Clearly, $A(R)$ is an $S_n$-torsor. The translation of Theorem 15.13 in terms of torsors shows that in the decomposition of $A(R)$ as a disjoint union of $\Gamma$-orbits

$$A(R) = \Gamma(\alpha_1) \cup \cdots \cup \Gamma(\alpha_t) \qquad \text{for some } \alpha_1, \ldots, \alpha_t \in A(R),$$

each $\Gamma(\alpha_i)$ is a $G_i$-torsor for

$$G_i = \{\sigma \in S_n \mid \alpha_i^\sigma \in \Gamma(\alpha_i)\} = \{\sigma \in S_n \mid \Gamma(\alpha_i)^\sigma = \Gamma(\alpha_i)\}.$$

**Proposition 15.14.** *Each $\Gamma(\alpha_i)$ is a group of arrangements of $R$ (in the technical sense of Definition 14.4, p. 222). For $i = 1, \ldots, t$ the group of permutations $\Pi(\Gamma(\alpha_i)) \subset \mathrm{Sym}(R)$ is the image $\Gamma_R$ of the action $\Gamma \to \mathrm{Sym}(R)$, and the group $\Pi'(\Gamma(\alpha_i)) \subset S_n$ is $G_i$.*

**Proof.** Let $\gamma_1(\alpha_i), \gamma_2(\alpha_i), \gamma_3(\alpha_i) \in \Gamma(\alpha_i)$ for some $\gamma_1, \gamma_2, \gamma_3 \in \Gamma$. The permutation $\binom{\gamma_1(\alpha_i)}{\gamma_2(\alpha_i)} \in \mathrm{Sym}(R)$ results from the action of $\gamma_2 \circ \gamma_1^{-1}$ on $R$, hence it maps $\gamma_3(\alpha_i)$ to $\gamma_2 \circ \gamma_1^{-1} \circ \gamma_3(\alpha_i)$, and this arrangement lies in $\Gamma(\alpha_i)$ because $\gamma_2 \circ \gamma_1^{-1} \circ \gamma_3 \in \Gamma$. It follows that $\Gamma(\alpha_i)$ is a group of arrangements of $R$, and it is clear from the definitions that $\Pi(\Gamma(\alpha_i)) = \Gamma_R$ and $\Pi'(\Gamma(\alpha_i)) = G_i$. $\qquad\square$

Now, let $E$ be an étale $F$-algebra of dimension $n$. By applying the construction above to the $\Gamma$-set of roots of $E$, we obtain an $S_n$-torsor $A(R(E))$. Fix some arrangement $\varpi \in A(R(E))$. Proposition 15.14 shows that its orbit $\Gamma(\varpi)$ is a group of arrangements of the roots of $E$, with $\Pi(\Gamma(\varpi)) = \Gamma_{R(E)}$.

**Theorem 15.15.** *If $P \in F[X]$ is a separable polynomial and $E = F[X]/(P)$, then $\Gamma(\varpi)$ is the Galois group of $P(X) = 0$ containing $\varpi$.*

**Proof.** Let $n = \deg P = \dim E$. As in Example 15.10, we identify $R(E)$ with the set $\{r_1, \ldots, r_n\}$ of roots of $P$ in $F_s$. Let $\varphi \in F[x_1, \ldots, x_n]$ be a polynomial as in Lemma 14.8 (p. 226), which takes $n!$ different values when evaluated on the arrangements of $R(E)$. Following the construction of the Galois group of $P(X) = 0$ in §14.2, we consider the polynomial

$$\Omega(Y) = \prod_{\alpha \in A(R(E))} (Y - \varphi(\alpha)) \in F[Y].$$

Let $E_\Omega = F[Y]/(\Omega)$. From the characteristic property of $\varphi$, it follows that $\Omega$ is separable, hence $E_\Omega$ is an étale $F$-algebra, and the map $\alpha \mapsto \varphi(\alpha)$ defines a $\Gamma$-equivariant bijection from $A(R(E))$ onto the set of roots of $\Omega$ in $F_s$, so this map yields an isomorphism of $\Gamma$-sets

$$\varphi \colon A(R(E)) \xrightarrow{\sim} R(E_\Omega). \tag{15.7}$$

Since $\mathsf{A}\big(R(E)\big)$ is an $S_n$-torsor, it follows that $E_\Omega$ is an $S_n$-Galois étale $F$-algebra. Now, let $\Phi \in F[Y]$ be the irreducible factor of $\Omega$ that has $\varphi(\varpi)$ as a root, and let $\mathsf{G} \subset \mathsf{A}\big(R(E)\big)$ be the set of arrangements $\alpha$ such that $\varphi(\alpha)$ is a root of $\Phi$:

$$\Phi(Y) = \prod_{\alpha \in \mathsf{G}} \big(Y - \varphi(\alpha)\big).$$

By definition (see Theorem 14.9, p. 227), $\mathsf{G}$ is the Galois group of $P(X) = 0$ containing $\varpi$. Let also $E_\Phi = F[Y]/(\Phi)$. The isomorphism (15.7) restricts to an isomorphism

$$\varphi \colon \mathsf{G} \xrightarrow{\sim} R(E_\Phi).$$

Because $\Phi$ is irreducible, the algebra $E_\Phi$ is a field, hence $R(E_\Phi)$ is an indecomposable $\Gamma$-set. Therefore, $\mathsf{G}$ also is indecomposable, hence it must be the orbit of $\varpi$. This proves $\mathsf{G} = \Gamma(\varpi)$. $\qquad\square$

Note that since $\Phi$ is irreducible $E_\Phi$ is one of the (isomorphic) Galois field extensions of $F$ in the direct product decomposition of $E_\Omega$ as in Theorem 15.13. The Galois group $\mathrm{Gal}(E_\Phi/F)$ is canonically isomorphic to the subgroup $\Pi'\big(\Gamma(\varpi)\big) \subset S_n$, hence also (under the isomorphism $f_\varpi$ of Exercise 14.2, p. 268) to $\Pi\big(\Gamma(\varpi)\big) = \Gamma_{R(E)}$.

**Remark 15.16.** For another viewpoint on the Galois group of $E_\Phi$, recall from §14.2 that every $\varphi(\alpha) \in F_s$ for $\alpha \in \mathsf{G}$ is a Galois resolvent of $P(X) = 0$. It has the property that

$$F\big(\varphi(\alpha)\big) = F(r_1, \ldots, r_n) \subset F_s$$

(see Lemma 14.14, p. 231). To each $\alpha \in \mathsf{G}$ we may associate the $F$-algebra isomorphism of evaluation at $\varphi(\alpha)$:

$$g_\alpha \colon E_\Phi \xrightarrow{\sim} F(r_1, \ldots, r_n), \qquad f + (\Phi) \mapsto f\big(\varphi(\alpha)\big).$$

A group homomorphism $\Pi(\mathsf{G}) \to \mathrm{Aut}_F F(r_1, \ldots, r_n)$ is then defined by mapping the permutation $\binom{\alpha}{\beta}$ for $\alpha$, $\beta \in \mathsf{G}$ to the automorphism $g_\beta \circ g_\alpha^{-1}$, which carries $\varphi(\alpha)$ to $\varphi(\beta)$. This group homomorphism is injective because each permutation in $\Pi(\mathsf{G})$ has a unique representation of the form $\binom{\varpi}{\alpha}$ with $\alpha \in \mathsf{G}$, and if $\alpha \neq \varpi$ the automorphism $g_\alpha \circ g_\varpi^{-1}$ is not the identity since it maps $\varphi(\varpi)$ to $\varphi(\alpha)$. We thus see that

$$|\mathrm{Aut}_F F(r_1, \ldots, r_n)| \geq |\Pi(\mathsf{G})| = |\mathsf{G}| = \deg \Phi.$$

But $[F(r_1, \ldots, r_n) : F] = [E_\Phi : F] = \deg \Phi$, hence it follows from Corollary 15.7 that $F(r_1, \ldots, r_n)$ is a Galois extension of $F$. Moreover, the group homomorphism above is an isomorphism

$$\Pi(\mathsf{G}) \xrightarrow{\sim} \mathrm{Gal}(F(r_1, \ldots, r_n)/F).$$

**Exercises**

We use the same notation as in Appendix 2.

*Exercise 15.1.* For any étale $F$-algebra $E$, construct an étale $F$-algebra $\Delta(E)$ of dimension 2 with the property that $\Delta(E) \simeq F \times F$ if and only if $\Gamma$ acts on $R(E)$ by even permutations. (The algebra $\Delta(E)$ is the *discriminant algebra* of $E$.) [Hint: consider the orbits of $\mathsf{A}\big(R(E)\big)$ under the action of the alternating group.]

*Exercise 15.2.* For any étale $F$-algebra $E$ of dimension 4, construct an étale $F$-algebra $C(E)$ of dimension 3 in such a way that if $E = F[X]/(P)$ for some separable polynomial $P$ (of degree 4), then $C(E) \simeq F[Y]/(Q)$ for any cubic resolvent $Q$ of $P$. Show that the discriminant algebra of $C(E)$ is canonically isomorphic to the discriminant algebra of $E$.

*Exercise 15.3.* Let $E$ be an étale $F$-algebra of dimension 4 such that $C(E) \simeq F \times \Delta(E)$ and $\Delta(E)$ is a field. Show that $E$ contains a field $K$ with $[K : F] = 2$, and construct an étale $F$-algebra $\check{E}$ of dimension 4 containing $\Delta(E)$ such that $C(\check{E}) \simeq F \times K$, in such a way that $E \otimes \Delta(E) \simeq \check{E} \otimes K$ and $\check{E} \simeq E$. Show that $E \otimes \Delta(E)$ is a $G$-Galois algebra for $G$ a dihedral group of order 8.

# Selected Solutions

## Chapter 10

*Exercise 10.1.* For $i = 1$, 2, 3, the permutations of $x_1, \ldots, x_4$ which leave $v_i$ invariant are the same as those which leave $u_i$ invariant. Therefore, $u_i$ is a rational function in $v_i$ with symmetric coefficients. Denoting by $s_1$, $s_2$ the first two elementary symmetric polynomials (see equation (8.2), p. 93), it is easily checked that $u_i = s_2 - v_i$ for $i = 1$, 2, 3, and

$$w_1 = s_1^2 - 4v_2, \qquad w_2 = s_1^2 - 4v_1, \qquad w_3 = s_1^2 - 4v_3.$$

Therefore, if $R$ (resp. $Q$, resp. $S$) is the monic cubic polynomial with roots $v_1$, $v_2$, $v_3$ (resp. $u_1$, $u_2$, $u_3$, resp. $w_1$, $w_2$, $w_3$), then

$$R(X) = -Q(s_2 - X) = -\frac{1}{64}S(s_1^2 - 4X),$$

$$Q(X) = -R(s_2 - X), \qquad S(X) = -64\,P\big(\tfrac{1}{4}(s_1^2 - X)\big).$$

*Exercise 10.2.* In order to reproduce the notation of Theorem 10.5, p. 135, let

$$g_1 = x_1 + x_2, \qquad g_2 = x_1 + x_3, \qquad g_3 = x_2 + x_3,$$
$$f_1 = x_1x_2, \qquad f_2 = x_1x_3, \qquad f_3 = x_2x_3.$$

Then

$$a_0 = s_2, \qquad a_1 = s_1s_2 - 3s_3, \qquad a_2 = s_1^2s_2 - 5s_1s_3,$$

$$\theta(Y) = Y^3 - 2s_1Y^2 + (s_1^2 + s_2)Y - (s_1s_2 - s_3),$$
$$\psi(Y) = Y^2 + (g_1 - 2s_1)Y + (g_1^2 - 2s_1g_1 + s_1^2 + s_2),$$

and

$$f_1 = \frac{s_2g_1^2 - s_1s_2g_1 - 3s_3g_1 + s_2^2 - s_1s_3}{3g_1^2 - 4s_1g_1 + s_1^2 + s_2}.$$

This expression is not unique. Indeed, it is clear that $f_1 = s_3/X_3$ and $g_1 = s_1 - X_3$, hence $f_1 = s_3(s_1 - g_1)^{-1}$. This non-uniqueness stems from the fact that $\theta(g_1) = 0$, for it is easy to check that

$$\frac{s_2Y^2 - s_1s_2Y - 3s_3Y + s_2^2 + s_1s_3}{3Y^2 - 4s_1Y + s_1^2 + s_2}$$
$$= \frac{s_3}{s_1 - Y} - \frac{s_2\theta(Y)}{(3Y^2 - 4s_1Y + s_1^2 + s_2)(s_1 - Y)}.$$

*Exercise 10.3.* Let $\sigma\colon x_1 \mapsto x_2 \mapsto x_3 \mapsto x_1$. If $\sigma(f^3) \neq f^3$ then $f^3$, $\sigma(f^3)$ and $\sigma^2(f^3)$ are pairwise distinct. This contradicts the hypothesis that $f^3$ takes only two values. Therefore $\sigma(f^3) = f^3$ and it follows that $\sigma(f) = \omega f$ for some cube root of unity $\omega$. Comparing coefficients in $\sigma(f)$ and $f$, we obtain $A = \omega B = \omega^2 C$, whence

$$f = A(x_1 + \omega^2 x_2 + \omega x_3).$$

Moreover, $\omega \neq 1$ since $x_1$, $x_2$ and $x_3$ can be rationally expressed from $f$.

*Exercise 10.4.* By Theorem 10.5 (p. 135), it suffices to prove that $t(\omega^k)t(\omega)^{-k}$ is invariant under the permutations which leave $t(\omega)^n$ invariant, i.e., by $\tau\colon x_1 \mapsto x_2 \mapsto \cdots \mapsto x_n \mapsto x_1$ (and its powers) (compare Proposition 10.9, p. 141). This is clear, since $\tau\big(t(\omega^k)\big) = \omega^{-k}t(\omega^k)$.

*Exercise 10.5.* Identifying $\{0, 1, \ldots, n-1\}$ with $\mathbb{F}_n$, we can represent $\sigma_i$ and $\tau$ as follows:

$$\sigma_i(x) = ix, \qquad \tau(x) = x + 1 \qquad \text{for } x \in \mathbb{F}_n.$$

Then $\tau \circ \sigma_i(x) = ix + 1$ and $\sigma_i \circ \tau^k(x) = i(x + k)$, hence $\tau \circ \sigma_i = \sigma_i \circ \tau^k$ if $ik \equiv 1 \bmod n$. One easily checks that if $\sigma_i \circ \tau^j = \sigma_k \circ \tau^\ell$, then $i = k$ and $j = \ell$ (for $i = 1, \ldots, n-1$ and $j = 0, \ldots, n-1$), and it follows that $|\mathrm{GA}(n)| = n(n-1)$.

## Chapter 11

*Exercise 11.1.*

$$[\alpha \; \beta \; \gamma \; \delta \; \varepsilon] = a^\alpha b^\beta c^\gamma d^\delta e^\varepsilon + a^\delta b^\varepsilon c^\beta d^\gamma e^\alpha + a^\gamma b^\alpha c^\varepsilon d^\beta e^\delta$$
$$\text{v iii iv i ii} \quad + a^\beta b^\delta c^\alpha d^\varepsilon e^\gamma + a^\varepsilon b^\gamma c^\delta d^\alpha e^\beta,$$

$$[\alpha \; \varepsilon \; \delta \; \beta \; \gamma] = a^\alpha b^\varepsilon c^\delta d^\beta e^\gamma + a^\beta b^\gamma c^\varepsilon d^\delta e^\alpha + a^\delta b^\alpha c^\gamma d^\varepsilon e^\beta$$
$$\text{v iii iv i ii} \quad + a^\varepsilon b^\beta c^\alpha d^\gamma e^\delta + a^\gamma b^\delta c^\beta d^\alpha e^\varepsilon,$$

$$[\alpha \; \gamma \; \beta \; \varepsilon \; \delta] = a^\alpha b^\gamma c^\beta d^\varepsilon e^\delta + a^\varepsilon b^\delta c^\gamma d^\beta e^\alpha + a^\beta b^\alpha c^\delta d^\gamma e^\varepsilon$$
$$\text{v iii iv i ii} \quad + a^\gamma b^\varepsilon c^\alpha d^\delta e^\beta + a^\delta b^\beta c^\varepsilon d^\alpha e^\gamma,$$

$$[\alpha \; \delta \; \varepsilon \; \gamma \; \beta] = a^\alpha b^\delta c^\varepsilon d^\gamma e^\beta + a^\gamma b^\beta c^\delta d^\varepsilon e^\alpha + a^\varepsilon b^\alpha c^\beta d^\delta e^\gamma$$
$$\text{v iii iv i ii} \quad + a^\delta b^\gamma c^\alpha d^\beta e^\varepsilon + a^\beta b^\varepsilon c^\gamma d^\alpha e^\delta.$$

The sum of these four partial types is a partial type corresponding to the subgroup generated by the permutations $\tau \colon a \mapsto e \mapsto b \mapsto c \mapsto d \mapsto a$ and $\sigma \colon a \mapsto a, \; b \mapsto d \mapsto c \mapsto e \mapsto b$. If $a, b, c, d, e$ are numbered as $a = 0$, $b = 2$, $c = 3$, $d = 4$, $e = 1$, then the subgroup is $\mathrm{GA}(5) \subset S_5$ (see p. 140).

*Exercise 11.2.* Applying $a \mapsto b \mapsto c \mapsto d \mapsto e \mapsto a$ to both sides of $a^2 = b + 2$ (for instance), one gets $b^2 = c + 2$, a relation which does not hold.

*Exercise 11.3.* The relations between $a, b, c$ are the following:

$$a^2 = b + 2, \qquad b^2 = c + 2, \qquad c^2 = a + 2,$$
$$ab = a + c, \qquad bc = b + a, \qquad ca = c + b.$$

It is readily verified that these relations are preserved under $a \mapsto b \mapsto c \mapsto a$.

*Exercise 11.4.* Using the fact that 2 is a primitive root of 13, it follows from Proposition 12.18, p. 169 (see also Proposition 12.21, p. 173) that $2\cos\frac{2\pi}{13} \mapsto 2\cos\frac{4\pi}{13} \mapsto 2\cos\frac{8\pi}{13} \mapsto 2\cos\frac{10\pi}{13} \mapsto 2\cos\frac{6\pi}{13} \mapsto 2\cos\frac{12\pi}{13} \mapsto 2\cos\frac{2\pi}{13}$ preserves the relations among $2\cos\frac{2\pi}{13}, \ldots, 2\cos\frac{12\pi}{13}$.

## Chapter 12

*Exercise 12.3.* Periods with an even number of terms are sums of periods of two terms, which are real numbers, see p. 171.

*Exercise 12.4.* Let $\zeta = \zeta_k' (= \zeta'^{g'^k})$ for some $k = 0, \ldots, p - 2$ and let $g = g'^\ell$ for some integer $\ell$, which is prime to $p - 1$ since $g$ is a primitive root of $p$ (see Proposition 7.10, p. 86). A straightforward computation yields $\zeta_i = \zeta'_{\sigma(i)}$, where $\sigma\colon \{0,\ldots,p-2\} \to \{0,\ldots,p-2\}$ is defined by $\sigma(\alpha) \equiv \ell\alpha + k \bmod p - 1$, for $\alpha = 0, \ldots, p-2$. The same arguments as in Proposition 10.7 (p. 139) show that $\sigma$ is a permutation. Moreover, $\alpha \equiv i \bmod p - 1$ implies $\sigma(\alpha) \equiv \sigma(i) \bmod p - 1$, hence

$$\sum_{\alpha \equiv i \bmod e} \zeta_\alpha = \sum_{\beta \equiv \sigma(i) \bmod e} \zeta'_\beta.$$

*Exercise 12.5.* Let $ef = gh = p - 1$. If $K_g \subset K_f$, then

$$\sigma^e(\zeta_0 + \zeta_h + \cdots + \zeta_{h(g-1)}) = \zeta_0 + \zeta_h + \cdots + \zeta_{h(g-1)}.$$

In particular, $\sigma^e(\zeta_0) = \zeta_e$ is of the form $\zeta_{h\ell}$ for some $\ell$. It follows that $e = h\ell$, hence $h$ divides $e$ and $f$ divides $g$.

Let $\eta$ be a period of $f$ terms. From Proposition 12.23, p. 176, it follows that $K_f = K_g(\eta)$. Now, Proposition 12.25 (p. 177) shows that $\eta$ is a root of a polynomial of degree $k = g/f$ over $K_g$. It is not a root of a polynomial of smaller degree, since if $P(\eta) = 0$ with $P \in K_g[X]$, then $P(\sigma^h(\eta)) = P(\sigma^{2h}(\eta)) = \cdots = P(\sigma^{h(k-1)}(\eta)) = 0$. Therefore, $g/f$ is the degree of the minimum polynomial of $\eta$ over $K_g$ (see Remark 12.16, p. 168), and it follows from Proposition 12.15, p. 166, that $\dim_{K_g} K_f = g/f$.

## Chapter 13

*Exercise 13.1.* The fundamental theorem of algebra (Theorem 9.1, p. 109) shows that all the roots of any polynomial equation $P(X) = 0$, with $P \in \mathbb{R}[X]$ (resp. $\mathbb{C}[X]$), are in $\mathbb{C}$, which is a radical extension of $\mathbb{R}$ (resp. $\mathbb{C}$).

*Exercise 13.2.* Corollary 13.9 (p. 205) shows that the degree of a radical extension of height 1 is its dimension as a vector space over the base field, hence $p = q$. If $F$ contains a primitive $p$-th root of unity $\omega$, then there is an automorphism $\sigma$ of $R$ such that $\sigma(u) = \omega u$ and $\sigma(c) = c$ for all $c \in F$. We have $(\sigma(v)v^{-1})^p = 1$, hence $\sigma(v) = \omega^k v$ for some $k \in \{1,\ldots,p-1\}$. Then $\sigma(u^k v^{-1}) = u^k v^{-1}$, hence $u^k v^{-1} \in F$ and $a^k b^{-1}$ is a $p$-th power in $F$. If $F$ does not contain a primitive $p$-th root of unity, let $F' = F(\omega)$ and $R' = R(\omega) = F'(u) = F'(v)$. The dimension of $F'$ as a vector space over $F$ is at most $p - 1$, hence $a$ and $b$ are not $p$-th powers in $F'$. Similarly, if $a^k b^{-1}$ is a $p$-th power in $F'$, then $a^k b^{-1}$ is a $p$-th power in $F$, so to prove the claim we may argue with $F'$ instead of $F$.

*Exercise 13.3.* From Lagrange's result (Theorem 10.5, p. 135), it is known that $x_1$, $x_2$, and $x_3$ can be rationally expressed from $t = x_1 + \omega x_2 + \omega^2 x_3$,

where $\omega = \frac{1}{2}(-1 + \sqrt{-3})$. Explicitly,
$$x_1 = \tfrac{1}{3}\big(s_1 + t + (s_1^2 - 3s_2)t^{-1}\big).$$
A straightforward computation yields
$$t^3 = s_1^3 - \tfrac{9}{2}(s_1 s_2 - 3s_3) + \tfrac{3\sqrt{-3}}{2}\Delta,$$
where $\Delta = (x_1 - x_2)(x_1 - x_3)(x_2 - x_3) = \sqrt{D(s_1, s_2, s_3)}$, where $D(s_1, s_2, s_3)$
is the discriminant (see §8.2). Therefore, a radical extension of $\mathbb{Q}(s_1, s_2, s_3)$
containing $x_1$ can be constructed as follows:
$$R_0 = \mathbb{Q}(s_1, s_2, s_3),$$
$$R_1 = R_0\big(\sqrt{-3D(s_1, s_2, s_3)}\big),$$
$$R_2 = R_1\bigg(\sqrt[3]{s_1^3 - \tfrac{9}{2}(s_1 s_2 - 3s_3) + \tfrac{3}{2}\sqrt{-3D(s_1, s_2, s_3)}}\bigg).$$
The field $R_2$ is not contained in $\mathbb{Q}(x_1, x_2, x_3)$ since $\sqrt{-3D(s_1, s_2, s_3)}$ is not
in $\mathbb{Q}(x_1, x_2, x_3)$. In order to show that $\mathbb{Q}(x_1, x_2, x_3)$ does not contain any
radical extension of $\mathbb{Q}(s_1, s_2, s_3)$ containing $x_1$ (or $x_2$ or $x_3$), one can argue
as in §13.3: if $u \in \mathbb{Q}(x_1, x_2, x_3)$ has the property that some power $u^p$ (with
$p$ prime) is invariant under $\sigma: x_1 \mapsto x_2 \mapsto x_3 \mapsto x_1$, then $u$ is invariant
under $\sigma$. Indeed, from $\sigma(u^p) = u^p$, it follows that $\sigma(u) = \omega u$ for some $p$-th
root of unity $\omega$. Since $\sigma^3 = \mathrm{Id}$, one has $u = \sigma^3(u) = \omega^3 u$, hence $p = 3$. But
1 is the only cube root of unity in $\mathbb{Q}(x_1, x_2, x_3)$, so $\sigma(u) = u$.

In order to obtain a solution of the general equation of degree 4, one
first solves a resolvent cubic equation, for instance the equation with root
$v = (x_1 + x_2)(x_3 + x_4)$, i.e.,
$$X^3 - a_1 X^2 + a_2 X - a_3 = 0$$
with $a_1 = 2s_2$, $a_2 = s_2^2 + s_1 s_3 - 4s_4$, $a_3 = s_1 s_2 s_3 - s_1^2 s_4 - s_3^2$. (See Exercise 8.3,
p. 108.) Then, $v' = x_1 + x_2$ is obtained as a root of the quadratic equation
$$X^2 - s_1 X + v = 0.$$
(The other root is $x_3 + x_4$.) Then $x_1$ and $x_2$ are obtained as roots of
$$X^2 - v' X + \frac{(s_1 - 2v')s_4}{(s_1 - v')(s_2 - v) - s_3} = 0.$$
(Observe that the constant term is $x_1 x_2$, expressed as a function of $v'$,
$v$ and the symmetric polynomials.) Therefore, a radical extension of
$\mathbb{Q}(s_1, s_2, s_3, s_4)$ containing $x_1$ can be constructed as follows:
$$R_0 = \mathbb{Q}(s_1, s_2, s_3, s_4),$$
$$R_1 = R_0\big(\sqrt{-3D(a_1, a_2, a_3)}\big)$$
$$(= R_0\big(\sqrt{-3D(s_1, s_2, s_3, s_4)}\big), \quad \text{see Exercise 8.3, p. 108})$$
$$R_2 = R_1(t)$$

where

$$t = \sqrt[3]{a_1^3 - \tfrac{9}{2}(a_1 a_2 - 3a_3) + \tfrac{3}{2}\sqrt{-3D(a_1, a_2, a_3)}}.$$

Then

$$v = \frac{1}{3}\left(a_1 + t + (a_1^2 - 3a_2)t^{-1}\right) \in R_2.$$

Let then

$$R_3 = R_2\left(\sqrt{s_1^2 - 4v}\right);$$

then $v' = \tfrac{1}{2}(s_1 + \sqrt{s_1^2 - 4v}) \in R_3$ and

$$x_1 \in R_4 = R_3\left(\sqrt{v'^2 - 4(s_1 - 2v')s_4\left((s_1 - v')(s_2 - v) - s_3\right)^{-1}}\right).$$

*Exercise 13.4.* Since 3 is a primitive root of 7, Proposition 12.21 (p. 173) shows that there is an automorphism $\sigma$ of $\mathbb{Q}(\zeta_7)$ defined by $\sigma(\zeta_7) = \zeta_7^3$. Suppose $\mathbb{Q}(\zeta_7)$ is radical over $\mathbb{Q}$; then there is a tower of extensions

$$\mathbb{Q}(\zeta_7) = R_0 \supset R_1 \supset \cdots \supset R_h = \mathbb{Q}$$

where $R_i = R_{i+1}(u_i)$ with $u_i^{p_i} = a_i$ for some prime number $p_i$ and some element $a_i \in R_{i+1}$ which is not a $p_i$-th power in $R_{i+1}$.

We will prove that if $a_i$ is invariant under $\sigma^2$, then $u_i$ is invariant under $\sigma^2$ too. Therefore every element in $R_i$ is invariant under $\sigma^2$. By induction, it follows that every element in $R_0 = \mathbb{Q}(\zeta_7)$ is invariant under $\sigma^2$, a contradiction.

From $\sigma^2(a_i) = a_i$, it follows that $\sigma^2(u_i)^{p_i} = u_i^{p_i}$, whence $\sigma^2(u_i) = \omega u_i$ for some $p_i$-th root of unity $\omega \in \mathbb{Q}(\zeta_7)$. By Theorem 12.32 (p. 184) the cyclotomic polynomial $\Phi_{p_i}$ is irreducible over $\mathbb{Q}(\zeta_7)$ if $p_i \neq 7$. Therefore, the only prime numbers $p_i$ such that $\mathbb{Q}(\zeta_7)$ contains a $p_i$-th root of unity other than 1 are $p_i = 2$ or 7.

Now, Corollary 13.9 (p. 205) shows that $\dim_{R_{i+1}} R_i = p_i$. Therefore, $p_i = 7$ is impossible, since $\dim_{\mathbb{Q}} \mathbb{Q}(\zeta_7) = 6$ by Theorem 12.13 (p. 166). If $p_i = 2$, then $\omega = 1$ or $-1$. However, it is impossible that $\sigma^2(u_i) = -u_i$, since by applying $\sigma^2$ twice to both sides of this equation one gets $\sigma^6(u_i) = -u_i$, a contradiction since $\sigma^6 = \mathrm{Id}$. Therefore, the only possibility is that $\sigma^2(u_i) = u_i$, and the claim is proved. On the other hand, $\mathbb{Q}(\zeta_7, \zeta_3)$ is radical over $\mathbb{Q}$. To see this, let

$$\eta_0 = \zeta_7^{3^0} + \zeta_7^{3^2} + \zeta_7^{3^4} = \zeta_7 + \zeta_7^2 + \zeta_7^{-3},$$
$$\eta_1 = \zeta_7^{3^1} + \zeta_7^{3^3} + \zeta_7^{3^5} = \zeta_7^3 + \zeta_7^{-1} + \zeta_7^{-2}$$

be the periods of three terms of $\Phi_7 = 0$. It is readily checked that $\eta_0 + \eta_1 = -1$ and $\eta_0\eta_1 = 2$, hence $\eta_0, \eta_1 = \frac{1}{2}(1 \pm \sqrt{-7})$. Now, let $t = \zeta_7 + \zeta_3\zeta_7^2 + \zeta_3^2\zeta_7^{-3}$; then

$$\zeta_7 = \tfrac{1}{3}\big(\eta_0 + t + (2\eta_0 + 1)t^{-1}\big)$$

and $t^3 = 8 + 2\eta_0 + 3\zeta_3 + 6\zeta_3\eta_0 \in \mathbb{Q}(\zeta_3, \eta_0) = \mathbb{Q}(\sqrt{-3}, \sqrt{-7})$, hence $\mathbb{Q}(\zeta_3, \zeta_7) = \mathbb{Q}(\zeta_3, \eta_0, t) = \mathbb{Q}(\sqrt{-3}, \sqrt{-7}, t)$ is a radical extension of $\mathbb{Q}$.

*Exercise 13.5.* If $R = F(u)$ with $u^p = a$, an isomorphism $f$ is given by $f\big(P(X) + (X^p - a)\big) = P(u)$.

*Exercise 13.6.* The cube roots of 2 in $\mathbb{C}$ are $\sqrt[3]{2} \in \mathbb{R}$, $\frac{1}{2}(-1 + i\sqrt{3})\sqrt[3]{2}$ and $\frac{1}{2}(-1 - i\sqrt{3})\sqrt[3]{2}$. Since the last two are not in $\mathbb{R}$, the field $\mathbb{Q}(\sqrt[3]{2}) \subset \mathbb{R}$ contains only one cube root of 2. Since, by the preceding exercise, all the fields obtained from $\mathbb{Q}$ by adjoining a cube root of 2 are isomorphic, they all contain only one cube root of 2. Therefore,

$$\mathbb{Q}(\sqrt[3]{2}), \quad \mathbb{Q}\big(\tfrac{1}{2}(-1 + i\sqrt{3})\sqrt[3]{2}\big) \quad \text{and} \quad \mathbb{Q}\big(\tfrac{1}{2}(-1 - i\sqrt{3})\sqrt[3]{2}\big)$$

are pairwise distinct subfields of $\mathbb{C}$.

## Chapter 14

*Exercise 14.1.* For any $\sigma, \tau \in \Gamma$, we have $\binom{\sigma(\varpi)}{\tau(\varpi)} = \tau\sigma^{-1} \in \Gamma$, hence $\mathsf{G}$ is a group of arrangements with $\Pi(\mathsf{G}) = \Gamma$. Similarly, for $\sigma, \tau \in \Gamma$ we have $\varpi^\tau = (\varpi^\sigma)^{\sigma^{-1}\tau}$.

*Exercise 14.2.* For $\sigma_1, \sigma_2 \in \Pi(\mathsf{G})$ and $\tau_1 = f_\alpha(\sigma_1)$, $\tau_2 = f_\alpha(\sigma_2)$ we have $\sigma_1\sigma_2(\alpha) = \sigma_1(\alpha^{\tau_2}) = \sigma_1(\alpha)^{\tau_2} = \alpha^{\tau_1\tau_2}$, hence $f_\alpha(\sigma_1\sigma_2) = \tau_1\tau_2$, which shows that $f_\alpha$ is a homomorphism. Its kernel is trivial, hence it is an isomorphism since $|\Pi(\mathsf{G})| = |\Pi'(\mathsf{G})|$. For $\sigma \in \Pi(\mathsf{G})$ fixed, let $\tau = f_\alpha(\sigma)$ and $\nu = f_\beta^{-1}(\tau)$, so $\nu(\beta) = \beta^\tau$. Then $\nu = f_\beta^{-1} \circ f_\alpha(\sigma)$ and

$$\nu = \begin{pmatrix} \beta \\ \beta^\tau \end{pmatrix} = \begin{pmatrix} \alpha^\tau \\ \beta^\tau \end{pmatrix} \circ \begin{pmatrix} \alpha \\ \alpha^\tau \end{pmatrix} \circ \begin{pmatrix} \beta \\ \alpha \end{pmatrix}.$$

The claim follows because $\sigma = \binom{\alpha}{\alpha^\tau}$ and $\binom{\alpha^\tau}{\beta^\tau} = \binom{\alpha}{\beta}$.

*Exercise 14.3.* By Exercise 14.1 we have $\mathsf{G}_2 = \Pi(\mathsf{G}_2)(\alpha_2)$, hence if $\Pi(\mathsf{G}_1) = \Pi(\mathsf{G}_2)$ and $\alpha_2 = \alpha_1^\sigma$ it follows that

$$\mathsf{G}_2 = \Pi(\mathsf{G}_1)(\alpha_1^\sigma) = \Pi(\mathsf{G}_1)(\alpha_1)^\sigma = \mathsf{G}_1^\sigma.$$

Similarly, if $\Pi'(\mathsf{G}_1) = \Pi'(\mathsf{G}_2)$ and $\alpha_2 = \tau(\alpha_1)$, then

$$\mathsf{G}_2 = \alpha_2^{\Pi'(\mathsf{G}_2)} = \tau(\alpha_1)^{\Pi'(\mathsf{G}_1)} = \tau\big(\alpha_1^{\Pi'(\mathsf{G}_1)}\big) = \tau(\mathsf{G}_1).$$

*Exercise 14.4.* (a) $\Rightarrow$ (c) Let $\alpha \in \sigma(\mathsf{G}) \cap \mathsf{G}^\tau$. By Exercise 14.1 we have $\sigma(\mathsf{G}) = \Pi(\sigma(\mathsf{G}))(\alpha)$ and $\mathsf{G}^\tau = \Pi(\mathsf{G}^\tau)(\alpha)$. But $\Pi(\sigma(\mathsf{G})) = \Pi(\mathsf{G})$ because $\Pi(\mathsf{G})$ is a normal subgroup of $\mathrm{Sym}(R)$ (see Proposition 14.5, p. 223) and $\Pi(\mathsf{G}^\tau) = \Pi(\mathsf{G})$, hence $\sigma(\mathsf{G}) = \mathsf{G}^\tau$.

(c) $\Rightarrow$ (a) Let $\sigma \in \mathrm{Sym}(R)$. Choose $\alpha \in \sigma(\mathsf{G})$ and $\tau \in S_n$ such that $\alpha \in \mathsf{G}^\tau$. By (c) we have $\sigma(\mathsf{G}) = \mathsf{G}^\tau$, hence $\Pi(\sigma(\mathsf{G})) = \Pi(\mathsf{G}^\tau) = \Pi(\mathsf{G})$. By Proposition 14.5 it follows that $\sigma \circ \Pi(\mathsf{G}) \circ \sigma^{-1} = \Pi(\mathsf{G})$.

The proof of the equivalence (b) $\Longleftrightarrow$ (c) is similar.

*Exercise 14.5.* Suppose $\Pi'(\mathsf{G}_1) = \cdots = \Pi'(\mathsf{G}_t)$. Let $\alpha \in \mathsf{G}_1$, and for $i = 1$, ..., $t$ pick $\sigma_i \in \Pi(\mathsf{G})$ such that $\sigma_i(\alpha) \in \mathsf{G}_i$. By Exercise 14.1 we have $\mathsf{G}_i = \sigma_i(\alpha)^{\Pi'(\mathsf{G}_i)}$ hence $\mathsf{G}_i = \sigma_i(\alpha^{\Pi'(\mathsf{G}_1)}) = \sigma_i(\mathsf{G}_1)$.

*Exercise 14.6.* With the notation of (14.4) and (14.5), we have $|\mathsf{G}| = \deg \Phi$ and $n! = \deg \Omega$. By Corollary 14.18 all the irreducible factors of $\Omega$ have the same degree, hence $|\mathsf{G}|$ divides $n!$. If $P$ is irreducible, then by Proposition 14.35 the group $\Pi(\mathsf{G})$ acts transitively on the $n$ roots of $P$, hence $n$ divides $|\mathsf{G}|$ by Lagrange's theorem (Proposition 14.3).

*Exercise 14.7.* If the Galois resolvent equation $\Phi(Y) = 0$ does not remain irreducible when a root of $Q$ is adjoined to the base field, then Theorem 14.22 shows that $\mathsf{G}$ decomposes into a disjoint union of $p$ subgroups of the same order. This is impossible if $|\mathsf{G}|$ is not a multiple of $p$.

*Exercise 14.9.* This follows from Corollary 13.5 (p. 202) by induction on the number of irreducible factors of $P$.

## Chapter 15

For Exercise 15.1, see Knus *et al.* [47, §18.B] or Knus–Tignol [48, (3.11)]. For Exercises 15.2 and 15.3, see Knus–Tignol [48, §5.3, §6.2].

# Bibliography

Page references at the end of each entry indicate the places in the text where the entry is quoted. Note that a large number of references is available online, in particular on the following web sites:

- European Cultural Heritage Online: `http://echo.mpiwg-berlin.mpg.de/home`
- Bibliothèque nationale de France: `http://gallica.bnf.fr/`

[1] Abel, N. H. (1881). *Œuvres complètes (2 vol.)* (Grøndahl & Søn, Christiania), L. Sylow and S. Lie, eds. → *pp. 194, 203, 215, and 216.*

[2] Artin, E. (1948). *Galois Theory*, Notre Dame Math. Lectures (Notre Dame Univ. Press, Ind.). → *pp. xi and 274.*

[3] Artin, E. (1965). *Collected Papers* (Addison-Wesley, Reading, Mass.), S. Lang and J. Tate, eds. → *p. 274.*

[4] Ayoub, R. (1980/81). Paolo Ruffini's contributions to the quintic, *Arch. Hist. Exact Sci.* **23**, pp. 253–277. → *pp. 194, 203, and 209.*

[5] ben Musa, M. (2013). *The Algebra of Mohammed ben Musa*, Cambridge Library Collection (Cambridge Univ. Press, Cambridge), translated by Friedrich August Rosen, reprint of the 1831 edition. → *pp. 9 and 10.*

[6] Bosmans, H. (1908). Sur le "Libro de algebra" de Pedro Nuñez, *Biblioth. Mathem., Sér. 3* **8**, pp. 154–169. → *pp. 26 and 37.*

[7] Bourbaki, N. (1967). *Algèbre, chapitres 4 et 5* (Hermann, Paris). → *pp. xi and 39.*

[8] Bourbaki, N. (1981). *Algèbre, chapitres 4 à 7* (Masson, Paris). → *pp. 274, 283, 284, and 285.*

[9] Bourbaki, N. (2012). *Algèbre, chapitre 8* (Springer, Berlin Heidelberg New York). → *p. 287.*

[10] Boyer, C. (1968). *A History of Mathematics* (J. Wiley & Sons, New York, NY). → *p. 105.*

[11] Bühler, W. (1981). *Gauss. A Biographical Study* (Springer, Berlin). → *pp. 155 and 182.*

[12] Cajori, F. (1974). *A History of Mathematical Notations, vol. 1: Notations in Elementary Mathematics* (Open Court, La Salle, Ill.). → *pp. 25, 26, 27, 29, and 30.*

[13] Cardano, G. (1968). *The Great Art or The Rules of Algebra* (The M.I.T. Press, Cambridge, Mass.-London), translated from the Latin and edited by T. Richard Witmer. → *pp. 11, 15, 16, 19, 21, 22, and 40.*

[14] Carrega, J.-C. (1981). *Théorie des corps. La règle et le compas*, Coll. formation des enseignants et formation continue (Hermann, Paris). → *p. 191.*

[15] Chase, S., Harrison, D. and Rosenberg, A. (1968). Galois theory and Galois cohomology of commutative rings, *Mem. Amer. Math. Soc.* **52**, pp. 1–19. → *p. 274.*

[16] Chuquet, N. (1881). *Le Triparty en la Science des Nombres* (Imprimerie Sci. Math. Phys., Rome), edited by A. Marre. → *p. 27.*

[17] Cohn, P. M. (1977). *Algebra, vol. 2* (John Wiley & Sons, London-New York-Sydney). → *p. xi.*

[18] Cotes, R. (1722). *Harmonia Mensurarum, sive Analysis & Synthesis per Rationum & Angulorum Mensuras promotae: Accedunt Alia Opuscula Mathematica per Rogerum Cotesium* (Cantabrigiae), edited by R. Smith. → *pp. 75 and 300.*

[19] Cotes, R. (1722). *Theoremata tum Logometrica tum Trigonometrica Datarum Fluxionum Fluentes exhibentia, per Methodum Mensurarum Ulterius extensam*, in [18], pp. 111–249, edited by R. Smith. → *pp. 75 and 76.*

[20] de Moivre, A. (1707). Aequationum quarundam potestatis tertiæ, quintæ, septimæ, novæ, &c superiorum, ad infinitum usque pergendo, in terminis finitis, ad instar regularum pro cubicis quae vocantur *Cardani*, resolutio analytica, *Phil. Trans. Royal Soc.* **25**, pp. 2368–2371. → *p. 77.*

[21] Deligne, P. (1977). *Cohomologie étale*, Lecture Notes in Mathematics, Vol. 569 (Springer-Verlag, Berlin-New York), Séminaire de Géométrie Algébrique du Bois-Marie SGA 4½, Avec la collaboration de J. F. Boutot, A. Grothendieck, L. Illusie et J. L. Verdier. → *p. 284.*

[22] Descartes, R. (1683). *Geometria (2 vol.)* (Ex typographia Blaviana, Amstelodami), translated by F. van Schooten, third edition. → *pp. 53 and 301.*

[23] Descartes, R. (1954). *The Geometry* (Dover, New York, NY), translated from the French and Latin by D.E. Smith and M.L. Latham. → *pp. 21, 28, 29, 30, 37, 66, and 67.*

[24] Dieudonné, J. (1978). *Abrégé d'histoire des mathématiques 1700–1900 (2 vol.)* (Hermann, Paris). → *p. 114.*

[25] Douady, R. and Douady, A. (1979). *Algèbre et théories galoisiennes. Vol. 2* (CEDIC, Paris). → *p. 274.*

[26] Edwards, H. M. (1984). *Galois Theory*, Graduate Texts in Mathematics, Vol. 101 (Springer-Verlag, New York). → *pp. vii and xi.*

[27] Ehrhardt, C. (2012). *Itinéraire d'un texte mathématique* (Hermann, Paris). → *pp. vii and 272.*

[28] Euler, L. (2007). A double demonstration of a theorem of Newton, which gives a relation between the coefficients of an algebraic equation and the sums of the powers of its roots, arXiv:0707.0699v1, Translated from the

Latin by J. Bell. → *p. 39.*

[29] Gandz, S. (1937). The origin and development of the quadratic equations in Babylonian, Greek and early Arabic algebra, *Osiris* **3**, pp. 405–557. → *p. 5.*

[30] Gauss, C. F. (1799). *Demonstratio nova theorematis omnem functionem algebraicam rationalem integram unius variabilis in factores reales primi vel secundi gradus resolvi posse* (Apud C.G. Fleckeisen, Helmstadii), reprinted in Werke Bd III, Georg Olms, Hildesheim, 1981, pp. 1–30. → *pp. 110 and 194.*

[31] Gauss, C. F. (1801). *Disquisitiones Arithmeticae* (Apud G. Fleischer Iun. Lipsiae). → *pp. 155, 156, 157, 159, 160, 163, 166, 173, 181, 191, and 215.*

[32] Gauss, C. F. (1816). Demonstratio nova altera theorematis omnem functionem algebraicam rationalem integram unius variabilis in factores reales primi vel secundi gradus resolvi posse, *Comment. Soc. Regiae Sci. Gottingensis, Comment. Cl. Math.* **3**, pp. 107–134, reprinted in Werke Bd III, Georg Olms, Hildesheim, 1981, pp. 31–56. → *pp. 100, 101, and 110.*

[33] Girard, A. (1884). *Invention Nouvelle en l'Algèbre* (Muré Frères, Leiden), réimpression par D. Bierens De Haan. → *pp. 35, 36, 37, 38, 67, and 68.*

[34] Goldstine, N. (1977). *A History of Numerical Analysis from the 16th through the 19th Century, Studies in the History of Math. and Phys. Sciences,* Vol. 2 (Springer, New York, NY). → *pp. 30, 34, 40, and 105.*

[35] Hankel, H. (1874). *Zur Geschichte der Mathematik in Alterthum und Mittelalter* (Teubner, Leipzig). → *p. 17.*

[36] Hardy, G. and Wright, E. (1979). *An Introduction to the Theory of Numbers* (Clarendon Press, Oxford). → *p. 191.*

[37] Heath, T. (1956). *The Thirteen Books of Euclid's Elements* (Cambridge Univ. Press, Dover, New York, NY). → *p. 7.*

[38] Hudde, J. (1683). *Epistola prima, de Reductione Aequationum,* Vol. 1 of [22], pp. 406–506, translated by F. van Schooten, third edition. → *pp. 53 and 55.*

[39] Hudde, J. (1683). *Epistola secunda, de Maximis et Minimis,* Vol. 1 of [22], pp. 507–516, translated by F. van Schooten, third edition. → *p. 53.*

[40] Jacobson, N. (1975). *Lectures in Abstract Algebra III, Theory of Fields and Galois Theory, Graduate Texts in Mathematics,* Vol. 32 (Springer-Verlag, New York-Heidelberg), second corrected printing. → *pp. xi and 274.*

[41] Jacobson, N. (1985). *Basic Algebra I,* 2nd edn. (W. H. Freeman and Company, New York). → *pp. xi and 253.*

[42] Kaplansky, I. (1972). *Fields and Rings,* 2nd edn., Chicago Lectures in Mathematics (The University of Chicago Press, Chicago, Ill.-London). → *pp. xi and 283.*

[43] Karpinski, L. C. (1915). *Robert of Chester's Latin Translation of the Algebra of Al-Khowarizmi, Univ. of Michigan Studies, Humanistic series,* Vol. 11 (MacMillan, New York, NY). → *pp. 9 and 10.*

[44] Kiernan, B. (1971). The development of Galois theory from Lagrange to Artin, *Arch. Hist. Exact Sci.* **8**, pp. 40–154. → *p. 272.*

[45] Kline, M. (1972). *Mathematical Thought from Ancient to Modern Times* (Oxford University Press, New York). → *pp. 8, 11, and 114.*

[46] Knorr, W. R. (1975). *The Evolution of the Euclidean Elements, Synthese Historical Library,* Vol. 15 (D. Reidel Publishing Co., Dordrecht-Boston,

Mass.). → *pp. 5 and 6.*

[47] Knus, M.-A., Merkurjev, A., Rost, M. and Tignol, J.-P. (1998). *The Book of Involutions, American Mathematical Society Colloquium Publications,* Vol. 44 (American Mathematical Society, Providence, RI), with a preface in French by J. Tits. → *pp. 283, 285, 287, and 298.*

[48] Knus, M.-A. and Tignol, J.-P. (2003). Quartic exercises, *Int. J. Math. Math. Sci.* **2003**, 68, pp. 4263–4323. → *p. 298.*

[49] Lagrange, J.-L. (1770, 1771). Réflexions sur la résolution algébrique des équations, *Nouveaux Mémoires de l'Acad. Royale des sciences et belles-lettres, avec l'histoire pour la même année* **1, 2**, pp. 134–215, 138–253, Œuvres de Lagrange, vol. 3 (J.-A. Serret, éd.) Gauthier-Villars, Paris (1869), pp. 203–421. → *pp. 70, 96, 121, 129, 130, 133, 135, 138, and 221.*

[50] Lebesgue, H. (1955). L'œuvre mathématique de Vandermonde, *Enseignement Math. Sér. II* **1**, pp. 201–223. → *pp. 143 and 153.*

[51] Leibniz, G. W. (1899). *Der Briefwechsel von Gottfried Wilhelm Leibniz mit Mathematikern, Bd. 1* (Mayer & Müller, Berlin), herausg. von C.I. Gerhardt. → *p. 69.*

[52] Leibniz, G. W. (1962). *Mathematische Schriften, Bd V* (Georg Olms, Hildesheim), herausg. von C.I. Gerhardt. → *pp. 74 and 75.*

[53] Meo, M. (2004). The mathematical life of Cauchy's group theorem, *Historia Math.* **31**, 2, pp. 196–221. → *p. 270.*

[54] Morandi, P. (1996). *Field and Galois Theory, Graduate Texts in Mathematics,* Vol. 167 (Springer-Verlag, New York). → *p. xi.*

[55] Neumann, P. M. (2011). *The Mathematical Writings of Évariste Galois,* Heritage of European Mathematics (European Mathematical Society (EMS), Zürich). → *pp. vii, 216, 218, 219, 220, 226, 231, 232, 240, 241, 254, 271, and 272.*

[56] Newton, I. (1967). *The Mathematical Papers of Isaac Newton, vol. I: 1664–1666* (Cambridge Univ. Press), edited by D.T. Whiteside. → *pp. 38 and 39.*

[57] Newton, I. (1967). *The Mathematical Works of Isaac Newton, vol. 2,* The sources of science (Johnson Reprint Corp., New York, London), assembled with an introduction by D.T. Whiteside. → *p. 39.*

[58] Newton, I. (1971). *The Mathematical Papers of Isaac Newton, vol. IV: 1674–1684* (Cambridge Univ. Press, Cambridge). → *pp. 75 and 89.*

[59] Newton, I. (1972). *The Mathematical Papers of Isaac Newton, vol. V: 1683–1684* (Cambridge Univ. Press), edited by D.T. Whiteside. → *p. 39.*

[60] Nový, L. (1973). *Origins of Modern Algebra* (Noordhoff, Leyden). → *p. 272.*

[61] Romanus, A. (1593). *Ideae Mathematicae Pars Prima, sive Methodus Polygonorum* (Apud Ioannem Masium, Lovanii). → *pp. 30 and 32.*

[62] Rosen, M. (1981). Abel's theorem on the lemniscate, *Amer. Math. Monthly* **88**, 6, pp. 387–395. → *p. 215.*

[63] Rotman, J. (1998). *Galois Theory,* 2nd edn., Universitext (Springer-Verlag, New York). → *p. xi.*

[64] Ruffini, P. (1953–1954). *Opere Matematiche (3 vol.)* (Ed. Cremonese della Casa Editrice Perrella, Roma), edited by E. Bortolotti. → *pp. 193, 194, and 209.*

[65] Samuel, P. (1971). *Théorie algébrique des nombres* (Hermann, Paris). → *p. 110.*

[66] Serret, J.-A. (1866). *Cours d'algèbre supérieure (2 vol.)*, 3rd edn. (Gautier-Villars, Paris). → *pp. 209, 219, and 220.*

[67] Smith, D. E. (1959). *A Source Book in Mathematics (2 vol.)* (Dover, New York, NY). → *pp. 73, 77, 78, 79, 80, and 100.*

[68] Stedall, J. (2011). *From Cardano's Great Art to Larange's Reflections: filling a gap in the history of algebra*, Heritage of European Mathematics (European Math. Soc.). → *pp. vii, 30, 37, 39, 40, 53, 56, 68, and 120.*

[69] Stevin, S. (1958). *The Principal Works of Simon Stevin. Volume II B: Mathematics* (C.V. Swets en Zeitlinger, Amsterdam), D.J. Struik, ed. → *pp. 1, 26, 27, 28, and 40.*

[70] Stewart, I. (2004). *Galois Theory*, 3rd edn., Chapman & Hall/CRC Mathematics (Chapman & Hall/CRC, Boca Raton, Florida). → *pp. xi and 191.*

[71] Struik, D. J. (1969). *A Source Book in Mathematics, 1200–1800* (Harvard Univ. Press, Cambridge, Mass.). → *pp. 15, 19, 25, 27, 35, 36, and 37.*

[72] Szamuely, T. (2009). *Galois Groups and Fundamental Groups*, Cambridge Studies in Advanced Mathematics, Vol. 117 (Cambridge University Press, Cambridge). → *p. 274.*

[73] Tartaglia, N. (1546). *Quesiti et inventioni diverse* (Venice). → *p. 13.*

[74] Toscano, F. (2009). *La formula segreta* (Sironi, Milano). → *p. 15.*

[75] Tschirnhaus, E. W. (1683). Methodus anferendi omnes terminos intermedios ex data aequatione, *Acta Eruditorum (Leipzig)* , pp. 204–207. → *p. 69.*

[76] van der Poorten, A. (1979). A proof that Euler missed ... Apéry's proof of the irrationality of $\zeta(3)$, *Math. Intel.* **1**, pp. 195–203. → *p. 107.*

[77] van der Waerden, B. L. (1966, 1967). *Algebra, Heidelberger Taschenbücher*, Vol. 12, 23 (Springer, Berlin), 7th edition of "Moderne Algebra". → *pp. xi, 49, 182, and 273.*

[78] van der Waerden, B. L. (1972). Die Galois-Theorie von Heinrich Weber bis Emil Artin, *Arch. History Exact Sci.* **9**, 3, pp. 240–248. → *p. 272.*

[79] van der Waerden, B. L. (1975). *Science Awakening, vol. I*, 4th edn. (Noordhoff International Publishing, Leyden), translated from the Dutch by Arnold Dresden. → *pp. xi, 2, 7, and 8.*

[80] van der Waerden, B. L. (1985). *A History of Algebra, from al-Khwārizmī to Emmy Noether* (Springer-Verlag, Berlin). → *p. 11.*

[81] Vandermonde, A. (1771). Mémoire sur la résolution des équations, *Histoire de l'Acad. Royale des Sciences* , pp. 365–416. → *pp. 143, 144, 148, 149, 153, and 154.*

[82] Vieta, F. (1595). *Ad Problema quod omnibus Mathematicis totius orbis construendum proposuit Adrianus Romanus Francisci Vietae Responsum* (Apud Iametium Mettayer, Parisiis), reprinted in "Francisci Vietae Opera Mathematica". → *pp. 33 and 34.*

[83] Vieta, F. (1983). *The Analytic Art* (Kent State Univ. Press, Kent, Ohio), translated by T.R. Witmer. → *pp. 29, 34, 35, 40, 63, and 64.*

[84] Waring, E. (1991). *Meditationes Algebraicae* (Amer. Math. Soc., Providence, RI), translated by D. Weeks. → *pp. 13, 96, and 98.*

[85] Weber, H. (1898). *Lehrbuch der Algebra, Bd. I* (F. Vieweg u. Sohn, Braunschweig). → *pp. 70 and 182.*

[86] Weil, A. (1960). De la métaphysique aux mathématiques, *Sciences*, pp. 52–56(Œuvres Scientifiques–Collected Papers, vol. 2, Springer, New York, NY, 1979, pp. 408–412). → *p. 272.*

[87] Weil, A. (1974). Two lectures on number theory, past and present, *Enseignement Math.* **20**, pp. 87–110, (Œuvres Scientifiques–Collected Papers, vol. 3, Springer, New York, NY, 1979, pp. 279–302). → *p. 5.*

# Index

Abel, Niels-Henrik, 194–196, 215
  Abel's condition, 216, 265–268
  theorem on natural irrationalities,
    203–209
abelian group, 267
absolute Galois group (of a field), 284
accessory irrationality, 203, 212
action (of a group on a set), 221
al-Khowarizmi, Mohammed ibn
  Musa, 9–10, 22
Alembert, Jean Le Rond d', 109
algebra
  Arabic, 9–11, 22
  Babylonian, 2–5, 7, 22
  Greek, 5–8, 21
algebraically independent elements,
  101
alternating group $\text{Alt}(X)$, 244
alternating group $A_n$, 212, 254
Apéry, Roger, 107
arrangement, 219, 222
Artin, Emil, 273–274, 277

Bernoulli, Jacques, 105
Bernoulli, Nicholas, 75
Bézout, Etienne
  Bézout's method, 117–120,
    127–130, 194
  elimination theory, 56, 70, 120
Bolzano, Bernhard, 114
Bombelli, Rafaele, 19, 26, 27
Bourbaki, Nicolas, 274

Cardano, Girolamo, 13–15, 21–22,
  25–26, 40
  Cardano's formula, 15–19, 121–125
casus irreducibilis, 19, 105
Cauchy, Augustin-Louis, 111,
  194–195, 212, 218
characteristic (of a field), 54, 196
Chebyshev polynomials, 34, 77, 90–91
Chuquet, Nicolas, 27
complete solvability (by radicals), 246
congruence (modulo an integer), 156
constructible point, 186
coset, 134
Cotes, Roger, 75–77
  Cotes–de Moivre formula, 73,
    75–80, 82, 90–91
Cramer, Gabriel, 56
cyclic group, 86, 162
cyclotomic polynomial (or equation),
  87–88, 216
  Galois group, 230
  irreducibility, 162–165, 174,
    182–185
  solvability by radicals, 82, 148–153,
    178–181
cyclotomy, 82

Dedekind, Richard, 163, 182, 273, 277
degree
  of a field extension, 275
  of a multivariate polynomial, 96

of a polynomial, 42
of a radical extension of height one,
    196
Delambre, Jean-Baptiste, 193
derivative (of a polynomial), 53
Descartes, René, 21, 26, 28–30, 37, 40
    Descartes' method for quartic
        equations, 66
Diophantus of Alexandria, 8, 29
discriminant, 102, 107–108, 265
discriminant algebra, 290

Eisenstein, Ferdinand Gotthold Max,
    163, 164
elementary symmetric polynomial, 95
elimination theory, 56, 70, 118
equation
    cubic, 13–19, 63–65, 70–72, 105,
        119, 121–127, 146
    cyclotomic, *see* cyclotomic
        polynomial
    general, 94, 193
    of squared differences, 108
    quadratic, 1–10
    quartic, 21–24, 66, 119–120,
        127–128, 146
    rational, 67–68
    resolvent cubic, 24, 290
étale algebra, 283
Euclid, 5–8, 10, 22
    Euclid's algorithm for the GCD,
        27, 44
    Euclidean division property, 43, 88,
        97
Euler, Leonhard
    Euler's method, *see* Bézout's
        method
    on elimination theory, 56
    on Newton's formulas, 39
    on number theory, 92, 159, 190, 192
    on roots of unity, 73, 80
    on the fundamental theorem of
        algebra, 109
    on the summation of series, 105
exponent

of a root of unity, 85
of an element in a group, 162
of an integer modulo a prime, 160

Fermat, Pierre de, 159
    Fermat prime, 190
    Fermat's theorem, 159, 162, 192
Ferrari, Ludovico, 21
    Ferrari's method, 22–24, 127–128
Ferro, Scipione del, 13
Fior, Antonio Maria, 13
Foncenex, Daviet François de, 110
fundamental theorem
    of algebra, 36, 74, 79, 100, 101,
        109–116
    of Galois theory, 134, 274–282
    of symmetric functions or
        polynomials, 95–101

$GA(n)$ (group of affine
    transformations), 140, 256
Galois algebra, 286
Galois extension (of fields), 275
Galois group
    of a field extension, 275
    of a polynomial (or equation), 228
Galois resolvent, Galois resolvent
    equation, 232
Galois, Évariste, 216–265, 271–272
$\Gamma$-set, 284
Gauss, Carl Friedrich, 155, 194, 215
    on cyclotomic equations, 162–181
    on number theory, 156–162
    on regular polygons, 191
    on the fundamental theorem of
        algebra, 100, 101, 110, 114
general (generic) polynomial, 94
    Galois group, 229
Girard, Albert, 28, 35–38, 67
    Girard's theorem, 36, 70, 93, 110
greatest common divisor (GCD), 44
Grothendieck, Alexander, 274
group
    abelian, 267
    alternating $\mathrm{Alt}(X)$, 244
    alternating $A_n$, 212

cyclic, 162
early results, 131–134, 147, 212
Galois, *see* Galois group
of arrangements, 222
solvable, 251
symmetric Sym($X$), 220
symmetric $S_n$, 131

Harriot, Thomas, 30, 37
height (of a radical extension), 197
Hero, 8
Hippasus of Metapontum, 5
Hudde, Johann, 53

ideal (in a commutative ring), 111
idempotent, 286
index
of a normal decomposition, 241
of a subgroup, 134, 241
irrationality
accessory, 203, 212
natural, 203
irreducible polynomial, 48
isotropy group, 131, 221

Jacobson, Nathan, 274

Khayyam, Omar, 10
Klein, Felix, 252
Kronecker, Leopold, 49, 111, 163, 182, 203, 273

Lacroix, Sylvestre-François, 193
Lagrange, Joseph-Louis, 95, 110, 120–138, 143, 193, 218
Lagrange resolvent, 130, 139–141, 145, 179, 181, 200, 248, 260
Landau, Edmund, 163
leading coefficient, 42
Lebesgue, Henri, 143, 153
Legendre, Adrien-Marie, 193
Leibniz, Gottfried Wilhelm, 69, 73–75, 89, 105, 109
lemniscate, 215
Liouville, Joseph, 216, 220, 271

Mertens, Franz, 163
minimum polynomial, 168
modulo (modulus), 156
Moivre, Abraham de, 77–80, 82, 83, 109, 117, 148
de Moivre formula, *see* Cotes–de Moivre formula
monic polynomial, 42
multiplicity (of a root), 52

natural irrationality, 203
Newton, Isaac, 38, 40, 56, 73, 75
Newton's formulas, 38–40, 105
on the summation of series, 89–90
normal
decomposition of a group of arrangements, 241
subgroup in a group, 241
Nunes, Pedro, 26, 37

orbit (under a group action), 221
order
of a group, 131
of an element in a group, 162

Pacioli, Luca, 11
period (of a cyclotomic equation), 171
primitive
element (of a field extension), 232
idempotent, 286
root of a prime number, 157, 160
root of unity, 85
Pythagoras, 5

quotient (of Euclidean division), 43
quotient ring (by an ideal), 111

radical
expression (solution) by radicals, 197, 246
field extension, 197
rational function, 43
Recorde, Robert, 26
regular polygon, 82, 155, 185–191
relatively prime polynomials, 44

remainder (of Euclidean division), 43
resolvent
  cubic, 24, 290
  Galois, 232
  Lagrange, *see* Lagrange resolvent
resultant, 57
Romanus, Adrianus, *see* Van Roomen
root
  common root of two polynomials,
    56
  multiple root, 52, 103
  of a complex number, 79, 80
  of a polynomial, 51
  of an étale algebra, 284
  of unity, 81
  rational, 67
  simple root, 52
Ruffini, Paolo, 193–195, 203, 209, 218
ruler and compass constructions, 155,
  185–191, 215

Schur, Issai, 163
separable closure (of a field), 283
separable polynomial, 283
Serret, Joseph-Alfred, 219, 220
solvable group, 251
Stevin, Simon, 1, 26–28, 40
substitution, 219

Sylvester, James Joseph, 56
symmetric group $\text{Sym}(X)$, 220
symmetric group $S_n$, 131
symmetric polynomial (or rational
  function), 94

Tartaglia, Niccolò, 13–15, 17
torsor (principal homogeneous $\Gamma$-set),
  286
transitive action, 254
transposition, 213
Tschirnhaus, Ehrenfried Walther, 69
  Tschirnhaus' method, 69–72, 118,
    125–127

van der Waerden, Bartel Leendert,
  273
Van Roomen, Adriaan, 30–34
Vandermonde, Alexandre-Théophile,
  84, 95, 107, 120, 143–153, 169
Viète, François, 1, 29–30, 32–35, 40,
  63–65, 93
Vierergruppe, 252

Wantzel, Pierre Laurent, 189, 195,
  209
Waring, Edward, 95–99, 120
Weil, André, 5, 272

Printed in the United States
By Bookmasters